D1029797

NUMBER
THEORY

老馬識途

陳省身題

André Weil

NUMBER THEORY
An approach through history

From Hammurapi to Legendre

BIRKHÄUSER
Boston • Basel • Stuttgart

Library of Congress Cataloging in Publication Data

Weil, André, 1906–
  Number theory.

  Bibliography: p.
  Includes index.
  1. Numbers, Theory of—History. I. Title.
  QA241.W3418 1983   512'.7'09   83-11857
  ISBN 0-8176-3141-0

Weil, André:
Number theory: an approach through history; From
Hammurapi to Legendre/André Weil.—Boston;
Basel; Stuttgart: Birkhäuser, 1983.—250 S.
  ISBN 3-7643-3141-0 geb.

Printed in the United States of America.

ISBN 3-7643-3141-0
ISBN 0-8176-3141-0

Preface

The texts to be examined in this volume range from an Old Babylonian tablet, datable to the time of Hammurapi or thereabouts, to Legendre's *Essai sur la Théorie des Nombres* of 1798. In the main, the volume stops short of Gauss's *Disquisitiones* of 1801, even though it includes an episode from Legendre's later career and does not avoid relevant references to the discoveries of Gauss and his successors.

Until rather recently, number theory, or arithmetic as some prefer to call it, has been conspicuous for the quality rather than for the number of its devotees; at the same time it is perhaps unique in the enthusiasm it has inspired, an enthusiasm eloquently expressed in many utterances of such men as Euler, Gauss, Eisenstein, Hilbert. Thus, while this book covers some thirty-six centuries of arithmetical work, its bulk consists in nothing more than a detailed study and exposition of the achievements of four mathematicians: Fermat, Euler, Lagrange, Legendre. These are the founders of modern number theory. The greatness of Gauss lies in his having brought to completion what his predecessors had initiated, no less than in his inaugurating a new era in the history of the subject.

Our main task will be to take the reader, so far as practicable, into the workshop of our authors, watch them at work, share their successes and perceive their failures. Luckily this does not require any deep delving into archives and manuscript collections. As the bibliography will witness, almost all the mathematicians to be mentioned in this volume have had their complete works, and their surviving correspondence, excellently edited and published. I have also had the good fortune of being able to consult the original editions of Viète, Bachet, Fermat, Wallis, Legendre in the Rosenwald Rare Book Collection at the Institute for Advanced Study.

The method to be followed here is historical throughout; no specific knowledge is expected of the reader, and it is the author's fond hope that some readers at least will find it possible to get their initiation into number theory by following the itinerary retraced in this volume. Some appropriate background may be obtained from the author's *Number theory for beginners* (whose content, incidentally, is almost entirely taken from Euler), or also (at a more sophisticated level) from Chapter I of J.-P. Serre's *Cours d'Arithmétique*. More detailed background information, along with some supplementary material (not of strictly historical character) will be supplied in the Appendices following Chapters II, III, IV.

My warm thanks, and the reader's, are due to my friend S. S. Chern for the beautiful calligraphy adorning the frontispiece. Thanks are also due to the University Museum, University of Philadelphia, for permission to reproduce the photograph of the Tang horse in their collection; to O. Neugebauer for the photograph of the tablet PLIMPTON 322; and to the Archives de l' Académie des Sciences, Paris, for the photograph of Lagrange's portrait engraved by Delpech. Finally I wish to express my gratitude to the Birkhäuser Verlag, Boston, and more specifically to Klaus Peters and his team of collaborators for the keen interest they have taken in this project, their helpful hand on more than one occasion, and the high technical skill they have brought to bear upon all matters pertaining to the publication of this volume.

Princeton, 15 April 1983

Contents

Table of Illustrations

Abbreviations, Basic References and Notations

Arch. = ARCHIMEDES, Opera omnia cum commentariis Eutocii, it.ed. J. L. Heiberg, Teubner, 3 vol., 1910–1915.

D. Bern. = Die Werke von DANIEL BERNOULLI, ed. D. Speiser, Bd. 2, Birkhäuser 1982.

J. Bern. = Johannis BERNOULLI ... Opera Omnia ... Lausannae et Genevae, sumptibus Marci Michaelis Bousquet et sociorum, 4 vol., 1742–1743.

Col. = H. T. COLEBROOKE, Algebra with arithmetic and mensuration. From the sanscrit of Brahmegupta and Bháscara, London 1817 (Reprint, 1972).

Corr. = Correspondance mathématique et physique de quelques célèbres géomètres du XVIIIème siècle ... publiée ... par P.-H. Fuss, St.-Pétersbourg, 2 vol., 1843 (reprint, Johnson Reprint Corp., 1968).

Desc. = Œuvres de DESCARTES, pub. par Charles Adam et Paul Tannery, Paris, 11 vol., 1897–1909.

Dioph. = DIOPHANTI Alexandrini Opera omnia cum graecis commentariis, ed. Paulus Tannery, Teubner, 2 vol. 1893–1895.

Dir. = G. Lejeune DIRICHLET's Werke, herausg. v. L. Kronecker und L. Fuchs, Berlin, 2 vol., 1889–97.

Disq. = Disquisitiones arithmeticae, auctore D. Carolo Friderico GAUSS, Lipsiae 1801 (= vol. I of Gauss, Werke, Göttingen 1870).

Eucl. = EUCLIDIS Elementa, ed. I. L. Heiberg, Lipsiae, Teubner, 4 vol., 1883–1885 (and: Post I. L. Heiberg ed. E. S. Stamatis, Teubner, 4 vol., 1969–1973).

Eu. = Leonhardi EULERI Opera omnia, sub ausp. Soc. scient. Nat. Helv. .... Series I–IV A, 1911–   ).

Fag. = Opere Matematiche del Marchese Giulio Carlo de' Toschi di FAGNANO, pubbl. ... dai Soci V. Volterra, G. Loria, D. Gamboli, 3 vol., 1911–1913.

Fe. = Œuvres de Fermat, pub. par . . . Paul Tannery et Charles Henry, Paris, 4 vol., 1891–1912 (+ Supplément, pub. par M. C. de Waard, 1 vol. 1922).

Gal. = Le Opere di GALILEO Galilei, Edizione Nazionale, Firenze, G. Barbera, 20 vol., 1890–1909.

Gau. = Carl Friedrich GAUSS, Werke, Göttingen, 12 vol., 1870–1929.

Huy. = Œuvres complètes de Christian HUYGENS, pub. par la Soc. Holl. des Sc., La Haye, Martinus Nijhoff, 22 vol., 1888–1950.

Jac. = C. G. J. JACOBI's Gesammelte Werke, herausg. v. K. Weierstrass, Berlin, 7 vol., 1881–1891.

JEH. = J. E. Hofmann, Neues über FERMATs zahlen-theoretische Herausforderungen von 1657, Abh. d. preuss. Akad. d. Wiss. 1943–1944. Nr. 9, pp. 41–47.

Lag. = Œuvres de LAGRANGE, pub. par M. J.-A. Serret et M. Gaston Darboux, Paris, 14 vol., 1867–1892.

Leib. = LEIBNIZens mathematische Schriften, herausg. von C. I. Gerhardt, Zweite Abtheilung, die mathematischen Abhandlungen Leibnizens enthaltend, Halle, 3 vol., 1858–1863.

Leon. = Scritti di LEONARDO PISANO, matematico del Secolo decimoterzo, pubbl. da Baldassare Boncompagni, Roma, 2 vol., 1857–1862.

LVE. = LEONARD DE PISE, Le livre des nombres carrés, traduit . . . par P. Ver Eecke, Desclée de Brouwer et Cie, Bruges 1952 (French translation of *Leon.* II. 253–283).

Mers. = Correspondance du P. Marin MERSENNE, Religieux Minime, Pub. par Mme Paul Tannery et Cornélis de Waard, Paris, 14 vol., 1955–1980.

New. = The mathematical papers of Isaac NEWTON, ed. by D. T. Whiteside, Cambridge University Press, 8 vol., 1967–1981.

PkU. = Leonhard EULER, Pis'ma k učënym, edd, T. N. Klado, Ju. Ch. Kopelevič, T. A. Lukin (red. V. I. Smirnov), Moskva-Leningrad 1963.

Viète, *Op.* = Francisci VIETAE Opera Mathematica . . . Operâ atque studio Francisci à Schooten Leydensis, Matheseos Professoris, Lugduni Batavorum, ex officinâ

Bonaventurae et Abrahami Elzeviriorum, 1646 (reprint
Georg Holms Verlag 1970).

Wal. = Johannis WALLIS S.T.D., Geometriae Professoris
in Celeberrima Academia Oxoniensi, Opera Mathematica:
(I) Volumen Primum, Oxoniae, e Theatro Sheldoniano
1695; (II) De Algebra Tractatus Historicus et Practicus,
Anno 1685 anglice editus, nunc auctus latine . . . Operum
Mathematicorum Volumen alterum, Oxoniae . . . 1693.

(For additional bibliography, see end of volume.)

References and dates

References to the above works are mostly given by in-
dicating the volume (when necessary) and the page. Example:
*Fe.*II.194 means Fermat's *Œuvres* (in the standard Tannery-
Henry edition), vol.II, p.194.

Exceptionally references to Euclid and to Diophantus are
given by indicating the Book and the proposition or problem.
Example: *Eucl.*VII.2 means Euclid (in the standard Heiberg
edition), Book VII, prop.2; similarly *Dioph.*V.11 means
Diophantus (the Tannery edition), Book V, problem 11. In
the case of Diophantus, it has occasionally been found necessary to include the numbering in Bachet's edition of 1621
(or, what amounts to the same, in S. de Fermat's edition of
1670; cf. Bibliography); this is indicated thus: *Dioph.*V.11
= *Dioph.*V.14$_b$, meaning that prob. 11, Book V, of Diophantus (Tannery's edition) is numbered 14 in Bachet's
edition.

Dates are of particular importance in the description of
Fermat's and of Euler's work (Chap.II and III). In the case
of letters this usually raises no question; the difference between the Julian ("old style") and the Gregorian ("new style")
calendar, being of little moment, has mostly been disregarded. As to Euler, a more elaborate system has been found
necessary, as it was desirable to indicate, for each paper, its
number in G. Eneström's catalogue of Euler's writings (cf.
Bibliography) and its probable date of composition. E.g.

*Eu.*I-2.531–555 = E 271|1758 means Euler's memoir, no.271 of Eneström, vol.2 of Series I, pages 531 to 555, written presumably in 1758; when the reference is only to a portion of some memoir, it is given thus: *Eu.*I-20.81 in E 252|1752. The dates are those given by Eneström (usually the date of presentation, either to the Petersburg or to the Berlin Academy) unless an earlier date is indicated by Euler's correspondence. References to vol.1 of Series IV A, which is a repertory of all available Euler correspondence, are given by number: *Eu.*IV A-1, no.1887 means no.1887 of that volume.

Additional references will be found in the bibliography on pages 361–364.

Notations

Traditional algebraic notation has been used throughout; as it cannot be said to have been well established until late in the seventeenth century, it is anachronistic in the description of the work of earlier writers (including Fermat, who used Viète's notations). From Euler on, our notation will usually conform to that of the original authors, except that the congruence notation, which only goes back to Gauss, has been adopted throughout as a convenient abbreviation. We recall that the "congruence"

$$a \equiv b \ (\mathrm{mod}.m),$$

with the "modulus" m, means that $a - b$ is a multiple of m. Thus an integer a is a "quadratic residue" modulo m if there is b such that $a \equiv b^2 \ (\mathrm{mod}.m)$, and a "quadratic non-residue" otherwise; it is an n-th power residue if there is b such that $a \equiv b^n \ (\mathrm{mod}.m)$.

Also for brevity's sake, the matrix notation has been used occasionally (in Chapter III, § XIII, and Appendix III of Chap.IV); **Z** is used to designate the "ring" consisting of all integers (positive, negative and 0); **Q**, **R**, **C**, to designate the "fields" consisting respectively of all rational numbers, of all real numbers, and of all "imaginary" (or "complex") numbers $a + b\sqrt{-1}$ (where a, b are real); when p is any prime, $\mathbf{F}_p$ is the "field" of p elements, consisting of the congruence classes modulo p. If θ is a "quadratic irrationality", actually $\sqrt{N}$, where N is a non-square integer, or the "cubic root of unity" $\dfrac{-1 + \sqrt{-3}}{2}$, then $\mathbf{Z}[\theta]$ is the "ring" consisting of all elements $a + b\theta$ where a and b are integers, and $\mathbf{Q}(\theta)$ is the "field" consisting of all elements $r + s\theta$, where r and s are rational numbers.

For reasons of typographical convenience, the Legendre symbol

$$\left(\frac{n}{p}\right)$$

(invariably so written by Legendre, Dirichlet and most classical authors) has frequently been printed as (n/p). It is the symbol, defined whenever p is a prime and n an integer prime to p, which has the value $+1$ when n is a quadratic residue modulo p, and -1 otherwise.

Chapter One

Protohistory

§I.

According to JACOBI, the theory of elliptic functions was born between the twenty-third of December 1751 and the twenty-seventh of January 1752. On the former date, the Berlin Academy of Sciences handed over to EULER the two volumes of Marchese FAGNANO's *Produzioni Matematiche*, published in Pesaro in 1750 and just received from the author; Euler was requested to examine the book and draft a suitable letter of thanks. On the latter date, Euler, referring explicitly to Fagnano's work on the lemniscate, read to the Academy the first of a series of papers, eventually proving in full generality the addition and multiplication theorems for elliptic integrals.

One might similarly try to record the date of birth of the modern theory of numbers; like the god Bacchus, however, it seems to have been twice-born. Its first birth must have occurred at some point between 1621 and 1636, probably closer to the latter date. In 1621, the Greek text of DIO-PHANTUS was published by BACHET, along with a useful Latin translation and an extensive commentary. It is not known when FERMAT acquired a copy of this book (the same one, no doubt, into whose margins he was later to jot down some of his best discoveries), nor when he began to read it; but, by 1636, as we learn from his correspondence, he had not only studied it carefully but was already developing ideas of his own about a variety of topics touched upon in that volume.

From then on, "numbers", i.e., number theory, never ceased to be among Fermat's major interests; but his valiant efforts to gain friends for his favorite subject were not, on the whole, crowned with success. "There is no lack of better topics for us to spend our time on" was young HUYGENS' comment to WALLIS (*Huy*.II.211 = *Fe*.IV.121). At one time Fermat cherished the thought of devoting a whole book to number theory (*Fe*.I.305). On another occasion he tried to persuade PASCAL to collaborate with him in writing such a book (*Fe*.II.299–300); of course he realized that Pascal's gifts of exposition were far superior to his own. To our great loss, Pascal showed no interest and politely declined the suggestion (*Fe*.II.314); his views may well have been similar to those of Huygens. After Fermat's death in 1665, there was a great demand for a publication of his writings, hardly any of which had ever appeared in print. In 1670, his son Samuel published a reprint of Bachet's *Diophantus*, along with Fermat's marginal notes and an essay by the Jesuit father Jacques de BILLY on Fermat's methods for solving certain types of so-called diophantine equations. This was followed in 1679 by the publication, also by Samuel, of a volume of his father's *Varia Opera*, including a few letters of arithmetical content. But it took half a century for this to have any effect, and in the meanwhile number theory seemed to have died off.

As to its rebirth, we can pinpoint it quite accurately. In 1729, young EULER was in St. Petersburg, an adjunct of the newly founded Academy of Sciences; his friend and patron GOLDBACH was in Moscow. Their correspondence has been carefully preserved and was published by Euler's great-grandson in 1843. Goldbach, in his own amateurish way, was keenly interested in mathematics and particularly in "numbers"; it is in a letter to Euler that he stated the conjecture to which his name has remained attached. On the first of December 1729, Goldbach asked Euler for his views about Fermat's statement that all integers $2^{2^n} + 1$ are primes (cf. *infra*, Chap.III, §IV). In his answer, Euler expresses some doubts; but nothing new occurs until the fourth of June, when Euler reports that he has "just been reading Fermat" and that he has been greatly impressed by Fermat's assertion that every integer is a sum of four squares (and also of 3 triangular numbers, of 5 pentagonal numbers, etc.). After that day, Euler never lost sight of this topic and of number theory in general; eventually LAGRANGE followed suit, then LEGENDRE, then GAUSS with whom number theory reached full maturity. Although never a popular subject, it has been doing quite well ever since.

Thus an account of number theory since Fermat can do full justice to the inner coherence of the topic as well as to the continuity of its development. In contrast with this, the mere fact that Fermat initially had to draw his main inspiration from a Greek author of the third century, only lately come to light, points to the entirely different character of much of the earlier mathematical work as well as to the frequent disappearance and re-appearance of essential sources of knowledge in former times. About ancient mathematics (whether Greek or Mesopotamian) and medieval mathematics (Western or Oriental), the would-be historian must of necessity confine himself to the description of a comparatively small number of islands accidentally emerging from an ocean of ignorance, and to tenuous conjectural reconstructions of the submerged continents which at one time must have bridged the gaps between them. Lacking the continuity which seems essential to history, his work might better be described by some other name. It is not prehistory, since it depends on written sources; protohistory seems more appropriate.

Of course new texts may turn up; they do from time to time. Our knowledge of ARCHIMEDES was greatly enriched in 1906 by the discovery of a palimpsest in Istanbul. What survives of DIOPHANTUS consists of six chapters or "books", while thirteen books are announced in the introduction; some new material pertaining to Diophantus, re-worked or perhaps translated from the original text, has been found recently in an Arabic manuscript; more may yet be forthcoming. Important cuneiform texts may still be buried underground in Mesopotamia, or even more probably (according to Neugebauer) in the dusty basements of our museums. Arabic and Latin medieval manuscripts by the score await identification, even in well-explored libraries. Still, what hope is there of our ever getting, say, a full picture of early Greek geometry? In the third century B.C., EU-DEMOS (not himself a mathematician) wrote in four "books" a history of geometry, some fragments of which have been preserved. But what may have been the contents of his history of arithmetic, comprising at least two books, all but entirely lost? Even if part of it concerned topics which we might regard as algebra, some of it must have been number theory. To try to reconstruct such developments from hints and allusions found in the work of philosophers, even of those who professed a high regard for mathematics, seems as futile as would be an attempt to reconstruct Newton's *Principia* out of the writings of Locke and Voltaire, or his differential calculus from the criticism of Bishop Berkeley.

It will now be our purpose to describe briefly, without any claim to completeness, a few highlights from the scanty remains of number-theorists prior to the seventeenth century. I have tried to exclude what belongs more properly to algebra (for instance, the solution of linear equations and linear systems), but the distinction between the two topics is often far from clearcut.

§II.

Of all the topics occurring in ancient mathematics, perhaps the one which belongs most clearly to number theory concerns the basic multiplicative properties of positive integers; they receive a fairly full treatment in EUCLID's books VII, VIII and IX. It is generally agreed upon that much, if not all, of the content of those books is of earlier origin, but little can be said about the history behind them. Some facts concerning divisibility must have been known in Mesopotamia; any table of reciprocals in the sexagesimal system points clearly to the essential distinction between those integers which contain no other prime than 2, 3 and 5, and all others. That the strictly additive treatment of fractions in Egyptian mathematics was eventually supplemented in Greece by a multiplicative treatment of the ratios between integers indicates a basic change of attitude which, according to Paul Tannery's very plausible hypothesis, may well have had its source in musical theory. This may in turn have had some connection with early proofs for the irrationality of the simplest square roots, such as $\sqrt{2}$ and $\sqrt{5}$, but we do not know what those proofs were; if Aristotle, in the course of a discussion on the logical structure of proofs (*Analytica Priora* I.23), hints at one proof for $\sqrt{2}$, this hardly gives us a right to credit it to some hypothetical "Pythagoreans". The concept of a prime may have occurred quite early, together with that of a divisor and that of a common multiple for given integers; all we can say is that Plato, in his late work *The Laws* (737e–738a), mentions some properties of the number 5040, stressing the fact that it is a common multiple of numbers up to 10 (but so is 2520) and that it has 59 divisors, not counting 5040 itself; this points to some advanced knowledge about the factorization of integers among the mathematicians in Plato's Academy, but how much cannot be ascertained. Was there originally a relation between the so-called "Euclidean algorithm", as described in *Eucl*.VII.1–2, for finding the g.c.d. of two integers, and the theory of the same process (*Eucl*.X.2) as it applies to possibly incommensurable magnitudes? Has it not often happened that a mathematical process has been discovered twice, in different contexts, long before the substantial identity between the two discoveries came to be perceived? Some of the major advances in mathematics have occurred just in this manner.

Even in Euclid, we fail to find a general statement about the uniqueness of the factorization of an integer into primes; surely he may have been aware of it, but all he has is a statement (*Eucl*.IX.14) about the l.c.m. of any number of given primes. Finally, the proof for the existence of infinitely many primes (*Eucl.*IX.20) represents undoubtedly a major advance, but there is no compelling reason either for attributing it to Euclid or for dating it back to earlier times. What matters for our purposes is that the very broad diffusion of Euclid in later centuries, while driving out all earlier texts, made this body of knowledge widely available to mathematicians from then on.

§III.

Magical or mystical properties of numbers occur in many cultures. Somehow, either in Greece or earlier, the idea of perfection attached itself to those integers which are equal to the sum of their divisors. The last theorem in the arithmetical books of Euclid (*Eucl.*IX.36), and possibly, in their author's view, the apex of his number-theoretical work, asserts that $2^n(2^{n+1} - 1)$ is perfect if the second factor is a prime. The topic, along with some of its extensions (such as the pairs of "amicable" numbers), occurs sporadically in later work, perhaps because of the special appeal of the words designating those concepts. It is of little theoretical importance and would not have to be mentioned here, were it not for the fact that it did attract a good deal of attention among some of Fermat's contemporaries, such as Mersenne and Frenicle, including even Fermat himself, and played some part in his early investigations (cf. *infra*, Chap.II, §IV).

§IV.

Indeterminate equations of the first degree, to be solved in integers, must have occurred quite early in various cultures, either as puzzles (as exemplified by various epigrams in the Greek *Anthology*; cf. *Dioph.*, vol. II, pp. 43–72), or, more interestingly for the mathematician, as calendar problems. A typical problem of this kind may be formulated as a double congruence

$$x \equiv p \pmod{a}, x \equiv q \pmod{b},$$

or as the linear congruence $ax \equiv m \pmod{b}$, or as an equation $ax - by = m$ in integers. The general method of solution for this is essentially identical with the "Euclidean algorithm" for finding the g.c.d. of a and b (*Eucl.*VII.2) or also (in modern terms) with the calculation of the continued fraction for a/b; the relation between the two problems is indeed so close that whoever knows how to solve the one can hardly fail to solve the other if the need for it arises. Nevertheless, if we leave China aside, the first explicit description of the general solution occurs in the mathematical portion of the Sanskrit astronomical work *Āryabhaṭīya*, of the fifth–sixth century A.D. (cf. e.g. Datta and Singh, *History of Hindu Mathematics*, Lahore 1938, vol. II, pp. 93–99). In later Sanskrit texts this became known as the *kuṭṭaka* (= "pulverizer") method; a fitting name, recalling to our mind Fermat's "infinite descent". As Indian astronomy of that period is largely based on Greek sources, one is tempted to ascribe the same origin to the *kuṭṭaka*, but of course proofs are lacking.

Then, in 1621, Bachet, blissfully unaware (of course) of his Indian predecessors, but also of the connection with the seventh book of Euclid, claimed the same method emphatically as his own in his comments on *Dioph.*IV.41$_b$ (= IV, lemma to 36), announcing that it was to be published in a book of arithmetical "elements"; as this never appeared, he inserted it in the second edition of his *Problèmes plaisants et délectables* (Lyon 1624), which is where Fermat and Wallis found it; both of them, surely, knew their Euclid too well not to recognize the Euclidean algorithm there.

§V.

There is no need to try to enumerate here the many sources, both Mesopotamian and Greek, for the summation of arithmetic progressions (such as $\Sigma_1^N n$), geometric progressions (such as $\Sigma_0^N 2^n$) and arithmetic progressions of higher order (such as $\Sigma_1^N n^2$); so far as Greece is concerned, this cannot be separated from the consideration of "figurative" numbers. The most primitive kind of tabulation leads directly to not a few formulas of this kind (e.g. to

n^2 = the sum of the first n odd numbers) which can then also be verified by suitable diagrams. Such results must have become fairly universally known at a comparatively early date; invoking the name of Pythagoras adds little to our understanding of the matter.

The same may be said about "pythagorean" triangles, a phrase by which we will merely understand those triples of integers (a, b, c) which satisfy the condition

$$(1) \qquad\qquad a^2 + b^2 = c^2$$

and which therefore, if the geometrical theorem known to us under the name of Pythagoras is assumed, measure the sides of a right-angled triangle. Of course, if (a, b, c) is such, so is (b, a, c); allowing for this permutation, one finds that the general solution of (1) is given by

$$(2) \quad a = d \times 2pq, \quad b = d \times (p^2 - q^2), \quad c = d \times (p^2 + q^2),$$

where p, q are mutually prime, and $p - q$ is odd and > 0; d is then the g.c.d. of a, b, c. Taking $d = 1$, one gets (up to the same permutation as before) the general solution of (1) in mutually prime integers. The simplest such triple is of course $(3, 4, 5)$, which seems to have belonged to a kind of mathematical folklore from very early times.

A table of fifteen pythagorean triples has been preserved in the Old Babylonian tablet PLIMPTON 322, published by O. Neugebauer and A. Sachs (*Mathematical Cuneiform Texts*, New Haven 1945, pp. 38–41) and dated by them to between 1900 and 1600 B.C. Perhaps it was calculated as an aid to trigonometry; it must have been based on some formula, either (2) itself or more simply the relation

$$a^2 = (c+b)(c-b);$$

the values for a are all "regular" numbers (containing no prime factor other than 2, 3, 5), while the values of b^2/a^2, listed in decreasing order, range at fairly regular intervals from nearly 1 down to almost $\frac{1}{3}$.

It is, alas, all too easy to ascribe similar knowledge to "the Pythagoreans"; taking Proclus's word for it merely begs the question. It is then no less easy to assert, either that they

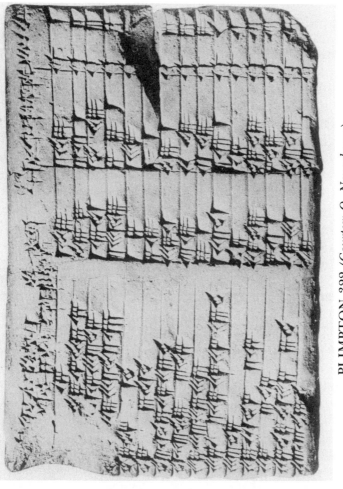

PLIMPTON 322 (*Courtesy O. Neugebauer*)

re-discovered it, or (more plausibly perhaps) that it came
to them through unknown channels from Mesopotamia.
Certain it is that the above formulas for constructing pythagorean triples were known to Euclid, who gave a proof
for them in a lemma to *Eucl*.X.29; five centuries later, Diophantus is so familiar with them that he has a technical term
($\pi\lambda\acute{\alpha}\sigma\sigma\varepsilon\iota\nu$, "to shape") for the construction of the triangle
$(2pq, p^2 - q^2, p^2 + q^2)$ from a pair of integers (p, q). Whatever
may have been the fate of that topic during the Middle
Ages, it is in Diophantine garb that it emerges again in
BOMBELLI's *Algebra* of 1572, whose Book III is almost
entirely based on Bombelli's perusal of a manuscript of
Diophantus in the Vatican library. The topic is then taken
up again by VIÈTE, at first in his *Notae priores* (copies of
which circulated long before they were printed in 1631)
and then in his *Zetetica* of 1593; both are also largely based
on Diophantus, whom Viète could have read in Xylander's
Latin translation of 1575 but had more probably studied
in a Greek manuscript in the royal library in Paris. Viète's
work, which exhibits more originality than Bombelli's in its
handling of Diophantine material, will be discussed in greater
detail further on.

§VI.

Whether one is interested only in pythagorean triangles,
or more generally in sums of two squares, there inevitably
comes a point where the algebraic identity

$$(3) \qquad (x^2 + y^2)(z^2 + t^2) = (xz \pm yt)^2 + (xt \mp yz)^2$$

has to play a role. In particular, because of the double sign
in the right-hand side, it appears whenever one wishes to
construct numbers which can be written as the sum of two
squares in more than one way. For $z = t = 1$, one gets the
special case

$$(4) \qquad 2(x^2 + y^2) = (x + y)^2 + (x - y)^2,$$

which, dressed up in geometric garb, appears in *Eucl*.II.9–10.

The identity (3) must have been familiar to Diophantus, as shown by the following passage (*Dioph*.III.19):

"*It is in the nature of 65 that it can be written in two different ways as a sum of two squares, viz., as 16 + 49 and as 64 + 1; this happens because it is the product of 13 and 5, each of which is a sum of two squares.*"

Since Diophantus has several references to a lost book of "porisms" (i.e., auxiliary theorems), and since the case $y = t = 1, x - z = \pm 1$ of (3), i.e., the identity

$$(x^2 + 1)(z^2 + 1) = (xz + 1)^2 + 1$$

or rather, equivalently:

$$x^2 z^2 + x^2 + z^2 = (xz + 1)^2$$

for $x - z = \pm 1$, seems implicit in one such reference (*Dioph*.V.5; cf. *Dioph*.III.15), it is perhaps not too far-fetched to imagine that (3) may have been one of his porisms. This must have been Bachet's view when he included (3) as Prop.III.7 of his own Porisms in his *Diophantus*. However that may be, while it would seem unreasonable, in view of the above-quoted passage, to deny a knowledge of (3) to Diophantus, no explicit statement of that identity appears to be known until we find one, with an elaborate proof for it, in Fibonacci's *Liber Quadratorum* of 1225 (*Leon*.II.257–260 = *LVE*.prop.IV). Fibonacci claims no credit for it, and seems rather to treat it as a result, well-known to specialists but deserving wider diffusion.

In contrast with Fibonacci's more popular *Liber Abaci*, his *Liber Quadratorum* fell into deep oblivion; it was with great difficulty that a copy of it was located by prince Boncompagni, who published it in 1856. As noted above, Bombelli, in his *Algebra* of 1572, took up many of the problems he had copied out of Diophantus; in particular, problem CXXVI of his Book III is translated from *Dioph*.III.19 and includes the passage quoted above about 65 being twice a sum of two squares; but he has no comment to indicate whether he understood what was involved. The decisive step was taken by Viète, who applied (3) to the construction of two new rectangular triangles from two given ones, the operation being called by him "*synaeresis*" or "*diaeresis*" according to the sign used in (3) (cf. *Notae priores*, prop.XLVI = *Op.* p. 34; *Zet.*IV.2 = *Op.* p. 62–63). Incidentally, Viète, far from confining himself to pythagorean triples and to integers, is well aware that his constructions and identities hold for arbitrary quantities. He also does not fail to point out (*Notae priores*, prop.XLVIII) the relation between (3) and the addition and subtraction formulas for trigonometric functions, a relation which becomes clear to us if we put

$$x = r \cos \alpha, \quad y = r \sin \alpha, \quad z = s \cos \beta, \quad t = s \sin \beta,$$

and hence:

$$xz \pm yt = rs \cos (\alpha \mp \beta), \quad xt \mp yz = rs \sin (\alpha \mp \beta);$$

and he introduces the term of "angle-duplication" for the case $(x, y) = (z, t)$. As this shows, Viète was interested in algebra and trigonometry much more than in number theory (cf. also his *Theoremata ad Angul.Sect.* = *Op.* pp. 287–304).

§VII.

A few more words about Fibonacci and his *Liber Quadratorum* will not be out of place here. Fibonacci, also known as Leonardo Pisano, was a well-traveled, well-informed and undoubtedly well-read mathematician. As a member of the international merchant community into which he was born, he was at home in the Mediterranean world from East to West. Born in Pisa, presumably around 1170, he had lived in North Africa and in Constantinople and had at least visited Sicily, Provence, Syria and Egypt during the latter part of the twelfth century, gathering information wherever he went from the most learned mathematicians of the day, before settling down to an enviable position as mathematical expert and teacher in his native city. When the emperor Frederic II sojourned there, Leonardo was of course brought to him and introduced into the court circle, at that time a meeting point for Latin, Arabic and Greek culture; it is hardly fanciful to surmise that Leonardo himself had some degree of familiarity with all three languages. In the fashion of the times, he held disputations in Frederic's presence with the learned men of the imperial entourage. On one occasion he was challenged to find three squares in arithmetic progression with the difference 5, i.e., in modern algebraic shorthand, to solve

$$y^2 - x^2 = z^2 - y^2 = 5$$

in rational numbers, or, what amounts to the same:

$$Y^2 - X^2 = Z^2 - Y^2 = 5T^2$$

in integers. This became the object of the *Liber Quadratorum*.

Problems about squares in arithmetic progression are quite old; because of identity (4), i.e., because of *Eucl*.II.9–10, finding such squares is equivalent to finding pythagorean triangles. In fact, if X^2, Y^2, Z^2 are in arithmetic progression, we have

$$Y^2 = \tfrac{1}{2}(X^2 + Z^2) = U^2 + V^2$$

with $U = \tfrac{1}{2}(X+Z)$, $V = \tfrac{1}{2}(Z-X)$, and the difference $Y^2 - X^2 = Z^2 - Y^2$ has the value $\tfrac{1}{2}(Z^2 - X^2) = 2UV$, which is four times the area of the triangle (U, V, Y). There is nothing to tell us that this was known to Leonardo's challenger, or to Leonardo himself. But is it mere coincidence that the problem "Find a pythagorean triangle of area $5m^2$" occurs in a Byzantine manuscript of the eleventh or twelfth century, now in the Old Palace library in Istanbul? It has been published by Heiberg (*Bibl.Math.*(III) 8 (1907–08), p. 122); Leonardo could have seen it in Constantinople. The writer of that problem knows that the area of a pythagorean triangle is a multiple of 6; "therefore", he says, "we must take for m^2 a multiple of 6"; taking $m = 6$, he writes down, without any further explanation, the triangle (9, 40, 41), of area $180 = 5 \times 6^2$. The corresponding squares in arithmetic progression would be 31^2, 41^2, 49^2, which already occur as such in *Dioph*.III.7.

On the other hand, Leonardo applies to his problem an

attractive and apparently original method, which consists in treating systematically each integral square n^2 as the sum of the first n odd integers, this being illustrated at each step by suitable diagrams. His first main result (*Leon*.II.265–271 = *LVE*.prop.XI–XII) amounts to saying that $Y^2 - X^2 = Z^2 - Y^2 = D$ has a solution in integers X, Y, Z if and only if D is of the form

$$D = 4d^2pq(p^2 - q^2) = 4d^2pq(p+q)(p-q)$$

where d, p, q are as in (2) above; of course this is four times the area of the triangle given by (2). Then he shows that D must be a multiple of 24, and finally he solves his problem by taking $d = 1, p = 5, q = 4$, corresponding to $D = 5 \times 12^2$ and to the same squares $31^2, 41^2, 49^2$ as above. In conclusion, he observes that other integers besides 5 have the same property (e.g. 7, as appears by taking $p = 16, q = 9$), but that many do not, and he asserts that D can never be a square, giving for this a totally inadequate reason; that this is indeed so was to be one of Fermat's major discoveries (cf. Chap.II, §X).

§VIII.

Just as the identity (3) plays a major role in any discussion of sums of two squares, the identities

(5) $(x^2 - Ny^2)(z^2 - Nt^2) = (xz \pm Nyt)^2 - N(xt \pm yz)^2$

(6) $(x^2 + Ny^2)(z^2 + Nt^2) = (xz \pm Nyt)^2 + N(xt \mp yz)^2$

where N is a positive integer, have a similar role to play in problems depending upon the quadratic forms $X^2 \mp NY^2$. Of course, from a modern algebraic standpoint, they do not really differ from (3), but this could not have been fully realized until the eighteenth century. For us, perhaps the easiest way of verifying (5) and (6) is to write

$$(x + y\sqrt{N})(z \pm t\sqrt{N}) = (xz \pm Nyt) \pm (xt \pm yz)\sqrt{N},$$

or the same identity with $\sqrt{-N}$ substituted for $\sqrt{N}$, and then to multiply both sides by the conjugate quantity obtained by substituting $-\sqrt{\pm N}$ for $\sqrt{\pm N}$; this was first pointed out by Euler in his *Algebra* of 1770 (*Eu*.I-1.422, Art.175). Since Euclid devotes the whole of his Book X to the theory of quadratic irrationalities, one feels tempted to imagine that he, or one of his successors, might have considered some such derivation for (5), or at least for special cases of (5). Not only is Euclid himself well aware of the relation

$$(\sqrt{r}+\sqrt{s})(\sqrt{r}-\sqrt{s}) = r - s,$$

but even the identity

(7) $$\frac{1}{\sqrt{r} + \sqrt{s}} = \frac{\sqrt{r}}{r - s} - \frac{\sqrt{s}}{r - s}$$

may be regarded as the essential content of prop. 112 of that book. Unfortunately, Euclid's motivation in Book X seems to have been the wish to construct a general framework for the theory of regular polygons and polyhedra, and not, as modern mathematicians would have it, an algebraic theory of quadratic fields. So we are left to speculate idly whether, in antiquity or later, identities involving square roots may not have been used, at least heuristically, in arithmetical work. Certain it is that, as late as the eighteenth century, Euler and Lagrange found it appropriate to congratulate each other on the original idea of introducing imaginary square roots into number theory (cf. *infra*, Chap.III, §XIV).

One is on somewhat firmer ground in assuming that problems of the type of $x^2 - Ny^2 = \pm m$, for given positive integers N and m, must have occurred rather early in Greek mathematics, presumably in connection with the problem of obtaining good rational approximations for $\sqrt{N}$ when N is not a square. Clearly, if $x^2 - Ny^2 = \pm m$ and if x and y are large in comparison with m, the ratio of x to y gives a good approximation for $\sqrt{N}$, as is apparent from the identity

$$\frac{x}{y} - \sqrt{N} = \frac{1}{y}\frac{x^2 - Ny^2}{x + y\sqrt{N}}$$

which is a special case of (7), i.e., of *Eucl.*X.112. Thus, when

Eutocius, in his commentary on Archimedes (*Arch.*III.234 and 246), wishes to verify the validity of Archimedes' approximations 265:153 and 1351:780 for $\sqrt{3}$, he merely writes:

$$265^2 - 3 \times 153^2 = -2, \ 1351^2 - 3 \times 780^2 = 1.$$

Now (5) gives an easy method for constructing such solutions of $x^2 - Ny^2 = \pm m$ as soon as one knows one of them and a solution of $p^2 - Nq^2 = \pm 1$, or, under suitable circumstances, $= \pm 2$ (cf. *infra*, §IX). For instance, Archimedes' approximate values for $\sqrt{3}$ are best explained by assuming that he was applying systematically the formula

$$(5x + 9y)^2 - 3 \times (3x + 5y)^2 = -2 \times (x^2 - 3y^2),$$

i.e., the special case $N = 3, z = 5, t = 3$ of (5) (cf. W. Knorr, *Arch.f.Hist.of Ex.Sc.* 15 (1975), pp. 137–138). Here one should note that, if $x^2 - 3y^2 = -2$, both x and y must be odd and $5x + 9y$ and $3x + 5y$ are even; the above relation becomes then

$$\left(\frac{5x + 9y}{2}\right)^2 - 3\left(\frac{3x + 5y}{2}\right)^2 = -\tfrac{1}{2}(x^2 - 3y^2).$$

In fact, Archimedes' approximations for $\sqrt{3}$ are all obtained by starting with $5^2 - 3 \times 3^2 = -2$ and applying alternately the substitutions

$$(x, y) \rightarrow \left(\frac{5x + 9y}{2}, \frac{3x + 5y}{2}\right)$$

and

$$(x, y) \rightarrow (5x + 9y, 3x + 5y).$$

Another example of the same method, presumably of earlier date but first attested in the second century A.D. by Theon of Smyrna, is given by the so-called "side and diagonal numbers", i.e., the successive solutions of $x^2 - 2y^2 = \pm 1$ derived from the obvious one $x = y = 1$ by iterating the substitution

$$(x, y) \rightarrow (x + 2y, x + y).$$

The effectiveness of this procedure depends upon the identity

$$(x+2y)^2 - 2 \times (x+y)^2 = -(x^2 - 2y^2);$$

this is the special case $N = 2$, $z = t = 1$ of (5), but it is also equivalent to (4), i.e., to $Eucl.$II.9–10. The name given to those numbers emphasizes the relationship with the ratio $\sqrt{2}$ of the diagonal of the square to its side.

We again find at least traces of the above identities in Diophantus, where cases $y = 1$, $t = 1$ or 2, $xt - z = \pm 1$ of both (5) and (6) seem implicit in $Dioph.$V.3 and $Dioph.$V.4; in the former passage there is an explicit reference to the Porisms, so that we are again left to surmise that those Porisms may have included many such identities, perhaps even (5) and (6); that they contained quite sophisticated algebraic identities is clear from the references in $Dioph.$V.5 and V.16. While this must remain conjectural, it is a fact that we find an explicit statement of (5) in the work of the Indian astronomer and mathematician Brahmagupta in the seventh century, of course in connection with problems of the form $x^2 - Ny^2 = \pm m$ (*Col.*, p. 363; cf. Datta and Singh, *loc.cit.*, vol.II, pp. 146–148). Here, too, the question of a Greek origin for (5) may be raised but cannot be answered.

§IX.

In 1773, Lessing, a well-known German literary figure and librarian of the famous Wolfenbüttel library, published a Greek "epigram", i.e., a short poem (in 22 distichs) which he had recently discovered in one of the manuscripts under his care. The manuscript describes it as a problem sent by Archimedes to the mathematicians in Alexandria.

Many mathematical epigrams are known. Most of them state problems of little depth; not so Lessing's find; there is indeed every reason to accept the attribution to Archimedes, and none for putting it in doubt. Only the question of its interpretation must remain open.

We are dealing here (*Arch.*II.528–534) with a problem

in 8 unknown integers; we shall denote them by $x, y, z, t,$ x', y', z', t'. The first 8 distichs state, in fairly standard form, a set of three equations:

$$x - \left(\frac{1}{2} + \frac{1}{3}\right)y = y - \left(\frac{1}{4} + \frac{1}{5}\right)z = z - \left(\frac{1}{6} + \frac{1}{7}\right)x = t;$$

this, of course, bespeaks a considerable degree of familiarity with systems of linear equations, but otherwise deserves no comment. Up to a factor, the solution is

$$x_0 = 2226, \quad y_0 = 1602, \quad z_0 = 1580, \quad t_0 = 891.$$

Next we have in 5 distichs the four equations:

$$x' = \left(\frac{1}{3} + \frac{1}{4}\right)(y+y'), y' = \left(\frac{1}{4} + \frac{1}{5}\right)(z+z'),$$

$$z' = \left(\frac{1}{5} + \frac{1}{6}\right)(t+t'), t' = \left(\frac{1}{6} + \frac{1}{7}\right)(x+x').$$

Here things are a little more complicated, but the scholiast or his source (of unknown date, of course) is able to obtain a solution. Since x', y', z', t' must be integers, the conclusion is that we must have:

$$(x, y, z, t) = n \times 4657 \times (x_0, y_0, z_0, t_0)$$

with some integer n; x', y', z', t' can be calculated correspondingly. In the numerical solution given by the scholiast, we have (unnecessarily, of course) $n = 80$.

So far this was linear algebra; "if you can solve the problem up to this point", says the author, "no one will call you ignorant, but this does not yet make you an expert; come and tell me the following"; whereupon he states, in two distichs, that $x + y$ has to be a square, and, in two more distichs, that $z + t$ has to be a triangular number; "if you do that", he concludes, "you win the prize for supreme wisdom". He may well say so; it can be shown that the smallest solution is of the order of magnitude of 10^{103275}.

In terms of the integer n, in fact, we can write $x + y = 4An$, where A turns out to be squarefree, so that we must have $n = AU^2$, where U is an integer. On the other hand, if $z + t$ is a triangular number, i.e., of the form $\frac{1}{2}m(m+1)$, then $8(z+t) + 1$ must be a square V^2; as we can put $z + t = Bn$, we get $V^2 - NU^2 = 1$ with $N = 8AB$. The problem thus amounts to a special case of "Pell's equation".

How are we to understand this? At the very least, it must have been meant to indicate that Archimedes was interested in such equations; perhaps also that he had some insight into the structure of the set of solutions; this, in turn, would presuppose a knowledge of the identity (5). At best, it could mean that he had devised a general method of solution for equations of the type $x^2 - Ny^2 = 1$; this would have had to depend, explicitly or not, upon the construction of the continued fraction for $\sqrt{N}$, i.e., upon the Euclidean algorithm as it applies to such a square root. Tannery seems to have favored the latter alternative and to have entertained the hope that the solution would some day re-appear with the lost books of Diophantus; so far no evidence for this has been forthcoming. Equations of the type $x^2 - Ny^2 = 1$ do occur in Diophantus (e.g. in $Dioph$.V.9 and 11), but it is a rational solution that is asked for, even when accidentally a solution in integers is obtained (e.g. in $Dioph$.V.9, where N is of the form $m^2 + 1$, giving $y = 2m, x = 2m^2 + 1$).

With the Indians we are on firm ground again, in spite of the huge gaps in our knowledge; some of the main texts have been known in the West, in English translation, since Colebrooke published his *Algebra with Arithmetic and Mensuration, from the Sanscrit of Brahmegupta and Bháscara* (London 1817). In the work of Brahmagupta, dating back to the seventh century, we find a whole section (Chap.XVIII, §7; *Col*.pp.363–372; cf. Datta and Singh, *loc.cit.* pp.146–161) devoted to equations $Ny^2 + m = x^2$, where N is a positive integer (tacitly assumed not to be a square), m is a positive or negative integer, and a solution (x, y) in integers is asked for; one should note that Brahmagupta and his successors are quite conversant with positive and negative numbers and 0, and that they state in full the rules for handling them. Also the technical vocabulary introduced by Brah-

ALGEBRA,

WITH

ARITHMETIC AND MENSURATION,

FROM THE

SANSCRIT

OF

BRAHMEGUPTA AND BHASCARA.

TRANSLATED BY

HENRY THOMAS COLEBROOKE, Esq.

F. R. S.; M. LINN. AND GEOL. SOC. AND R. INST. LONDON; AS. SOC. BENGAL,
AC. SC. MUNICH.

LONDON:

JOHN MURRAY, ALBEMARLE STREET.

1817.

magupta in connection with those equations has been adopted, with minor variations, by all his successors; the type of problem in question is known to them as *vargaprakṛti* ("square-natured"); N is the *guṇaka* ("qualifier") or *prakṛti* ("nature"); m is the *kṣepa* ("additive"); y and x are the "first", "minor" or "junior" root, and the "second", "major" or "senior" root, respectively. It is in that section (verses 64–65) that we find the identity (5), given in the form of two laws of composition

$$((x, y; m), (z, t; n)) \rightarrow (xz \pm Nyt, xt \pm yz; mn),$$

where for brevity we write $(x, y; m)$ for any triple of integers satisfying $x^2 - Ny^2 = m$, the "qualifier" N remaining the same once for all. In the subsequent literature, these laws became known as the *bhāvanā* ("production") rules; the *bhāvanā* is called positive or negative according as the sign is $+$ or $-$, and equal or unequal according as $(x, y) = (z, t)$ or not. Thus these terms correspond exactly to Viète's *synaeresis*, *diaeresis* and angle-duplication (cf. *supra*, §VI).

As Brahmagupta explains, the *bhāvanā* can then be used in various ways in order to derive new solutions from known ones; for instance, composition with a triple $(p, q; 1)$ will generate an indefinite number of solutions for a given "additive" m when one is known. Similarly, when one has obtained, e.g. by composing a triple $(x, y; m)$ with itself, a triple $(X, Y; M)$ with $M = m^2$, this gives a rational solution $(X/m, Y/m)$ for the equation with the additive 1, and it gives a triple $(X/m, Y/m; 1)$ if X/m and Y/m are integers. More generally, if $M = \mu m^2$, one can derive, from any triple $(X, Y; M)$, a triple $(X/m, Y/m; \mu)$ provided $X/m, Y/m$ are integers (Datta and Singh, *loc.cit.*, pp.150–151).

These remarks, in conjunction with the *bhāvanā* rules, already enable Brahmagupta to solve Pell's equation $x^2 - Ny^2 = 1$ in a variety of cases (for instance $N = 92$ and $N = 83$), and to give rules for its solution whenever a triple $(p, q; m)$ is known with $m = -1$, $m = \pm 2$ or $m = \pm 4$. For $m = -1$ or $m = \pm 2$, it is in fact enough to compose $(p, q; m)$ with itself once, the solution being

$$(p^2 + Nq^2, 2pq; 1) \text{ resp. } (\tfrac{1}{2}(p^2 + Nq^2), pq; 1);$$

similarly, we have the solution

$$(\tfrac{1}{4}(p^2 + Nq^2), \tfrac{1}{2}pq; 1)$$

if $m = \pm 4$ and p is even. On the other hand, if $m = \pm 4$ and p is odd, one has to compose $(p, q; m)$ twice with itself to obtain a triple $(P, Q; \pm 1)$, then, if the sign is $-$, to compose $(P, Q; -1)$ with itself; for these two cases, Brahmagupta gives explicit formulas, with polynomials in p, q of degree 3 in one case and 6 in the other.

With this, one is still far short of the general solution; since Colebrooke it has been known that such a one is to be found in the work of Bhāskara (twelfth century; *Col.* pp. 170–184; cf. Datta and Singh, *loc.cit.* pp.161–172), but an almost identical presentation has now been found, attributed to an otherwise unknown author Jayadeva, in a commentary of the eleventh century (cf. K. S. Shukla, *Gaṇita* 5 (1954), pp.1–20). Both authors describe the method as "the cyclic process" (*cakravāla*, from *cakra*, a wheel). Its true originator remains unknown.

As is the case with many brilliant discoveries, this one can be seen in retrospect as deriving quite naturally from the earlier work. For a given N, assume that we have a triple $(p, q; m)$ where m, in some sense, is small; one wishes to derive from this another such triple. To this end, one constructs a triple $(x, y; M)$ with $M = mm'$ and m' small; composing it with the first one, we get a triple $(X, Y; m^2m')$. If now X and Y are multiples of m, this gives us a triple $(p', q'; m')$ with $p' = X/m$, $q' = Y/m$, and we can proceed in the same manner, hoping eventually to reach a triple $(u, v; 1)$. In the *cakravāla*, this is achieved by taking $y = 1$ and therefore $M = x^2 - N$, where x remains to be suitably chosen. This gives

$$X = px + Nq, \quad Y = p + qx.$$

We may assume that q was prime to m; were this not so, the relation $p^2 - Nq^2 = m$ would show that p and q have a g.c.d. $d > 1$, and m would be a multiple of d^2, so that the triple

$(p, q; m)$ could be replaced by $(p/d, q/d; m/d^2)$. We can now, by the *kuṭṭaka* method (cf. *supra*, §IV), determine x so that $Y = p + qx$ is a multiple of m. Writing

$$q^2M = q^2x^2 - Nq^2 = q^2x^2 - p^2 + m$$
$$= m \times \left(\frac{qx+p}{m} \times (qx-p) + 1 \right),$$

we see that M is a multiple of m, since q is prime to m. Now the relation $X^2 = NY^2 + mM$ shows that X^2 is a multiple of m^2, so that X is a multiple of m. Writing $X = mp'$, $Y = mq'$, we get the triple $(p', q'; m')$.

To make m' small, observe that we can choose x, within its congruence class modulo m, such that $x < \sqrt{N} < x + |m|$; if then $\sqrt{N} + x$ were <0, we would have $2\sqrt{N} < |m|$; if therefore we assume $|m| < 2\sqrt{N}$, this cannot happen, and we get

$$0 < N - x^2 = (\sqrt{N}-x)(\sqrt{N}+x) < 2|m|\sqrt{N}$$

and therefore also $|m'| < 2\sqrt{N}$. Now we can apply the same procedure to $(p', q'; m')$, obtaining a triple $(p'', q''; m'')$ and so on. As the integers $m, m', m'', \ldots$ are bounded, they must repeat themselves; perhaps this was the origin of the name given to the "cyclic" method.

This calls for several observations. Firstly, the facts for which we have just given proofs may have been known to the Indians only experimentally; there is nothing to indicate whether they had proofs for them, or even for part of them. Secondly, in order to carry out the *cakravāla*, one must have a starting point; invariably they choose as such the triple $(p_0, 1; m_0)$ for which p_0^2 is the closest square to N, above or below; it is easy to see that $|m_0|$ is then $<2\sqrt{N}$. That being so, the *kuṭṭaka* is of course not needed at the first step, since $q_0 = 1$; but it is also not needed in any further operation, since, if notations are as above, we have $p' - q'x = -qm'$, so that at the next step one has to take $x' \equiv -x \pmod{m'}$. Strangely enough, this does not seem to have been noticed by any of our Indian authors (nor even by their later commentators, down to the sixteenth century); they make no mention of it, and invariably refer to the *kuṭṭaka* for the choice of x, even though their abundant numerical evidence could easily have convinced them that this was unnecessary. Thirdly, the Indian prescription for choosing x within its congruence class modulo m is not quite what we have indicated above, since the rule is to make $N - x^2$ "small" (i.e., in actual practice, as small as possible), but, as the context shows, in absolute value; in other words, one should substitute $x_1 = x + |m|$ for x if $x_1^2 - N$ happens to be less than $N - x^2$. It can be shown that this has merely the effect of abbreviating the procedure somewhat when that is the case. Finally, we are told to iterate the process only till we find an "additive" m with one of the values ± 1, ± 2, ± 4, and then to make use of the *bhāvanā*, i.e., of Brahmagupta's procedure for that case. Actually this is no more than a shortcut, since it can be shown that the *cakravāla*, applied in a straightforward manner, would inevitably lead to a triple $(p, q; 1)$ as desired; while this shortcut is quite effective from the point of view of the numerical solution, it destroys the "cyclic" character of the method, which otherwise would appear from the fact that the "additives" m, m', m'', ... would repeat themselves periodically, corresponding to the periodicity of the continued fraction of $\sqrt{N}$ (cf. Chap.III, §XII).

For the Indians, of course, the effectiveness of the *cakravāla* could be no more than an experimental fact, based on their treatment of a great many specific cases, some of them of considerable complexity and involving (to their delight, no doubt) quite large numbers. As we shall see, Fermat was the first to perceive the need for a general proof, and Lagrange the first to publish one. Nevertheless, to have developed the *cakravāla* and to have applied it successfully to such difficult numerical cases as $N = 61$ or $N = 67$ had been no mean achievement.

§X.

When Fermat embarked upon his career as a number-theorist, he had little to go by (apart from Euclid) except

Diophantus, in Bachet's edition, and Viète's masterly presentation of a large portion of Diophantus in his *Zetetica* of
1593.

As he had later occasion to observe, neither Viète nor
even Diophantus were really number-theorists. "Number
theory", he wrote in his challenge of 1657 to the English
mathematicians (*Fe*.II.335) "deals with integers; Diophantus,
on the other hand, deals with rational numbers, and Viète,
by extending Diophantus's work to continuous quantities,
has made it clear that it does not really belong to number
theory". Fermat then goes on to assign the work of Viète
to "geometry", and classifies that of Diophantus as "close
to geometry".

From our modern point of view, things look somewhat
differently. Firstly, since so much of Diophantus, and even
more of Viète, remains valid over arbitrary fields, we would
classify this primarily as algebra; of course Viète's algebra,
both in its notations and in its content, is far more advanced,
and much closer to ours, than that of Diophantus. Secondly,
the distinction between rational numbers and integers is
not as clearcut as the above quotation from Fermat would
suggest; as he knew very well, it does not apply when one
is dealing with homogeneous equations. Thirdly, as will be
seen presently, there is much, in Diophantus and in Viète's
Zetetica, which in our view pertains to algebraic geometry.
Moreover, modern developments have led to a better understanding of the analogies (already dimly perceived on
some occasions by Leibniz and Euler) between function-
fields and number-fields, showing that there is sometimes
little difference between solving a problem in rational numbers and solving it in a field of rational functions.

All these points of view acquire significance when one
tries to classify the problems and methods which occur in
Diophantus. His work, as it has been transmitted to us,
consists of not quite two hundred problems, each of which
asks for the solution, in positive rational numbers, of a given
set of algebraic equations. Diophantus has notations for one
unknown quantity at a time, and for its powers (with positive

1540   1603

FRANÇOIS VIÈTE

MAITRE DES REQUÊTES DE L'HOTEL
MEMBRE DU CONSEIL PRIVÉ DU ROI
INVENTEUR DE L'ALGÈBRE MODERNE

(D'après le Frontispice, par Rabel, de l'édition de 1630.)

or negative exponents), but not for more; if, in the course of a problem, an auxiliary problem has to be discussed, the same notation is used for the new auxiliary unknown, and is then discarded. The data frequently include one or more constants; as these cannot be designated by letters, arbitrary numerical values have to be assigned to them; moreover, the solution is usually expected to contain as many parameters as is compatible with the problem, or at least with the method of solution, but a numerical value has to be assigned to each at the moment of its appearance; this may be changed later if it turns out that it has to fulfil some condition, not satisfied by the initial choice (an obvious adaptation from the old *regula falsi*, well-known from its early use in linear problems). Occasionally a partial problem is discussed, sometimes as a lemma, with a parameter which later must play the role of the unknown; such explicitly parametric solutions are called "indefinite" (cf. e.g. *Dioph*.IV.19, where this term is defined). Of course Diophantus is aware of the existence of irrational quantities; occasionally, when an initial choice of some parameters would lead to such a value for the unknown, it is discarded by the words "the unknown becomes irrational" (cf. e.g. *Dioph*.IV.9).

It is not easy to characterize the arrangement of the problems into books, or within each book; it is also unnecessary, since we cannot be sure that this goes back to Diophantus (or whoever was the original author or compiler of our collection). As to the method, it consists (as d'Alembert put it, with his customary lucidity, in the article DIOPHANTE of the *Encyclopédie* of 1750) in "making such substitutions on the unknowns as to reduce the problem to a linear equation in one unknown" (*"manier tellement les inconnues ou l'inconnue, que le quarré et les plus hautes puissances de cette inconnue disparaissent de l'équation, et qu'il ne reste que l'inconnue au premier degré"*; *Encyclopédie*, tome IV, p. 1014). Some types of questions, however, occur with great frequency and will be examined now. It will be best to describe them first in modern terms (cf. Chap.II, §§XIV–XV and Appendix II).

Eventually, most problems are reduced to finding a point

with positive rational coordinates on an algebraic curve of genus 0 or 1, given by one or more equations. As explained above, the coefficients in those equations have to be stated numerically, but quite often they include parameters, and it is understood that the method of solution should give values to the unknowns which can be rationally expressed in terms of these. Of course it is easy to verify that this is so as soon as some modern algebraic notation, Viète's for instance, becomes available.

It is now possible to distinguish two main cases:

I. *Curves of genus* 0: the curve must be such that at least one rational point on it, in affine or in projective space, is obvious or easy to find. There would then be no problem, in the eyes of Diophantus, if this "visible" solution was in affine space and satisfied all side-conditions (for instance inequalities) required for it. If that is not so, then it becomes relevant to observe that, on such curves, rational points are everywhere dense. This is made explicit by Viète (*Zet*.IV.5 = *Op*. pp. 65–67) in the special case of the circle $X^2 + Y^2 = a^2 + b^2$; Diophantus seems aware of it, but he has no general way of expressing it, except by the use of a special word (παρισότης, which Fermat, following Xylander and Bachet, translates by *adaequatio* or *adaequalitas*) to designate his way of approximating a given number by a rational solution for a given problem (cf. e.g. *Dioph*.V.11 and 14). One typical case of a curve of genus 0, of frequent occurrence in Diophantus, is given by the plane conic $AX^2 + BX + C = U^2$, where either A, C or (as in the second lemma to *Dioph*.VI.12) $A + B + C$ is a square. One puts then $U = pX + q$ if $A = p^2$ or $C = q^2$; signs have of course to be adjusted according to the requirements of each case.

II. *Curves of genus* 1: the curve must have at least one "visible" rational point and one "visible" rational pair; this is so, for instance, for an equation

$$AX^4 + BX^3 + CX^2 + DX + E = U^2$$

provided A or E is a square (cf. e.g. *Dioph*.VI.10). More often, one meets with a "double equation" $P(X) = U^2$,

$Q(X) = V^2$, with P and Q of the form

$$P(X) = AX^2 + BX + C, \qquad Q(X) = A'X^2 + B'X + C'.$$

If P and Q have a root in common on the projective line, e.g. if $A = A' = 0$ or if $C = C' = 0$, this defines a curve of genus 0 in the (X, U, V)-space; otherwise it defines a curve of genus 1. In Diophantus, it always turns out that $P - Q$, or sometimes more generally $m^2P - n^2Q$ for suitable choices of m and n, has two rational zeros on the projective line; this is so e.g. if $A = A'$ or if $C = C'$. Then we can write $P - Q = LM$ with rational factors L, M, and the quadric $P - Q = U^2 - V^2$ has two rational systems of lines

$$U \pm V = tL, \qquad U \mp V = t^{-1}M.$$

The intersection of one such line with $P = U^2$ provides us then with a rational pair; if A and A', or C and C', are both squares, we have even two rational points. The general theory of curves of genus 1 makes it clear how further points may be derived from these. Diophantus achieves the same result by putting

$$U = \tfrac{1}{2}(tL + t^{-1}M)$$

and then adjusting the value of t so that $P = U^2$ has a rational solution (cf. e.g. *Dioph*.III.17 or IV.23). Also plane cubics fall within the scope of similar methods (cf. e.g. *Dioph*.IV.24). It should be noted, however, that Diophantus never goes beyond a single application of his process, and that this may fail to provide him with a solution satisfying required inequalities. We shall have to consider this point more closely when discussing Fermat's treatment of the same questions (cf. Chap.II, §XV).

§XI.

While the above sketch seems to cover adequately almost the whole of the extant work of Diophantus, one also meets there with a few puzzling hints, possibly suggesting a concern with quite different topics. One problem (*Dioph*.V.9) asks for a number $A = 2a + 1$ to be expressed as $x^2 + y^2$, with the side conditions $x > a, y > a$; the latter may be disregarded, in view of *Dioph.*II.9, or of the fact that rational points on a circle are everywhere dense if the circle has one rational point (cf. *supra*, §X). On the other hand, our text of *Dioph.*V.9, which makes no sense at this point and is obviously corrupt, specifies a condition for a (which, it says, should not be odd) and also seems to make some reference to prime divisors of A. Thus it implies at least that a is an integer (while ordinarily Diophantus assumes merely that the data are rational). In the solution of the problem, a is given the value 6, giving $A = 13 = 2^2 + 3^2$, and the solution is $x = \frac{257}{101}$, $y = \frac{258}{101}$. In *Dioph.*V.10, we have a similar question with other side-conditions and with $A = 9$. In *Dioph.*V.11 we have the corresponding problem with $A = 3a + 1$ and the equation $A = x^2 + y^2 + z^2$, and the text, which at this point is clear and unambiguous as it stands, merely prescribes that a be not of the form $8n + 2$, i.e., that A be not of the form $8n + 7$. To all this must be added the statement, in *Dioph.*VI.14, that 15 is not the sum of two squares and that "consequently" the equation $15x^2 - 36 = y^2$ cannot be solved in rational numbers; for these not altogether trivial statements, no explanation is given. Finally, as Bachet observed, the problems *Dioph.*IV.29 and 30 assume that any given integer (or, presumably, rational number) can be written as a sum of four squares; not surprisingly from our point of view, this is indeed so for the numerical data chosen in those problems.

One is thus left to wonder just how much Diophantus, or his predecessors, knew about the decomposition of numbers into sums of 2, 3 and 4 squares; it could be, of course, that their knowledge was merely experimental. Soon after taking up such questions, Fermat observed that the impossibility of some decompositions can be proved quite easily by using congruences modulo 4 or 8; one sees thus, for instance, that A cannot be written as $x^2 + y^2$ if it is of the form $4n + 3$, nor as $x^2 + y^2 + z^2$ if it is of the form $8n + 7$ (cf. Chap.II, §V). Such arguments are not found anywhere in Diophantus, but they cannot be said to be alien to the spirit of Greek arithmetic. Short of a chance discovery of hitherto unknown texts, the questions thus raised are unanswerable.

§XII.

For a mathematician of the sixteenth century, Diophantus was no easy text to decipher, and no small share of the credit for its re-discovery must go to Rafael BOMBELLI and to XYLANDER. Bombelli read and translated most of it in Rome, for his own use, about 1570, and then incorporated this into his Italian *Algebra* of 1572; Xylander was the first to attempt a complete translation, and published the fruit of his efforts in Basel in 1575. Of course the substantial difficulties of such undertakings were greatly increased by the comparatively poor condition of the available manuscript texts. As has been the case with virtually all Greek classical authors, these were all derived from a single codex (the so-called "archetype", now lost) marred by copying mistakes and omissions. Worst of all, in the case of Diophantus, were the numerical errors. Undoubtedly the copying had been done by professional scribes, not by mathematicians; but perhaps this was just as well; would-be mathematicians might have made things even worse.

The first translator, Holzmann, who hellenized his name as Xylander, was a humanist and a classical scholar who took up algebra as a hobby. Bombelli was a busy engineer, designing canals, desiccating marshes, and he dedicated to mathematics little more than those periods of leisure which were granted to him by his munificent employer and patron, the bishop of Melfi. When he inserted his Diophantine material into Book III of his *Algebra*, his intention seems to have been merely to enrich with it an earlier manuscript of his, composed at least ten years before. His own merits as an algebraist lay elsewhere, for instance in his treatment (quite a modern one) of complex numbers in his Book I, and in the theory of cubic and biquadratic equations in Book II. VIÈTE, on the other hand, while trained as a lawyer and employed as such for much of his life, came to regard himself, and to be regarded by his contemporaries, as a mathematician first of all. *"Ars nostra"* is the word he uses for mathematics in the preface of his *Isagoge* of 1591, the first in a series of publications which was to establish his fame as the greatest algebraist of his age. "Pure gold" is how he describes the discoveries he is about to disclose; "not alchemists' gold", he adds, "soon to go up in smoke, but the true metal, dug out from the mines where dragons are standing watch". Little did it matter to him, when he followed up the *Isagoge* with his *Zetetica* of 1593, that much of the ore for it had been supplied by Diophantus; his purpose was to demonstrate the power and scope of the new algebra; his use of Diophantus is directed wholly towards this end. In this, as Fermat rightly observed, whatever little number theory was present in Diophantus has been lost.

Thus it was still a huge task that awaited the future editor, translator and commentator of Diophantus, even after all the work done by his predecessors. As Fermat's son Samuel expressed it in his preface to the *Diophantus* of 1670 (echoing, no doubt, his father's sentiments, perhaps his very words):

"Bombelli, in his *Algebra*, was not acting as a translator for Diophantus, since he mixed his own problems with those of the Greek author; neither was Viète, who, as he was opening up new roads for algebra, was concerned with bringing his own inventions into the limelight rather than with serving as a torch-bearer for those of Diophantus. Thus it took Xylander's unremitting labors and Bachet's admirable acumen to supply us with the translation and interpretation of Diophantus's great work".

Claude Gaspar BACHET, sieur de Méziriac, was a country gentleman of independent means, with classical tastes, and no mathematician. Somehow he developed an interest for mathematical recreations and puzzles of the kind found in many epigrams of the Greek *Anthology* as well as in medieval and Renaissance mathematical texts, or nowadays in the puzzle columns of our newspapers and magazines. In 1612 he published in Lyon a collection of such puzzles under the title *Problèmes plaisants et délectables qui se font par les nombres.* As this indicates, he was thus led to number theory, and so to Diophantus. The latter must have occupied him for several years prior to its publication in 1621; after seeing it through the press he retired to his country estate, got married, and apparently gave up all mathematical activity, except that he prepared a second edition of his *Problèmes* of 1612, incorporating into it some of the materials he had intended for a treatise on arithmetic which never saw the light (cf. *supra*, §IV). Incidentally, the *Problèmes* won such applause that they have enjoyed successive editions even into the present century.

Samuel Fermat's praise of Bachet was by no means excessive. No mere philologist would have made sense out of ever so many corrupt passages in the manuscripts; Xylander had all too often failed to do so. Bachet never tires of drawing attention to the defects of Xylander's translation and comments, while naively extolling his own merits. He even ventures to speak disparagingly of Viète's algebraic methods, which he neither appreciated nor understood; this did not stop him from lifting two Porisms (III.11 and III.14) and some important problems about cubes (in his commentary to *Dioph.*IV.2) out of Viète's *Zetetica* (*Zet.*IV.12,15,18,20 = *Op.* pp. 71,73,74,75) without a word of acknowledgment. Nevertheless, his is the merit of having provided his successors and notably Fermat with a reliable text of Diophantus along with a mathematically sound translation and commentary. Even his lack of understanding for the new algebra may be said to have benefited number theory in the end. It led him, no doubt, to overload his commentary and the "porisms" with which he prefaced it with a huge ballast of ludicrously clumsy proofs. At the same time, since his chief interest was number theory and not algebra, he invariably laid the emphasis on those aspects of the text which were more properly arithmetical, and, prominently among these, on all questions regarding the decomposition of integers into sums of squares (cf. *supra*, §XI). He asked for the con-

ditions for an integer to be a sum of two or of three squares;
he extracted from Diophantus the conjecture that every
integer is a sum of four squares, and asked for a proof.

The curtain could rise; the stage was set. Fermat could
make his entrance.

PETRVS DE FERMAT SENATOR THOLOSANVS.

Chapter Two

Fermat and His Correspondents

§I.

Fermat's outwardly uneventful life is soon told. He was born into a prosperous middle-class family in Beaumont de Lomagne, a little town in the south of France, not far from Toulouse in the province of Languedoc. His mother Claire de LONG belonged to the "noblesse de robe", which means that her family had been ennobled through one or more of its members holding a magistrate's office. He was baptized Pierre FERMAT on 20 August 1601. Before 1631 he spent some time, perhaps some years, in Bordeaux. On 14 May 1631 he was received into the "parlement", i.e., the provincial High Court of Judicature, in Toulouse, with the title of councilor, an office which he held until his death; in virtue of it he was thereafter styled Monsieur de Fermat, and is so referred to by his contemporaries. On 1 June 1631 he was married to Louise de LONG, a distant cousin of his mother's. They had two sons and three daughters; the elder son Samuel became a magistrate, and, like his father, a councilor in the Toulouse High Court; his brother took orders and became a canon in the cathedral church in Castres; one daughter got married, two became nuns. Fermat's professional life was divided between Toulouse, his main residence, and Castres, which was the seat of the "Chambre de l'Edit" of the Toulouse High Court; this was the judicial instance concerned with the relations between the catholic and protestant communities of the province[1]. Fermat died in Castres on 12 January 1665 during one of his turns of office there.

Obviously Fermat had enjoyed an excellent classical education; he was well versed in Latin, Greek, Italian and Spanish, and generally praised for his skill in writing verse in several languages, a skill which he transmitted to his son Samuel. He collected manuscripts; his advice was eagerly sought on the emendation of Greek texts. With such interests, a seventeenth century gentleman might naturally think of a journey to Italy; in Fermat's case the active scientific life in that country, and the presence there of such men of science as Galileo, Cavalieri, Ricci, Torricelli, should have provided an additional attraction. Indeed some of his best friends did make the trip and visited the Italian scientists on the way: Carcavi in 1634, Beaugrand in 1635, Mersenne in 1644 and 1645. There is no indication that Fermat ever thought of following their example. In 1636, in one of his very first letters to Mersenne (*Fe.*II.14), he speaks somewhat doubtfully of "finding some occasion for spending three or four months in Paris", which would have given him, he says, the opportunity for writing up his ideas on geometrical topics; but this was not to happen. Late in life, in 1660, he expresses to Pascal the earnest wish to meet him; "as his health is little better than Pascal's" this would have to take place "somewhere halfway between Clermont and Toulouse" (*Fe.*II.450); Pascal, also for reasons of health (*Fe.*II.450–452), did not undertake the journey, which indeed, in those days, might have been a strenuous one for an invalid. In the same year, Fermat, hearing that Huygens is in Paris (*Fe.*II.452), assures him that he would go there to meet him if only his health would allow him to travel; obviously he

---

[1] A number of letters of Fermat's, touching upon his professional life, have been preserved (*Fe.*II, letters LI,LIV,LV,LIX,LXIV,LXV,LXVI, CXI, and *Fe.*IV.15–22; cf. *Fe.*III.505). Cf. also the excellent study by P. Chabbert, Fermat à Castres, *Rev.d'Hist.des Sc.* 20 (1967), pp. 337–348.

hopes that Huygens, with whom he had communicated through Carcavi on scientific matters (*Fe*.II.322,328,431, 446), would take the hint and come to visit him; if so, his expectation was not fulfilled.

Finally Fermat died without ever[2] having ventured farther from home than Bordeaux. As his letters show, his vacations were ordinarily spent "*à la campagne*", i.e., in his country house.

We have no data to tell us when Fermat became interested in mathematics, but this must have occurred at the latest during his stay in Bordeaux in the late twenties. There he met, among others, Etienne d'ESPAGNET, himself a magistrate and a magistrate's son; they soon became intimate friends (cf. e.g. *Fe*.II.71,105,136). D'Espagnet was some five years older than Fermat, was mathematically inclined, and one likes to think of him as having been the first to perceive Fermat's genius and having given him the kind of encouragement which few young men can quite do without. He had in his possession, and he communicated to Fermat, some of Viète's unpublished writings (*Fe*.II.133); presumably he had a more or less complete collection of Viète's works, which were not easily obtainable at that time; it is obvious that Fermat studied them closely, early in his career, long before they were made generally available by F. van Schooten's publication of Viète's *Opera Mathematica* in 1646. As to d'Espagnet, he remained Fermat's lifelong friend, keeping up, not only his personal relations with him, but also his keen interest in Fermat's mathematical progress (cf. *Fe*.II.221). In 1646, Mersenne had to get d'Espagnet to bring pressure on Fermat in order to "drag him away from Bergerac" for a three days' meeting in Bordeaux (*Fe*.Supp.136, note (1)).

Viète had died in Paris in 1603; in the twenties and thirties, some of his admirers were concerning themselves with the

---

[2] Or hardly ever; in 1631 he is listed as having a law degree from Orléans; but perhaps this could have been obtained *in absentiâ*.

FRANCISCI VIETÆ

OPERA

MATHEMATICA,

In unum Volumen congesta,

ac recognita,

Operâ atque studio

FRANCISCI à SCHOOTEN Leydensis,
Matheseos Professoris.

NON SOLUS

LVGDVNI BATAVORVM,

Ex Officinâ Bonaventuræ & Abrahami Elzeviriorum.

CIƆ IƆ C XLVI.

diffusion of his methods; their efforts were to culminate in the 1646 publication, referred to above. This small group included Jean de BEAUGRAND, "mathematician to H. H. Gaston d'Orléans" (whatever this may have meant), who later was to hold office as Court secretary ("secrétaire du Roy"; cf. *Fe*.IV.22, note (1), 2°); like most of the scientists or scientifically interested men in those days, he was in touch with the famous Father MERSENNE, no serious mathematician himself, but who nonetheless collaborated with him in annotating a 1631 edition of Viète's *Isagoge* and *Notae Priores*; the latter had never been published, and the former was out of print. In view of all this, when we find Beaugrand, in a 1630 letter to Mersenne (*Mers*.II.515), mentioning "*ces Messieurs de Bordeaux*", to whom he had sent a pretty little mechanical problem, it does not seem far-fetched to assume that this reference included Etienne d'Espagnet; it could well have included also Fermat if he was already known by name to Beaugrand at that time. Certain it is that, in the years that followed, and at least until 1638, a number of communications on mathematical topics passed between Fermat and Beaugrand (cf. *Fe*.II.4–5,20,28,72,94,105,106,133). During his trip to Italy in 1635, Beaugrand spoke glowingly of Fermat to the Italian scientists (cf. *Fe*.II.26, note (1)) and showed them some of Fermat's results (cf. *Fe*.Supp.99), possibly (according to contemporary hints) attributing some of the credit to himself (cf. *Fe*.II.26, note (1), *Gal*.XVI.328,345, and *Fe*.Supp.113,114).

Beaugrand died in 1640; it is not clear whether he ever met Fermat, and there are signs that the latter's warm regard for him (*Fe*.II.4–5) had cooled before 1640 (cf. *Fe*.II.111,133, and perhaps II.207). A more enduring friendship was the one between Fermat and Pierre de CARCAVI, who became his colleague at the Toulouse High Court in 1632 and remained there until 1636. In that year he transferred to Paris, where he formed a close acquaintance with the scientific circle gathered around Mersenne, Etienne Pascal and Roberval. Fermat's scientific correspondence with the members of that group begins with a letter to Mersenne a few days after Carcavi's arrival in Paris. From then on it never stopped until around 1662, when Fermat, perhaps for reasons of ill health (cf. *Fe*.II.450) allowed "his Geometry to fall into a deep sleep" (*Fe*.II.485). Clearly this correspondence, a not inconsiderable part of which has been preserved, provided Fermat with his main outside incentive for pursuing his mathematical work, since the closest he ever came to personal contact with a mathematician (apart from a hypothetical visit from Beaugrand) was the above-mentioned three days' meeting with Mersenne, which could hardly be so described. As to Carcavi, who took over, after Mersenne's death in 1648, the latter's role as an intermediary between contemporary men of science, he truthfully introduced himself to Huygens, in 1656, as "one not well versed in mathematics, but loving it passionately" (*Fe*.IV.118). Nor could Father LALOUVÈRE, the teacher of mathematics at the local Jesuits' college in Toulouse, have been of much use to Fermat; his unhappy excursion into integral calculus on the occasion of the famous prize competition of 1658 on the cycloid (cf. *Fe*.II.413,430) is gleefully made fun of by Pascal (no friend of the Jesuit order) in his *Histoire de la Roulette* of 1658. Nevertheless, we owe to him the one and only piece of writing of Fermat's that he ever allowed to be printed, the anonymous dissertation (*Fe*.I.199–254) on the rectification of curves which Lalouvère published in 1660 as an appendix to his book on the cycloid. In that respect, mention must be made also of the eight anonymous pages which J. E. Hofmann discovered in 1943, appended to a copy of Frenicle's exceedingly rare pamphlet of 1657 on "Pell's equation" and other topics[3]; this consists of a letter to Frenicle and one to Digby and must have been printed for private use without Fermat's knowledge, since apparently it was unknown even to his son at the time of the preparation of the *Varia Opera* of 1679.

---

[3] J. E. Hofmann, Neues über Fermats zahlentheoretische Herausforderungen von 1657, *Abh.d.preuss.Akad.d.Wiss.* 1943–1944, Nr.9, pp. 41–47.

DIOPHANTI
ALEXANDRINI
ARITHMETICORVM
LIBRI SEX,
ET DE NVMERIS MVLTANGVLIS
LIBER VNVS.

CVM COMMENTARIIS C. G. BACHETI V. C.
& obſeruationibus D. P. de FERMAT Senatoris Toloſani.

Acceſſit Doctrinæ Analyticæ inuentum nouum, collectum
ex varijs ciuſdem D. de FERMAT Epiſtolis.

OBLOQVITVR NVMERIS SEPTEM DISCRIMINA VOCVM

TOLOSÆ,
Excudebat BERNARDVS BOSC, è Regione Collegij Societatis Ieſu.

M. DC. LXX.

The question of the publication of Fermat's work was indeed raised on several occasions during his lifetime. As we have seen (Chap.I, §I), he once expressed a firm resolve of writing up his number-theoretical discoveries in book-form (*Fe.*I.305, Obs.XVIII). In 1654 he requested Carcavi's and Pascal's cooperation to a more ambitious plan: he wished them to help him prepare the bulk of his mathematical work for publication, while Pascal would have been in sole charge of the number theory, which Fermat despaired of ever writing up in full (*Fe.*II.299). This came to naught; but in 1656 Carcavi mentions the matter to Huygens (*Fe.*IV.120); he does so once more in 1659 (*Fe.*IV.126), in spite of Fermat's rather discouraging and discouraged conclusion to the summary of his arithmetical discoveries ("*mes resveries*", as he calls them) in his communication of August 1659 (*Fe.*II.436). Finally nothing was done before his death.

Actually, in those days, it was not quite a simple matter for a mathematician to send a work to the press. For the printer to do a tolerable job, he had to be closely supervised by the author, or by someone familiar with the author's style and notation; but that was not all. Only too often, once the book had come out, did it become the butt of acrimonious controversies to which there was no end. Should one wonder, then, if Fermat, whenever the question of publication arose, insisted on anonymity above all (*Fe.*II.106,299)? At the same time, it is clear that he always experienced unusual difficulties in writing up his proofs for publication; this awkwardness verged on paralysis when number theory was concerned, since there were no models there, ancient or modern, for him to follow.

Fermat never bothered to keep copies of his scientific communications to his correspondents (cf. e.g. *Fe.*II.218). Thus, after his death, it fell to the lot of his son Samuel to bring together, as best he could, the scattered remnants of his father's writings. He began with the *Diophantus* of 1670, a reprint of Bachet's *Diophantus* of 1621 where he inserted the full text of the notes jotted down by Fermat in the margins of his personal copy. To this volume the Jesuit father Jacques de BILLY, a former friend of Bachet's and a teacher of

VARIA OPERA
MATHEMATICA
D· PETRI DE FERMAT,
SENATORIS TOLOSANI.

Accefferunt felectæ quædam ejufdem Epiftolæ, vel
ad ipfum à plerifque doctifsimis viris Gallicè, Latinè,
vel Italicè, de rebus ad Mathematicas difciplinas,
aut Phyficam pertinentibus fcriptæ.

TOLOSÆ,

Apud JOANNEM PECH, Comitiorum Fuxenfium Typographum, juxta
Collegium PP. Societatis JESU.

M. DC. LXXIX.

mathematics in Dijon, contributed a 36-page essay, the *Doctrinae Analyticae Inventum Novum*, based on a series of letters from Fermat (now lost except for an early one of 1659: *Fe*.II.436–438); it describes in some detail Fermat's methods for handling diophantine equations of genus 1 (cf. Chap.I, §X, and *infra*, §XV), and naively extolls Fermat's merits far above those of Diophantus, Viète and Bachet. This was followed in 1679 by the volume of *Varia Opera*, the bulk of which consisted of Fermat's writings on geometry, algebra, differential and integral calculus, together with a number of letters to and from Mersenne, Roberval, Etienne Pascal, Frenicle, Blaise Pascal, Carcavi, Digby, Gassendi. Not a few letters, among them some important ones on "numbers", were not included, obviously because their recipients failed to send them to Samuel; thus they remained unknown until our times.

Now, with the publication of Fermat's complete works in 4 volumes by P. Tannery and Ch. Henry (Paris 1891–1912; *Supplément* by C. de Waard, 1922), to which must be added a few fragments in Mersenne's *Correspondance* and J. E. Hofmann's publication mentioned above, it is to be presumed, barring further discoveries, that we possess the entire Fermat *corpus,* tantalizingly incomplete as it must appear.

§II.

In fulfilment of our program, only Fermat's work on "numbers" will be considered here; we shall leave aside, not only his extensive writings on geometry and the differential and integral calculus, but also his important work on probabilities (contained in his correspondence with Pascal, *Fe*.II.288–312, and with Huygens, *Fe*.II.320–331) and his contributions to algebra, viz., elimination-theory (*Fe*.I.181–188) and his treatment of Adriaen van Roomen's cyclotomic equation (*Fe*.I.189–194).

Perhaps binomial coefficients pertain more to algebra than to number theory, even though Fermat, describing an early result of his about them (*Fe*.I.341, Obs.XLVI), com-

ments that "there can hardly be found a more beautiful or
a more general theorem about numbers". The theorem in
question concerns the integers denoted, in modern notation,
by

$$\binom{n+m-1}{m}$$

and described by Fermat and his contemporaries as "triangular" (for $m = 2$), "pyramidal" (for $m = 3$), "triangu-
lotriangular" (for $m = 4$), etc.; they are defined by the recurrent relation

$$\binom{n+m}{m+1} = \binom{n+m-1}{m} + \binom{n-1+m}{m+1} = \sum_{\nu=1}^{n} \binom{\nu+m-1}{m}.$$

Undoubtedly Fermat was aware of the role played by these
numbers in the binomial formula; he may have known about
it, if not from Stifel (whom he never read, as he says
in a different context: Fe.II.188), at least from his study of
Viète; cf. e.g. the latter's Ang.Sect., published in 1615
($= Op.Math.287-304$). Fermat's theorem is no other than
the formula

$$n \cdot \binom{n+m-1}{m-1} = m \cdot \binom{n+m-1}{m}.$$

There is of course no way of dating Fermat's marginal
notes in his Diophantus; but this one must surely have an-
tedated his somewhat complacent mention of the same result
(as "une tres-belle proposition") to Roberval in 1636 (Fe.II.84),
then again (as "propositionem pulcherrimam") to Mersenne
(Fe.II.70; for that letter we adopt the date 1638 proposed
by J. Itard, Rev.Hist.Sc. 2 (1948), pp. 95–98). This theorem
was to stand him in good stead in 1654 in his work on
probabilities; in connection with the latter topic, it was obtained independently in that same year by Pascal, who,
strangely enough (since he belonged to Roberval's circle)
had not been aware of Fermat's prior discovery (cf. Fe.II.308,
and Fe.II.70, note (1)). In the thirties, Fermat used it pri-

marily in order to find theorems on sums of the form

$$S_m(N) = \sum_{n=1}^{N} n^m$$

or more generally $\Sigma_1^N (an + b)^m$. Apparently (cf. *Fe*.II.84) he did this by writing, as above

$$\binom{N+m}{m+1} = \sum_{n=1}^{N} \binom{n+m-1}{m}$$

and then using his theorem to write

$$\binom{n+m-1}{m} = \frac{1}{m!} n(n+1) \dots (n+m-1)$$

$$= \frac{1}{m!} (n^m + A_1 n^{m-1} + \cdots + A_{m-1} n)$$

with numerical coefficients $A_1, \dots, A_{m-1}$. This gives

$$S_m(N) + A_1 S_{m-1}(N) + \cdots + A_{m-1} S_1(N)$$

$$= \frac{1}{m+1} N(N+1) \dots (N+m)$$

from which one can obtain the formulas for the sums $S_m(N)$ by induction on m. Fermat, while noting (*Fe*.II.68,84) that the case $m = 2$ had been known to Archimedes and the case $m = 3$ to Bachet, gives the formula for $m = 4$ as a sample, adding that the general case can be treated by the same method. The same approach was rediscovered by Jacob BERNOULLI (and is described in his posthumous *Ars Conjectandi* of 1713), leading him to the definition of the "Bernoulli numbers" and "Bernoulli polynomials", whose importance for number theory did not begin to appear until later at the hands of Euler. As to Fermat, after stating the above results to Roberval in 1636, he merely adds (*Fe*.II.85) that he had applied them to his calculation of the integral $\int x^m dx$ (or rather, in his language, to the quadrature of the "parabola" $y = x^m$; cf. *Fe*.II.73,94,95), in conscious imitation of one of the methods which Archimedes had used in the case $m = 2$ (cf. also *Fe*.II.83–84, and *Fe*.I.342, Obs.XLVIII).

§III.

In communicating his theorem on binomial coefficients to Roberval and to Mersenne, Fermat says nothing about the proof. On the other hand, his statement of it in the margin of the *Diophantus* ends up with the words *"demonstrationem margini inserere nec vacat, nec licet"* ("I have no time, nor space enough, for writing down the proof in this margin") quite similar to those (*"hanc marginis exiguitas non caperet"*, "the margin is too narrow for this proof") which conclude his notorious statement about the equation $X^n + Y^n = Z^n$. This coincidence, as may be noted here in passing, makes it plausible that the latter was also made rather early in Fermat's career. At any rate there cannot be much doubt about Fermat's proof for the theorem on binomial coefficients; if, as he asserts, he had worked out such a proof, it could only have been one by induction. It is of course a proof by induction that Pascal spells out for it, with almost pedantic clarity, in the *Triangle Arithmétique*, and Bernoulli does the same in the *Ars Conjectandi*. Fermat himself has at least one proof clearly worked out by induction in his *Apollonius* (Lib.I, prop.VII; *Fe*.I.26–27), a geometrical treatise written largely before 1630 and completed no later than 1636 (*Fe*.II.100).

In those days, and even much later, the word "induction" was used in an altogether different sense; it indicated no more than what is now understood by "conjecture". To verify "by induction" a statement $P(n)$, depending upon an integer n, meant merely to verify it for some (and sometimes very few) initial values of n. Of course it could happen that the procedure used, say, in order to verify $P(4)$ was to derive it from $P(3)$ in such a manner as to make it obvious that the same argument would serve to derive $P(n)$ from $P(n-1)$. In that case, which occurs not seldom in Euclid (e.g. in *Eucl*.I.45, VII.14, VIII.6, to give a few random examples) a mathematician has the right to regard such a proof as conclusive, even though it is not couched in the conventional terms which would be used for it to-day. For such proofs, it became customary to use the term of "complete" (or "mathematical") induction[4], while "incomplete" induction came to mean the heuristic argument described above.

Induction of the "incomplete" kind was used quite systematically by Wallis in his *Arithmetica Infinitorum* of 1656 and applied to a wide variety of questions, including the theory of binomial coefficients; eventually his results led to Newton's binomial series, and in the next century to Euler's theory of eulerian integrals and the gamma function. What has to be emphasized here, however, is the criticism offered by Fermat in 1657 to Wallis's use of "induction":

"One might use this method if the proof of some proposition were deeply concealed and if, before looking for it, one wished first to convince oneself more or less of its truth; but one should place only limited confidence in it and apply proper caution. Indeed, one could propose such a statement, and seek to verify it in such a way, that it would be valid in several special cases but nonetheless false and not universally true, so that one has to be most circumspect in using it; no doubt it can still be of value if applied prudently, but it cannot serve to lay the foundations for some branch of science, as Mr. Wallis seeks to do, since for such a purpose nothing short of a demonstration is admissible" ["*On se pourroit servir de cette méthode, si la démonstration de ce qui est proposé étoit bien cachée et qu'auparavant de s'engager à la chercher on se voulut assurer à peu près de la vérité; mais il ne s'y faut fier que de bonne sorte et on doit y apporter les précautions nécessaires. Car on pourroit proposer telle chose et prendre telle règle pour la trouver qu'elle seroit bonne à plusieurs particuliers et néanmoins seroit fausse en effet et non universelle. De sorte qu'il faut être fort circonspect pour s'en servir, quoiqu'en y apportant la diligence requise, elle puisse être fort utile, mais non pas pour prendre pour fondement de quelque science ce qu'on en aura déduit, comme fait le sieur Wallis; car pour cela on ne se doit contenter de rien moins que d'une démonstration . . .*": Fe.II.351–352].*

Predictably, Wallis was displeased, and, in a letter to Digby

---

[4] In French, "récurrence" (for "complete" induction) avoids the ambiguity due to the double meaning of "induction".

(*Comm.Epist.*, letter XVI = *Wal.*II.782–783), haughtily declined to answer the charge. As his results have largely been substantiated by later developments, Fermat's criticism may, in retrospect, strike us as too harsh, even though he sought to soften it by some high praise for Wallis (cf. also *Fe.*II.337,343). What concerns us here is that it seems to express Fermat's mature views about proofs. Similarly, writing to Clerselier in 1662, he states that "the essence of a proof is that it compels belief" ("*la qualité essentielle d'une démonstration est de forcer à croire*": *Fe.*II.483). This is not to deny, of course, that he was liable to error; he was so, perhaps more than others, in view of his fatal habit of seldom writing down his proofs in full and hardly ever keeping copies of his communications to his friends. Nevertheless, in view of the above quotations, when Fermat asserts that he has a proof for some statement, such a claim has to be taken seriously.

§IV.

Man's love for numbers is perhaps older than number theory; perhaps it assisted in its birth; it has ever been its faithful companion, not untainted by superstition at times (cf. Chap.I, §III). As noted above (Chap.I, §IX) the Indians' love for large numbers may have supplied a motivation for their treatment of Pell's equation. Bachet had been a number-lover before turning to Diophantus (cf. Chap.I, §XII).

Perhaps none of Fermat's contemporaries deserves to be called a number-theorist; but several were number-lovers. One of these was Mersenne, who was interested in "perfect" and "amicable" numbers; so were P. Bruslart de SAINT-MARTIN and A. Jumeau de SAINTE-CROIX, both older than Fermat, and B. FRENICLE de BESSY, who was some years younger and was to become a member of the first Academy of Sciences when it was founded by Colbert in 1666. Frenicle was once praised by Fermat, not without a touch of irony perhaps, as "this genius who, without the succor of Algebra, penetrates so deeply into the knowledge of numbers" (*Fe.*II.187).

Early in his career, in his Bordeaux period (cf. *Fe*.II.196), Fermat had been attracted by "magic squares"; these are square matrices of positive integers satisfying certain linear conditions. Stifel, Cardano, Bachet had mentioned this topic. In 1640, Fermat and Frenicle discovered through Mersenne their common interest in it; this was the beginning of a correspondence between them (cf. *Fe*.II.182–197), which, after a somewhat inauspicious beginning (*Fe*.II.182–185), soon turned to more substantial topics and, in spite of some occasional friction (cf. *Fe*.II.260) was continued, rather desultorily at times, throughout the next two decades.

It is difficult for us to take magic squares seriously, in spite of Fermat's professed enthusiasm for it (*Fe*.II.194) and of E. Lucas's intriguing suggestion (*Fe*.IV.190) that they may have led to the discovery of the fundamental identity for sums of four squares (cf. *infra*, §XIV, and Chap.III, §XI). An even more popular topic with the number-lovers of those times turned out to be far more fruitful at the hands of Fermat; it is the one which grew out of Euclid's theorem on "perfect numbers", and which was invariably referred to, in Fermat's time, under the rather unsuitable name of "aliquot parts". Somehow the term "aliquot part", of uncertain origin, came to denote any divisor, including 1 but excluding n, of an integer n, and the plural "the aliquot parts" to denote the sum $s(n)$ of such divisors. Thus "perfect" numbers are those which satisfy $s(n) = n$, and "amicable" numbers are the pairs n, m satisfying $s(n) = m, s(m) = n$. In addition to these, Mersenne and others started looking for "submultiple" numbers, i.e., solutions of $s(n) = a.n$, where a is 2, 3, or some other small integer. Even DESCARTES, around 1638, was drawn into the search for such numbers, and, with his usual arrogance, soon became the self-proclaimed champion at this game, while at the same time assuring Frenicle (a rival contender for the title) that it was less than a year since he had even heard of such matters. It was a basic tenet of Cartesian philosophy that any man, by dint of *proper* self-discipline and by *proper* use of his reasoning faculty, could construct within his own mind the whole system of everything there is to know about the material and immaterial world. The emphasis being on the word *proper*, it turned out, of course, that Descartes, and only he, had ever been capable of such a feat, and consequently that others could well learn from him, but he had learnt nothing from others and never had had to. He stoutly denied, for instance, that his algebra owed anything to Viète, directly or indirectly, and, however unlikely such assertions may appear, there is no way of checking their degree of veracity.

Actually, not a little ingenuity is required in order to obtain all the "perfect", "amicable" and "submultiple" numbers which turn up in the letters exchanged at that time between Fermat, Mersenne, Frenicle and Descartes. Fermat had started considering such questions long before 1636 (*Fe*.II.20); in 1636 he pronounced them to be very difficult, thought he had a general method for solving them all, and told Roberval that he was planning a small treatise on the subject (*Fe*.II.93); in 1643 he still speaks complacently of "excellent discoveries" about aliquot parts, but his interest in them seems to be waning (*Fe*.II.255); already by that time diophantine problems and sums of squares occupy the foreground.

To gauge the difficulty of the problem for some of Fermat's contemporaries, one must remember that, as late as 1640 (*Fe*.IV.69), Mersenne was asking Saint-Martin how to find the number of divisors of 49000 and their sum without enumerating them one by one; Euclid, and perhaps even the mathematicians around Plato (cf. Chap.I, §II) might have found this a naive question. Of course a good algebraic notation is a great help in such matters; with such notations as Descartes and Fermat had at their disposal, it is almost trivial to verify that, if p is a prime, then $s(p^n) = (p^n - 1)/(p - 1)$ for all n (a result which, in substance, is contained in *Eucl*.IX.35). If, for our convenience, we denote by $S(n)$ the sum of the divisors of n including n, i.e., $S(n) = s(n) + n$, then it is equally easy to show that $S(mn) = S(m)S(n)$ when m, n are mutually prime. Those two facts are at the bottom of Euclid's proof for his theorem on perfect numbers: $2^{n-1}(2^n - 1)$ is perfect if the second factor is a prime[5]. In the same way one may verify the rule for amicable numbers, sent by Descartes to Mersenne in 1638: $2^{n+1}(18 \times 2^{2n} - 1)$ and $2^{n+1}(3 \times 2^n - 1)(6 \times 2^n - 1)$ are amicable if the second factor of the former, and the second and third factors of the latter, are primes (*Desc.*II.93–94); undoubtedly this rule must have been known also to Fermat, and it was in fact much older (cf. *Fe.*IV.185). To quote just another example among those constructed at that time, here is one for which $s(n) = 5n$, split into its prime factors:

$$n = 2^{36} \times 3^8 \times 5^5 \times 7^7 \times 11 \times 13^2 \times 19 \times 31^2 \times 43$$
$$\times 61 \times 83 \times 223 \times 331 \times 379 \times 601 \times 757$$
$$\times 1201 \times 7019 \times 112303 \times 898423 \times 616318177.$$

It was proposed by Frenicle to Fermat, through Mersenne, in 1643 (*Fe.*II.255) and, from Fermat's response, it appears that he must have known of it already. As can readily be seen, the whole point, in constructing such numbers, lies in ascertaining whether some large numbers, particularly such of the form $2^n - 1$, $3^n - 1$, are prime, and, if not, splitting them into their prime factors. For instance, in 1640, Frenicle asked Fermat, also through Mersenne, for a perfect number between 10^{20} and 10^{22} (*Fe.*II.185); implicitly the question concerned "euclidean" perfect numbers, i.e., those of the form $2^{n-1}(2^n - 1)$, since no others were known at the time (nor indeed now). The inequality

$$10^{20} < 2^{n-1}(2^n - 1) < 10^{22}$$

gives $34 \leqslant n \leqslant 37$, and $2^n - 1$ cannot be a prime unless n is so, since $2^a - 1$ divides $2^{ab} - 1$ for all a and b. Therefore the point in Frenicle's question was really to find out whether $2^{37} - 1$ is a prime. It is not, and there is no perfect number between 10^{20} and 10^{22} (or at any rate no "euclidean" one), as Fermat answers soon enough (*Fe.*II.194); fortunately for

---

[5] That these are the only even perfect numbers was proved by Euler; the proof is not difficult and may be recommended to the reader as an exercise.

his self-esteem, he had detected, just in time, a numerical error in his calculations, which had nearly made him fall into the trap carefully laid for him by Frenicle (*Fe*.II.199). What he found was the decomposition

$$2^{37} - 1 = 137438953471 = 223 \times 616318177.$$

If, at the same time, one notices the prime factor decompositions

$$616318177 + 1 = 2 \times 7^3 \times 898423,$$

$$898423 + 1 = 2^3 \times 112303,$$

$$112303 + 1 = 2^4 \times 7019,$$

$$7^8 - 1 = 2^6 \times 3 \times 5^2 \times 1201,$$

$$3^9 - 1 = 2 \times 13 \times 757,$$

one has all that is needed in order to check Frenicle's "submultiple" number quoted above.

What is far more interesting for us is the way in which Fermat proceeded in order to factorize $2^{37} - 1$. His procedure, as he told Mersenne about June 1640 (*Fe*.II.198), depended upon the following three propositions:

(I) If n is not a prime, $2^n - 1$ is not a prime.
(II) If n is a prime, $2^n - 2$ is a multiple of $2n$.
(III) If n is a prime, and p is a prime divisor of $2^n - 1$, then $p - 1$ is a multiple of n.

As mentioned above, (I) is trivial in the sense that it is the special case $x = 2^a$ of the identity

$$x^b - 1 = (x-1)(x^{b-1}+x^{b-2}+\cdots+1).$$

That Fermat should mention this as a discovery merely shows how primitive his algebraic knowledge was at that time. As to (II) and (III), they are of course typical cases of what is known now as "Fermat's theorem". For a general formulation, we have to take up his letter to Frenicle of 18 October 1640, where it is stated as "*la proposition fondamentale des parties aliquotes*"; it is as follows (*Fe*.II.209):

   "Given any prime p, and any geometric progression 1, a, a^2, etc., p must divide some number $a^n - 1$ for which n divides $p - 1$; if then N is any multiple of the smallest n for which this is so, p divides also $a^N - 1$."

   Actually it is not hard to discover this "by induction", i.e., experimentally, for instance in the case $a = 2$, the decisive one for perfect numbers. If Frenicle knew of it, which seems to have been the case, one may assume that he had proceeded in that manner; so did Leibniz, in the years between 1676 and 1680[6]; so did Euler in 1731 and subsequent years: "*tentando comprobari potest*", "its truth can be verified by experiment", he writes in his very first paper on number theory (*Eu*.I-2.1 in E26|1732; cf. Chap.III, §IV). One can hardly doubt that Fermat had arrived at it in the same way; but already in 1640 he says he has a proof; "he would send it to Frenicle", he writes, "if he did not fear being too long". What could this proof be?

   There are two classical formulations for Fermat's theorem, and correspondingly two proofs for it. Writing it as $a^p \equiv a$ (mod p), as Fermat did for $a = 2$ in his letter to Mersenne, one proves it additively, writing, for $a = 2$:

$$2^p = (1+1)^p = 1 + \binom{p}{1} + \binom{p}{2} + \cdots + \binom{p}{p-1} + 1$$

and observing that all the binomial coefficients in the right-hand side are multiples of p; this was indeed contained in what Fermat had proved about binomial coefficients, not later than 1636 (cf. above, §II). The same result can then be extended to $a > 2$, either by "mathematical induction", writing

$$(a+1)^p = a^p + \binom{p}{1}a^{p-1} + \cdots + \binom{p}{p-1}a + 1$$
$$\equiv a^p + 1 \quad (\text{mod } p)$$

as Euler did in 1742, or else by using the "multinomial

---

   [6] On Leibniz's unpublished excursions into number theory, cf. the excellent article by D. Mahnke, *Bibl.Math.* (3) 13 (1912–13), 26–61.

formula"

$$(1 + 1 + \cdots + 1)^p = 1 + 1 + \cdots + 1 + \sum_{q+r+\cdots+s=p} \frac{p!}{q!r! \cdots s!}$$

as Leibniz did (cf. D. Mahnke, *loc.cit.*, p. 49).

On the other hand, if one writes the theorem as Fermat did in his letter to Frenicle, then it is a special case of the fact that the order of a group is a multiple of the order of any subgroup. One must then firstly observe that, if one divides by a prime p the numbers in the geometric progression $1, a, a^2, \ldots$, with a prime to p, the remainders must repeat themselves, so that, for some $r \geq 0$ and some $n > 0$, one must have $a^{n+r} \equiv a^r$, i.e., $a^r(a^n - 1) \equiv 0 \pmod{p}$; this gives $a^n \equiv 1 \pmod{p}$. Take for n the smallest number for which this is so; then one can arrange the $p - 1$ congruence classes prime to p modulo p into sets $\{b, ba, \ldots, ba^{n-1}\}$, each consisting of n elements; as these sets are easily seen to be disjoint, n must divide $p - 1$. This is the multiplicative proof obtained by Euler around 1750, of which he says that it is the better one since it can be generalized to $a^{\phi(m)} \equiv 1 \pmod{m}$ for any m (cf. Chap.III, §VI). In view of the way in which Fermat changed the formulation of his theorem from (II) to (III) and then to his statement to Frenicle, one may guess that his final proof was the second one.

As to the use to which this can be put, Fermat explains it quite clearly. Take for instance $2^{37} - 1$ as above. If it has a prime divisor p, then 37 must divide $p - 1$; as p is odd, it must be sought among the primes of the form $74n + 1$; after rejecting the first candidate 149, one tries the next one 223 and succeeds (*Fe.*II.199).

Fermat's enquiry does not stop there. Factorizing numbers of the form $a^n - 1$, one notices at once, for $n = 2m$, that one has also to factorize the numbers $a^m + 1$; so Fermat asks whether, for any a and any prime p, there always is m such that p divides $a^m + 1$. The answer, of course, is negative, and Fermat has the criterion for this to happen: if, and only if, the smallest n for which p divides $a^n - 1$ is even, then there is m such that p divides $a^m + 1$, and the smallest such m is $n/2$ (*Fe*.II.209–210).

For $a = 2$, another question arises for Fermat, also an important one from the point of view of "aliquot parts": when is $2^m + 1$ a prime? It cannot be so if m has an odd divisor $d > 1$; for, if $m = ed$, then, putting $N = 2^e$, one sees at once that $N + 1$ divides $2^m + 1 = N^d + 1$. The remaining case is $m = 2^r$; then $2^m + 1$ is indeed a prime for $r = 0, 1, 2, 3, 4$. Writing to Frenicle in 1640, Fermat enumerates such numbers up to $r = 6$:

$$3, 5, 17, 257, 65537, 4294967297, 18446744073709551617,$$

and conjectures that all are primes. It is hard to believe that he did not try to apply, at least to the sixth one $2^{32} + 1$, the method he had used to factorize $2^{37} - 1$; it shows that any prime divisor of $2^{32} + 1$ is of the form $64n + 1$, which leaves the candidates $193, 257, 449, 577, 641$, etc.; 641 divides $2^{32} + 1$; in fact, this is how Euler proceeded, almost a century later, incidentally re-discovering Fermat's theorem on that occasion (cf. Chap.III, §IV). What is even more surprising is that Frenicle, who had also factorized $2^{37} - 1$, did not at once point out the error, as (judging from the general tone of their correspondence; cf. *Fe*.II.185) he would have been only too pleased to do; on the contrary, Frenicle expressed agreement (*Fe*.II.208). Fermat persisted in his conjecture to the end of his days, usually adding that he had no full proof for it (cf. *Fe*.II.309–310, where he mentions that it is important in the matter of "aliquot parts"; cf. also *Fe*.II.405,434). One may imagine that, when he first conceived the conjecture, he was so carried away by his enthusiasm that he made a numerical error, and then never checked his calculation again. As to $2^{64} + 1$, it has the prime factor 274177, but this was undoubtedly beyond Fermat's range, and even beyond Frenicle's, though the latter was the more stubborn calculator of the two. Even to verify that 616318177 is a prime must have sorely tried Fermat's patience, if indeed he took the trouble to do it.

§V.

In the years from 1636 to 1640, Fermat turned his attention increasingly towards two types of questions: diophantine problems on the one hand, and sums of squares on the other.

Also the latter, of course, were suggested by Diophantus, or at least by Bachet's comments on Diophantus (cf. Chap.I, §XI). We see them appear in 1636, at the beginning of the correspondence between Fermat and Mersenne. One of the first questions to occur in those letters amounts to asking whether an integer which is the sum of 2 (resp. 3) rational squares is also the sum of 2 (resp. 3) integral squares; as previously observed (Chap.I, §XI), Diophantus sometimes seems to take this for granted, and so does Bachet in his commentary (cf. e.g. his comment on *Dioph.*V.9 = V.12$_b$ and on *Dioph.*V.11 = V.14$_b$). Fermat's letter to Mersenne of 15 July 1636 (*Fe.*II.29–30) implies that on that date he thinks he has proved it, while on 2 September 1636 he merely says that he is working at it ("*j'y travaille*"; *Fe.*II.58). An elementary proof for this has been published by L. Aubry in 1912 (cf. Chap.III, App.II); it is one that Fermat could have understood. It is idle, of course, to ask whether he could have found it; had he done so, occasions would not have been lacking for him to mention it to his correspondents; but the matter never turns up again in his letters.

Obviously, in those years, Fermat was still a raw novice, so far as "numbers" were concerned. On 16 September 1636, he submits to Roberval a problem formulated in the terminology of the tenth book of Euclid; it amounts to showing that

$$x^2 + y^2 = 2(x+y)z + z^2$$

has no non-trivial solution in rational numbers. On that day, he says he cannot prove it (*Fe.*II.62–63). A few days later, on 22 September 1636, he has found a proof, but has had to work very hard on it ("*elle m'a donné grandissime peine*"; *Fe.*II.74). His proof, which also Roberval had found in the

meanwhile (*Fe*.II.81,83), is to put $t = z + x + y$, to rewrite
the equation as

$$2x^2 + 2xy + 2y^2 = t^2$$

and then observe that, taking x, y, t to be integers without
a common divisor, t must be even, therefore x and y cannot
both be even, so that $x^2 + xy + y^2$ must be odd. In other
words, the equation has no non-trivial solution even as a
congruence modulo 4. One could also observe that the original equation can be written as

$$3z^2 = (x-z)^2 + (y-z)^2$$

so that the question is to show that 3, which is not a sum of
two squares in integers, is also not the sum of two rational
squares; this is a special case of the question mentioned to
Mersenne a short time before. It could equally well be
answered by considering the congruence $3z^2 \equiv u^2 + v^2$
(mod 3).

Actually, in virtue of "Hasse's principle" (cf. Chap.IV,
Appendix I), if F is any indefinite quadratic form with integral coefficients, and if $F(x) = 0$ has no non-trivial solution
in integers (or, what amounts to the same, in rational numbers), then there is a modulus m such that the congruence
$F(x) \equiv 0$ (mod m) has no non-trivial solution. In other words,
in the case of a single homogeneous quadratic equation, an
impossibility of this kind can always be verified by a congruence argument; this applies in particular to the question
of representing a number by a sum of squares. It is no
wonder, therefore, that Fermat, at the beginning of his
career as a number-theorist, should have started with problems of that kind. That he did not yet regard congruence
arguments as essentially trivial only shows his lack of experience at that time; on one occasion this must have caused
him some embarrassment. In 1638, writing to Mersenne
(*Fe*.II.66), he announces *inter alia,* with some self-satisfaction,
that no integer of the form $8n - 1$ can be a sum of less
than four squares, not only in integers, but even in fractions;
beside this, he apparently had stated in an earlier letter the

similar fact about numbers of the form $4n - 1$ and sums of two squares. Mersenne, as was his wont, passed this on to Descartes, who at that time thought he had reasons for being displeased with Fermat. So far as integral squares are concerned, Bachet had already (*loc.cit.*) given proofs based on nothing more than congruences modulo 8 resp. modulo 4. That this can easily be extended to fractions was disdainfully pointed out by Descartes; pretending that such a trivial matter was beneath his notice, he even left it to be carried out by an allegedly unschooled young man, actually his former pupil Jean Gillot (*Desc.*II.179,195; cf. ibid.II.91–93, and F. Rabelais, *Pantagruel,* Livre II, Chap.XVIII).

It is striking that, in the letter of 1638 to Mersenne, Fermat's statement about numbers of the form $8n - 1$, with which he was unwittingly laying himself open to Descartes's sarcasms, follows immediately the celebrated statement about every integer being "the sum of three triangular numbers, of four squares, of five pentagonal numbers, and so on"; it is the same statement which, perhaps at the same time, he inscribed in his *Diophantus* (*Fe.*I.305, Obs.XVIII); he was to repeat it to Pascal in 1654 (*Fe.*II.313) and to Digby in 1658 (*Fe.*II.404). Also in his letter of 1638, Fermat challenges Sainte-Croix to find two cubes whose sum is a cube, or two fourth powers whose sum is a fourth power, with the implication that he already knew or suspected that there are none. We shall never know for sure when, or whether, Fermat proved all these results. But clearly, if he was still a novice in number theory in 1638, his genius had begun to shine forth.

§VI.

Leaving sums of squares aside for a moment, we turn to what has been known since Euler as the theory of quadratic residues.

As we have seen, Fermat's theorem arose from the wish to find out when a prime p divides a number $a^n - 1$; a companion to that question, which turns up repeatedly in his correspondence with Frenicle in 1640 and 1641, is to find out when p divides $a^n + 1$ (*Fe*.II.209–210,220,226,232). In modern terms, call $n_p(a)$ the order of a modulo p, i.e., the smallest $n > 0$ such that p divides $a^n - 1$. By Fermat's criterion (cf. above, §IV), p divides some integer $a^m + 1$ if and only if $n_p(a)$ is even. On the other hand, as has been known since Euler, a is a quadratic residue modulo p if and only if $(p - 1)/n_p(a)$ is even; in view of the law of quadratic reciprocity, the primes p with that property are those which occur in certain arithmetic progressions of the form $4ax + b$; as Euler found out, this can rather easily be discovered "by induction" (cf. Chap.III, §VIII). If one considers only primes of the form $4n - 1$, then $n_p(a)$ is even when $(p - 1)/n_p(a)$ is odd, and *vice versa;* thus, for such primes, there is a simple criterion in terms of quadratic residues for p to have the property sought for by Fermat; for the others, there is nothing of the kind. Small wonder, then, that Fermat judged the problem a very hard one (*Fe*.II.210)! Characteristically perhaps, when he tried to explore it experimentally, he immediately made trivial numerical errors which nullified his attempt. In a letter sent to Mersenne in 1641 and intended for Frenicle (*Fe*.II.220–221), he asserts (A) that primes of the form $12n \pm 1$ divide no integer of the form $3^m + 1$; (B) that every prime $12n \pm 5$ divides some $3^m + 1$; (C) that primes $10n \pm 1$ divide no $5^m + 1$; (D) that every prime $10n \pm 3$ divides some $5^m + 1$; he specifies 11, 13, 23, 37 as examples for (A), 5, 17, 19 for (B), 11, 19 for (C) and 7, 13, 17 for (D). He even promises "for another occasion" the general rules for the prime divisors of $a^m + 1$ for any a. Actually the primes of type (A) resp. (B) are those for which 3 is a quadratic residue resp. non-residue; those of type (C) resp. (D) are those for which 5 is a quadratic residue resp. non-residue; Fermat's assertions are true for primes of the form $4n - 1$, but not necessarily for the others; in fact, 37 is a counterexample for (A), since it divides $3^9 + 1$; also, 41 divides $5^{10} + 1$ and provides a counterexample for (C). Thus Fermat missed his chance of guessing the quadratic reciprocity law; as we know from hindsight, he had asked the wrong question.

It was easier, no doubt, to deal with the quadratic character of -1, although even this, at that time, had to be regarded as a major discovery. Writing to Roberval in 1640, Fermat asserts $(Fe.II.204)$ that, if a and b are mutually prime, no prime divisor of $a^2 + b^2$ can be of the form $4n - 1$. Alternatively, if an integer has a prime divisor of the form $4n - 1$ and has no square divisor, then it cannot be a sum of two squares ("not even in fractions"). Writing to Frenicle a little later $(Fe.II.210)$, he has a rather cryptic statement which, when taken apart, is a direct consequence of the theorem he had sent to Roberval; neither Frenicle nor Fermat, at that stage, seemed willing to disclose their best results to one another. When in the same letter Fermat tells Frenicle that a prime of the form $x^2 + 2$ cannot divide a number $a^2 - 2$ $(Fe.II.211)$, one is left to wonder whether he is just testing Frenicle's knowledge or whether he is being deliberately misleading; such a prime, in fact, would obviously be of the form $4n - 1$ and would divide $a^2 + x^2 = (a^2 - 2) + (x^2 + 2)$.

As to Fermat's proof, it must have been the same which Euler was to discover in 1742 (cf. Chap.III, §Vb). Assume that $p = 4n - 1$ is a prime and divides $a^2 + b^2$, with a prime to b; both a and b must then be prime to p. Using modern notation for brevity, we can write $a^2 \equiv -b^2 \pmod{p}$; putting $m = 2n - 1, p - 1 = 2m$, we get $a^{2m} \equiv -b^{2m} \pmod{p}$. But, by Fermat's theorem, we have $a^{2m} \equiv b^{2m} \equiv 1 \pmod{p}$; this is a contradiction.

§VII.

We are so used to the fact that the congruence classes modulo a prime p make up a field, viz., the so-called prime field $\mathbf{F}_p$, and to all the algebraic properties that this implies, that we do not feel the need to analyze it any more. Actually Bachet's solution of the equation $ax - my = 1$ (cf. Chap.I, §IV) implies that every integer a prime to m has an inverse modulo m; for m prime this shows that $\mathbf{F}_m$ is a field. The same fact may also be regarded as a consequence of Fermat's theorem, since the congruence $a^n \equiv 1 \pmod{p}$ implies that a has the inverse a^{n-1} modulo p. From this we deduce at once, for instance, that if p divides $a^2 + b^2$, with a and b prime to p, then -1 is a quadratic residue modulo p; in fact, if a' is an inverse of a modulo p, the congruence $x^2 \equiv -1 \pmod{p}$ has the solution $x = a'b$. Thus the result sent by Fermat to Roberval amounts in modern language to the determination of the quadratic residue character of -1 modulo primes of the form $4n - 1$. At some time Fermat must have become aware of this; some of his later results could hardly be explained otherwise (cf. infra, §XII).

The case $p = 4n + 1$ must have given him more trouble, just as it gave more trouble to Euler and gives more trouble to the historian attempting to unravel the mysteries of Fermat's number theory. Actually, not only does every prime $p = 4n + 1$ divide some sum $a^2 + b^2$, but every prime of that form can itself be written as $a^2 + b^2$, in one and only one way. The first to publish a proof for these facts was Euler; they are both stated by Fermat, as early as 1640 (Fe.II.213; cf. Fe.I.314, Obs.XXVI, and Fe.I.293, Obs.VII). In all known proofs, one has to begin by proving the first part, i.e., by showing that -1 is a quadratic residue of $p = 4n + 1$; this must also have been the case with Fermat's proof (cf. Fe.II.432, and infra, §VIII). If we consult Euler again (Corr.I.494; cf. Chap.III, §VI) we see that Fermat could have proceeded as follows.

Take a prime $p = 4n + 1$, and, for x, y prime to p, put $a = x^n$, $b = y^n$. We have

$$(a^2 - b^2)(a^2 + b^2) = x^{4n} - y^{4n}.$$

By Fermat's theorem this is a multiple of p, so that p divides either $a^2 - b^2 = x^{2n} - y^{2n}$ or $a^2 + b^2$. Thus, if p divides no number $a^2 + b^2$ with a, b prime to p, it must divide all numbers $x^{2n} - y^{2n}$; for $y = 1$, this gives $x^{2n} \equiv 1 \pmod{p}$ for $1 \leqslant x \leqslant p - 1$.

As we know now, a congruence of degree d cannot have

more than d solutions in the field $\mathbf{F}_p$; in particular, the congruence $x^{2n} \equiv 1 \pmod{p}$ cannot have $p - 1 = 4n$ solutions. For us, this simple remark is enough to clinch the matter. For the sake of convenient reference, we formulate as a lemma the relevant portion of the result in question:

Lemma 1. Let p be a prime; then there is no $m < p - 1$ such that $a^m - 1$ is a multiple of p for all integers a prime to p.

Did Fermat take this for granted, perhaps on the basis of a certain amount of numerical evidence? Of course he knew that an equation $f(x) = 0$ of degree m can have no more than m solutions; this is proved for instance in Descartes' *Géométrie* of 1636 (*Desc.*VI.444–445), the proof being the one that we would give for it now: if $f(x) = 0$ has the roots $x_1, x_2, ..., x_n$, one can, by successive division of $f(x)$ by $x - x_1, x - x_2, ..., x - x_n$, bring down the degree to $m - n$; here it is understood that the underlying field is the field of real numbers. In 1772, Euler realized that this is equally valid for the field $\mathbf{F}_p$ (cf. Chap.III, §VI); the steps involved are quite easy. But may we credit Fermat, in 1640, with a degree of abstract thinking that Euler reached only towards the end of his long career? Probably not.

In 1749 Euler had proceeded differently (*loc.cit.*); as a true disciple of Leibniz and Bernoulli, he used the difference operator D defined by

$$(Df)(x) = f(x+1) - f(x);$$

it is easily seen by induction on m that, for any m, $D^m f(x)$ is a linear combination of $f(x), f(x+1), ..., f(x+m)$ with integral coefficients. Thus, if, for some $m < p - 1$, p divides $a^m - 1$ for $1 \leq a \leq p - 1$, it must also divide $D^m f(1)$ for $f(x) = x^m - 1$; but, as Euler knew, and as is easily seen by induction on m, we have $D^m f = m!$; as p is a prime, and $m < p - 1$, this is not a multiple of p.

Another proof for the quadratic character of -1 was offered by Lagrange in 1771 (*Lag.*III.425–438), depending, not upon lemma 1, but upon "Wilson's theorem", i.e., upon the congruence $(p-1)! \equiv -1 \pmod{p}$ for any prime p. If that congruence is known or assumed, one merely has to write, for $p = 2m + 1$:

$$(p-1)! = (1 \times 2 \times \cdots \times m).(p-1)(p-2) \cdots (p-m)$$
$$\equiv (m!)(-1)^m m! \equiv (-1)^m (m!)^2 \pmod{p}$$

to conclude that, if m is even, p divides $(m!)^2 + 1$. It is by no means impossible that Fermat could have, experimentally at least, come across the conclusion of Wilson's theorem; Leibniz very nearly did so around 1682 (cf. D. Mahnke, *loc.cit.*, p. 42). Is it fanciful to imagine that Fermat could then have proceeded in the manner outlined above?

We know so much more about Leibniz than about Fermat, because apparently Leibniz jotted down every thought that ever came into his head and preserved every scrap of paper that bore his handwriting (cf. D. Mahnke, *loc.cit.*, p. 29–30). Such were not Fermat's habits.

§VIII.

At this point, if we allowed ourselves the use of modern concepts, we would introduce the Gaussian ring $\mathbf{Z}[i]$; this consists of the "complex integers" $a + bi = a + b\sqrt{-1}$, where a and b are ordinary integers; as will be explained in Appendix I, one has in this ring a "euclidean algorithm", from which it follows that every number in it can be uniquely decomposed into prime factors. Then, knowing that every ordinary prime $p = 4n + 1$ divides some sum $a^2 + b^2 = (a+bi)(a-bi)$, while no prime $p = 4n - 1$ does, one concludes that the latter remain prime in $\mathbf{Z}[i]$, while the former do not; if then, for $p = 4n + 1$, $r + si$ is a prime factor of p, p has to be equal to $r^2 + s^2$.

Bombelli, in Book I of his *Algebra* of 1572, had given an exposition of the theory of the numbers $a + bi$, perfectly sound even from a modern point of view (cf. above, Chap.I, §XII); but there is no sign that Fermat had studied Bombelli, and "imaginary numbers" seem always to have remained beyond his horizon. Even in the next century, after Euler had dealt with those numbers for many years in connection with problems in analysis, it is only quite late in life that he introduced them into his number theory (cf. Chap.III, §XIII). When Fermat, writing to Mersenne on Christmas day 1640, told him that every prime $p = 4n + 1$ is in one and only one way a sum of two squares (*Fe*.II.213; cf. *Fe*.I.293, Obs.VII), he must of course have proceeded differently. Fortunately he gave us a faint indication of his method in his communication of 1659 to Huygens (*Fe*.II.432; cf. *infra*, §X). There he says that he had used his "method of descent", showing that if it were not so for some prime it would also not be true for some smaller prime "and so on until you reach 5". This may not have seemed quite enlightening to Huygens; we are in a better position, because Euler, in the years between 1742 and 1747, constructed a proof precisely of that kind (cf. Chap.III, §IX); it is such that we may with some verisimilitude attribute its substance to Fermat. Actually it preserves, in a different arrangement, the various moments of the one we outlined above using the Gaussian ring. It is more or less as follows.

Take a prime $p = 4n + 1$; we already know that it divides some sum $a^2 + b^2$ with a, b prime to p. Let r be the remainder of the division of a by p; if $r \leqslant 2n$, put $a' = r$; otherwise we have $0 < p - r \leqslant 2n$, and we put $a' = p - r$. Construct b' similarly from b, and replace a, b by a', b'. After this is done, p still divides $a^2 + b^2$, and a, b are > 0 and $< p/2$, so that $a^2 + b^2$ is $< p^2/2$ (this is the step which corresponds to the use of the euclidean algorithm for $\mathbf{Z}[i]$). The same will hold if we divide a, b by their g.c.d., so that we may assume that they are mutually prime. Write now $N = a^2 + b^2$; all the prime divisors of N except p are then $< p/2$ and must be either 2 or of the form $4m + 1$; we must show that, if all are sums of two squares, so is p. Here, as Euler found, one could try to proceed in two different manners.

Firstly, if all the prime divisors of N/p are such, so is N/p, in virtue of the fundamental identity

(1) $$(a^2 + b^2)(x^2 + y^2) = (ax \pm by)^2 + (ay \mp bx)^2$$

(cf. Chap.I, §VI). Therefore, writing $N/p = x^2 + y^2$, we have

$$p = \frac{a^2 + b^2}{x^2 + y^2} = \left(\frac{ax \pm by}{x^2+y^2}\right)^2 + \left(\frac{ay \mp bx}{x^2+y^2}\right)^2,$$

so that p is a sum of two rational squares. This brings us back to the question which Fermat had raised but apparently not solved in 1636 (cf. above, §V, and Chap.III, Appendix II).

Following Euler (*Corr*.I.416–417,419), we shall now bring the proof to completion by a different route. For the sake of brevity, in this § and the next, a pair of integral squares with the sum N will be called a "representation" of N. If two integers N, N' have representations $N = a^2 + b^2$, $N' = x^2 + y^2$, the right-hand side of the identity (1) defines two representations of NN', which will be said to be derived from those of N and of N' by "composition". Now Euler's main step can be formulated as follows:

Lemma 2. For any $N = a^2 + b^2$, let $q = x^2 + y^2$ be a prime divisor of N. Then N/q has a representation $u^2 + v^2$ such that the representation $N = a^2 + b^2$ is one of those derived by composition from it and from $q = x^2 + y^2$.

Apply (1) to N and q. We have

(2) $$Nq = (ax \pm by)^2 + (ay \mp bx)^2.$$

But q divides the number

$$Ny^2 - b^2 q = a^2 y^2 - b^2 x^2 = (ay - bx)(ay + bx)$$

and therefore one of the numbers $ay \mp bx$. Consequently, if the sign is suitably chosen in (2), q divides the left-hand side and the second term in the right-hand side, hence also the first term. With the same choice of sign, we can now write

(3) $$ax \pm by = qu, \qquad ay \mp bx = qv;$$

dividing (2) by q^2, we get $N/q = u^2 + v^2$. Now (3), solved for a and b, gives

(4) $\qquad a = ux + vy, \qquad b = \pm(uy - vx).$

This proves the lemma.

If now, as before, p is the greatest prime divisor of $N = a^2 + b^2$, with a, b mutually prime, and all prime divisors of N except p are known to be of the form $x^2 + y^2$, we can apply lemma 2 to any one of these, say q, then to any prime divisor other than p of N/q, etc., until we obtain for p itself an expression as a sum of two squares. This completes the proof of Fermat's statement, as obtained by Euler more than a century later. Except for minor details, Fermat's proof could have been much the same.

§IX.

As we have seen, the question whether a number N can be written as a sum of two squares, either in fractions or in integers, originates from Diophantus (cf. Chap.I, §§XI–XII); so does the question whether a number N can be the hypotenuse of a right-angled triangle "in numbers", i.e., whether N^2 is a sum of two integral squares; so does also the question whether a number can be a sum of two squares in more than one way (cf. Chap.I, §VI).

Perhaps Fermat's first mention of such matters is in his marginal note on *Dioph*.V.12$_b$ (*Fe*.I.313, Obs.XXV), which is merely to the effect that $3n$ cannot be a sum of two squares unless 3 divides n; this just amounts to saying that -1 is not a quadratic residue modulo 3. It is immediately followed (*Fe*.I.314, Obs.XXVI) by a statement of "the true criterion" for $N = 2m + 1$ to be a sum of two squares: the necessary and sufficient condition is that m should be even, i.e., N of the form $4n + 1$, and that, after N has been divided by its greatest square divisor, the only prime factors left in the quotient should themselves be of the form $4n + 1$. This is contained in the results discussed above in §VIII.

Already in 1640, however, when he takes up these questions in his correspondence (*Fe*.II.213–214; cf. *Fe*.II.221–

222 and *Fe.*I.293–297, Obs.VII), Fermat does not stop at the question whether $N = x^2 + y^2$ has a solution; he asks for the number of those solutions, and for a way to find them. Partly, no doubt, for traditional reasons, he lays special stress on the corresponding questions for "hypotenuses", i.e., on the case when N is a square.

In counting the number of "representations" $N = x^2 + y^2$ of an integer N, we will pay no attention to the signs of x and y, nor to their order; this agrees in substance with Fermat's usage; of course he would always have taken x, y to be natural integers. A representation $N = x^2 + y^2$ will be called "proper" if x and y are mutually prime. If a representation $N = x^2 + y^2$ is not proper, i.e., if x, y have a g.c.d. $d > 1$, one can write $x = dx', y = dy', N = d^2N'$, and (x', y') is a proper representation of N'. Thus all questions about representations can be reduced, if one wishes, to similar questions about proper representations; this is implicitly understood in Fermat's treatment. In particular, if N has a proper representation, its odd prime factors must be all of the form $4n + 1$; at the same time, N cannot be a multiple of 4, since $x^2 + y^2$ cannot be a multiple of 4 unless x, y are even. If now $N = 2N'$, and N' is odd, then every representation $N' = x'^2 + y'^2$ of N' gives a representation $N = x^2 + y^2$ of N with $x = x' + y', y = x' - y'$; conversely, if $N = x^2 + y^2$, x and y must both be odd and we can put $x' = (x+y)/2, y' = (x-y)/2, N' = x'^2 + y'^2$ (cf. (4), Chap.I, §VI). These formulas determine a one-to-one correspondence between the representations of N and those of N'; if one is proper, so is the other. Thus it is enough to discuss representations for odd integers.

The first point on which Fermat rightly insists is that every prime of the form $4n + 1$ has a unique representation. Actually this follows at once from the result we have formulated as lemma 2 in §VIII; taking $N = q$ in that lemma, we get $N/q = 1 = u^2 + v^2$, so that (u, v) is $(\pm 1, 0)$ or $(0, \pm 1)$; (4) shows then that (a, b) defines, in the sense explained above, the same representation of q as (x, y); obviously it is a proper one.

From the same lemma it follows now that any representation of an integer N can be obtained by "composition" from a factoring of N into its prime factors. If, following Fermat, we call "primitive" the representations of primes by sums of two squares, one may say that all representations can be derived from the primitive ones by a systematic process, viz., by composition. This is indeed specified by Fermat for the case of "hypotenuses", i.e., for representations of squares (*Fe*.II.222); undoubtedly he was aware of it also in the general case. As he could hardly have arrived at such a conclusion by mere trial and error, this gives an additional argument for assuming that he was in possession of some such result as lemma 2.

As to the number of representations of an integer, Fermat has the following statements (*Fe*.II.213–215; cf.*Fe*.I.293–297, Obs.VII):

(I) If p is a prime of the form $4n + 1$, and $r \geqslant 1$, then $N = p^r$ has exactly $r/2$ representations if r is even and $(r+1)/2$ if r is odd.

(II) If m is any integer, take all the distinct primes of the form $4n + 1$ dividing m; let $\alpha, \beta, ..., \gamma$ be their exponents in m; then m is exactly h times a hypotenuse, i.e., m^2 has exactly h representations, with h given by

$$2h + 1 = (2\alpha+1)(2\beta+1) ... (2\gamma+1).$$

The letter to Mersenne containing (I) and (II) is dated 25 December 1640. Only a few months earlier (*Fe*.II.202–204), he had only been able to tell Roberval that a prime $p = 4n - 1$ cannot divide the sum of two mutually prime squares (i.e., that -1 is a non-residue modulo $p = 4n - 1$), adding that "nothing he had ever found about numbers had pleased him as much as the proof of that result". What a giant stride forward in less than half a year!

In making the statements quoted above, Fermat does not specify that he has proofs for them; nevertheless one may plausibly try to reconstruct his reasoning. As we have said, it is enough to deal with proper representations. If we go back to lemma 2 and its proof, we see from formula (4) that, if (a, b) is a proper representation for N, (u, v) must be one for N/q. This has a converse, which we also formulate as a lemma:

Lemma 3. Let $q = x^2 + y^2$ be an odd prime divisor of N; assume that N/q is >2 and has the proper representation $N/q = u^2 + v^2$. Let (a, b) and (a', b') be the two representations for N, arising from (u, v) and (x, y) by composition. Then these two representations of N are distinct; they are both proper if q does not divide N/q; if q divides N/q, one is proper and the other is not.

In fact, we have

$$a = ux + vy, \quad b = uy - vx; \quad a' = ux - vy, \quad b' = uy + vx.$$

Clearly $a \neq \pm a'$, and it is easily seen that $a \neq \pm b'$. We have

$$ax + by = uq, \qquad ay - bx = vq,$$

and similar formulas for a', b'; as (u, v) is proper, this shows that no prime, other than q, can divide a and b, or a' and b'. If q divided both a and a', it would divide $a + a'$, $a - a'$, hence u and v; as this is not so, either (a, b) or (a', b') must be proper. If q divides, say, a and b, q^2 divides $N = a^2 + b^2$, and q divides N/q; therefore, if it does not, both (a, b) and (a', b') are proper. Finally, if q divides N/q, we write, just as in the proof for lemma 2

$$bb' = u^2y^2 - v^2x^2 = \frac{N}{q}y^2 - v^2q;$$

this is a multiple of q, so that q divides b, hence also a (as shown by $N = a^2 + b^2$) or b', hence also a'. This completes the proof.

There cannot be much doubt that Fermat must have proceeded more or less along such lines, even though he would of course not have expressed his ideas in this fashion and might well have contented himself with a careful analysis of typical numerical cases. Probably he might have begun with the case of a power $P = p^r$ of a prime $p = 4n + 1$; applying lemma 3, or some similar result, to such a power, one finds at once that it has one and only one proper representation; this, in substance, is equivalent to Fermat's statement (I). If now N is a product of ρ powers $P_1, P_2, \ldots$, P_ρ of distinct primes, the lemma shows that it has $2^{\rho-1}$ proper representations, arising by composition from those of the factors P_i, since each additional factor multiplies the number of proper representations by 2; (II) follows easily from this fact. One could also obtain a formula for the number of representations, proper or not, of an arbitrary integer, but this would not be quite as simple as (II).

One consequence of the above results must be pointed out here, because of the role it must have played in some of Fermat's later work on diophantine equations (cf. *infra.*, §XVI); we shall formulate it as a lemma.

Lemma 4. Let N^3 be a cube with a proper representation $N^3 = a^2 + b^2$. Then N has a representation $u^2 + v^2$ such that $a = u^3 - 3uv^2$, $b = v^3 - 3vu^2$.

As N^3 has a proper representation, N must be odd and must be a product of powers $P_1, \ldots, P_\rho$ of ρ distinct prime factors of the form $4n + 1$; then so is N^3, and both have $2^{\rho-1}$ proper representations, all arising by composition from those of the factors P_i resp. P_i^3. Consequently, each proper representation for N^3 must arise by composition from one, and only one, of N; in particular, N must have a representation (u, v) such that, when it is first composed with itself and then with the resulting representation of N^2, one gets the given one (a, b) for N^3. If the signs for u, v are chosen properly, one obtains the formulas in the lemma. One would get the same result more easily by making use of the ring $\mathbf{Z}[i]$ of Gaussian integers (cf. Appendix I). There is of course a similar result for any power N^r with $r \geqslant 2$, and it can be proved in the same way. The case $r = 2$ is nothing else than Euclid's lemma on pythagorean triangles (cf. Chap.I, §V) and does not require the elaborate machinery needed for lemma 4.

Thus the theory of the representations of numbers by sums of squares may be regarded as essentially complete, so far as "composite" (i.e., non-prime) integers are concerned.

There remains the problem of actually finding what Fermat calls the primitive representations, i.e., those of the primes $p = 4n + 1$. In 1654 he told Pascal that he had found a systematic method for this (*"une règle générale"*: *Fe*.II.313; cf. *infra*, §XIII). In 1641, however, he knew of no such method, and had nothing better to offer than trial and error (*"tâtonner"*: *Fe*.II.223).

It is noteworthy that Fermat, "writing hurriedly" (as he says) on Christmas day 1640, still finds time to point out the analogy between numbers of representations by the form $X^2 + Y^2$, and similar numbers for $X^2 - Y^2$, or rather, as he formulates it, for the number of times that N can be written as $x - y$ with xy a square. This condition gives $x = du^2, y = dv^2, N = d(u^2 - v^2)$, where d is the g.c.d. of x and y, so that u, v are mutually prime. For N odd, the number of solutions is indeed given by the same formula as in (II), except that all the prime divisors of N are to be taken into account, and not merely those of the form $4n + 1$. The same question is mentioned on another occasion (*Fe*.II.256–258, of doubtful date), in connection with a method for factoring large integers. Take for N a large odd integer; let n be the largest integer $<\sqrt{N}$. The traditional way of factoring N is to try to divide it successively by all primes from 3 up to n. Instead of this, Fermat tries to write it as $N = x^2 - y^2$; as x must be $>n$, this can be done by writing the sequence

$$(n+1)^2 - N, (n+2)^2 - N, (n+3)^2 - N, ...$$

until one finds a square in it. As successive differences of consecutive numbers in that sequence are $2n + 3, 2n + 5$, etc., the calculation can be carried out rather easily, as shown by Fermat's example

$$N = 2027651281, \qquad n = 45029,$$

where he finds

$$N = 45041^2 - 1020^2 = 46061 \times 44021.$$

The procedure is quite effective if N has divisors close to

$\sqrt{N}$. As a means of testing whether a given large number is or is not a prime, Fermat, writing to Huygens in 1659, had to acknowledge that it is far from perfect (*Fe*.II.435), in spite of many shortcuts which he says he has discovered. Of course some progress has been made in this since Fermat; but it is diverting to observe that supposedly "unbreakable" codes are being constructed nowadays on the assumption that the factoring of very large numbers is beyond the power of our most sophisticated computers.

§X.

In the tantalizingly brief account of his number-theoretical work sent to Huygens *viâ* Carcavi in 1659 (*Fe*.II.431–436 = *Huy*.II.458–462) Fermat has the following to say about his early work:

"As ordinary methods, such as are found in the books, are inadequate to proving such difficult propositions, I discovered at last a most singular method . . . which I called the *infinite descent*. At first I used it only to prove negative assertions, such as: "No number of the form $3n - 1$ can be written as $x^2 + 3y^2$ ", "There is no right-angled triangle in numbers whose area is a square". . . . To apply it to affirmative questions is much harder, so that, when I had to prove that "Every prime of the form $4n + 1$ is a sum of two squares", I found myself in a sorry plight (*"en belle peine"*). But at last such questions proved amenable to my method . . ."

Was it pure negligence that Fermat, in introducing his infinite descent to Huygens, included the first statement about numbers $3n - 1$? Are we faced there with a copying error? Our text is based, not on an autograph of Fermat, but on a copy made by Huygens. By the time it was written, Fermat had long known (and surely Huygens knew very well) that the statement in question can be proved by a trivial congruence argument modulo 3. There had been a time when Fermat had not been aware of this (cf. above, §V); should one imagine that the first proof he had worked out for that statement (and perhaps for the companion statement about sums of two and four squares; cf. §V) had been one by infinite descent, before he realized that a congruence argument would suffice? However that may be, the passage quoted above indicates that his proof for the second statement must have preceded his work of 1640 about sums of two squares (cf. above, §IX). This is in part confirmed by his letter to Mersenne (*Fe*.II.65, presumably to be dated 1638; cf. above, §II), where he challenges Sainte-Croix to find a right-angled triangle whose area is a square, a cube equal to the sum of two cubes, etc.; clearly he must already have known, or at least suspected, that these are problems without solutions. In 1640 he sends the same (and another impossible one) to Frenicle through Mersenne in order to test Frenicle's theoretical knowledge in such matters; "if he says that there is no solution up to this or that limit", Fermat tells Mersenne, "be assured that he proceeds by tables". There is no sign that Frenicle gave any answer.

In his *Traité des triangles rectangles en nombres*, posthumously published in 1676 (and reprinted in 1729, *Mém.Acad.Sc.*t.V, pp. 127–206), Frenicle did include a proof that the area of a pythagorean triangle can neither be a square nor twice a square. In view of Fermat's statement to Huygens (*Fe*.II.436), one may safely assume that this was based on a communication from Fermat; the lack of an acknowledgment, in a manuscript which the author did not see through the press, does not imply that Frenicle was claiming credit for those proofs. In that same *Traité*, Frenicle gives "induction", i.e., experimentation, as his only reason for assuming that every prime of the form $4n + 1$ is a sum of two squares (cf. *infra*, §XI(ii)); all we know about him indicates that he was an experimentalist, in contrast with Fermat whose foremost interests were theoretical.

Writing to Huygens about his proof by descent for the theorem on the area of pythagorean triangles, Fermat merely described the method by saying that "if the area of such a triangle were a square, then there would also be a smaller one with the same property, and so on, which is impossible";

he adds that "to explain why would make his discourse too long, as the whole mystery of his method lies there" (*Fe*.II.432). Fortunately, just for once, he had found room for this mystery in the margin of the very last proposition of Diophantus (*Fe*.I.340–341, Obs.XLV); this is how it goes.

Take a pythagorean triangle whose sides may be assumed mutually prime; then they can be written as $(2pq, p^2 - q^2, p^2 + q^2)$ where p, q are mutually prime, $p > q$, and $p - q$ is odd (cf. Chap.I, §V). Its area is $pq(p+q)(p-q)$, where each factor is prime to the other three; if this is a square, all the factors must be squares. Write $p = x^2, q = y^2, p + q = u^2, p - q = v^2$, where u, v must be odd and mutually prime. Then x, y and $z = uv$ are a solution of $x^4 - y^4 = z^2$; incidentally, v^2, x^2, u^2 are then three squares in an arithmetic progression whose difference is y^2 (cf. Chap.I, §VII, and *Fe*.II.65, problem 4°). We have $u^2 = v^2 + 2y^2$; writing this as $2y^2 = (u+v)(u-v)$, and observing that the g.c.d. of $u + v$ and $u - v$ is 2, we see that one of them must be of the form $2r^2$ and the other of the form $4s^2$, so that we can write $u = r^2 + 2s^2$, $\pm v = r^2 - 2s^2, y = 2rs$, and consequently
$$x^2 = \tfrac{1}{2}(u^2 + v^2) = r^4 + 4s^4.$$
Thus $r^2, 2s^2$ and x are the sides of a pythagorean triangle whose area is $(rs)^2$ and whose hypotenuse is smaller than the hypotenuse $x^4 + y^4$ of the original triangle. This completes the proof "by descent".

Frenicle follows this proof faithfully, with little more than verbal changes (*loc.cit.* pp.173–175). As to triangles whose area is twice a square, he proceeds as follows (*ibid.*, pp.175–176). As above, one can write $p + q = u^2, p - q = v^2$, and either $p = x^2, q = 2y^2$, or $p = 2x^2, q = y^2$. As u, v are odd, and $2p = u^2 + v^2$, p must be odd, and so we can write $p = x^2, q = 2y^2$. Then $4y^2 = (u+v)(u-v)$; as the g.c.d. of $u + v$ and $u - v$ is 2, we can write $u + v = 2r^2, u - v = 2s^2$, $u = r^2 + s^2, v = r^2 - s^2$, and finally
$$x^2 = \tfrac{1}{2}(u^2 + v^2) = r^4 + s^4.$$
Then the triangle (r^2, s^2, x) has the area $2(rs/2)^2$, and the proof is complete.

OBSERVATIO D. P. F.

AREA trianguli rectanguli in numeris non potest esse quadratus, huius theorematis à nobis inuenti demonstrationem quam & ipsi tandem non sine opera & laboriosâ meditatione deteximus, subiungemus. Hoc nempe demonstrandi genus miros in arithmeticis suppeditabit progressus, si area trianguli esset quadratus darentur duo quadratoquadrati quorum differentia esset quadratus: Vnde sequitur dari duo quadrata quorum & summa, & differentia esset quadratus. Datur itaque numerus compositus ex quadrato & duplo quadra.s aequalis quadrato, ea conditione vt quadrati, cum componentes faciant quadratum. Sea si numerus quadratus componitur ex Quadrato & duplo alterius quadrati eius latus similiter componitur ex quadrato & duplo quadrati vt facillime possumus demonstrare.

Vnde concluditur latus illud esse summam laterum circa rectum trianguli rectanguli & vnum ex quadratis illud componentibus efficere basem & duplum quadratum aequari perpendiculo.

Illud itaque triangulum rectangulum conficitur à duobus quadratis quorum summa & differentia erunt quadrati. At isti duo quadrati minores probabuntur primis quadratis primò suppositis quorum tam summa quam differentia faciunt quadratu. Ergo si dentur duo quadrata quorum summa & differentia faciant quaaratum, dabitur in integris summa duorum quadratorum eiusdem natura priore minor. Eodem ratiocinio dabitur & minor istâ inuenta per viam prioris & semper in infinitum minores inuenientur numeri in integris idem praestantes: Quod impossibile est, quia dato numero quouis integro non possunt dari infiniti in integris illo minores. Demonstrationem integram & fusius explicatam inserere margini vetat ipsius exiguitas.

Hâc ratione deprehendimus & demonstratione confirmauimus nullum numerum triangulum praeter vnitatem aequari quadratoquadrato.

SCHOLIVM.

Hinc patet, idem ostendi posse de quadratoquadrato ipsius H quod ostensum est de quadratoquadrato ipsius K. Quare verum est in quolibet triangulo rectangulo quadratum aream duplicata additum quadratoquadrato cuiuslibet lateris circa rectum efficere quadratu. Vt in dato hypothesi 144. cum quadratoquadratis ipsorum 3. & 4. scilicet quadratis 225. & 420. sed & conditionis adiectae necessitas per .ligcoram sic demonstrabitur. Data area cuiuslibet trianguli rectanguli put a 6. ponatur vnum laterum circa rectum 1. N. erit alterum ¹². Vt autem sit triangulum rationale oportet, vt summa quadratorum, put a 1 Q. + ¹. aequetur quadrato, & omnia deducto in 1 Q. fiet 1 QQ + 144. aequalis quadrato. Vnde patet quadratoquadratum cuiuslibet lateris circa rectum adscito quadrato area duplicata, debere conficere quadratum.

Porrò hanc ipsam quaestionem tractans Franciscus Vieta Zetetico 16. libri quarti, duas alias ei prafigit conditiones. Prima est. Oportet vt area addendo aliquem quadratoquadratum, fiat quadratoquadratus, vel vt ducendo aream in aliquem quadratum, & producum addendo alicui quadratoquadrato, fiat quadratoquadratus. Et haec conditio, sufficiens quidem est, vt demonstrat Vieta, sed an sit necessaria merito quis ambigat. Sanè arbitror cùm eius necessitatem demonstrare non posset summi vir ingenij, secundam excogitasse. Secunda conditio est. Oportet vt datus area numerus, sit cubus suo multatus latere, vel vt idem per quadratum aliquem multiplicatus, sit cubus suo multatus latere. Et haec conditio non solùm sufficiens est, vt ostendit Vieta, sed etiam necessaria vt demonstrabimus, ne tanti viri commentum labare videatur. Quamuis tutius sit hanc conditionem ita proponere. Oportet vt area numerus sit cubus suo multatus latere, latusue suo multatum cubo. Vel vt eo per aliquem quadratum multiplicato vel diuiso, fiat cubus suo multatus latere, latusue suo multatum cubo. Huius autem rei ratio est, quia quodlibet triangulum rectangulum, potest concipi simile alteri triangulo, quod formatum sit ab vnitate, & ab alio quouis quadrato, cuius quadrati si ponas latus 1 N. fiet trianguli hypotenusa 1 Q. + 1. alterum laterum circa rectum 2 N. alterum 1 Q – 1 vel 1 – Q prout 1 N. supponitur maior vel minor vnitate. Quare fit area 1 C – 1 N. vel 1 N – 1 C. ita si fingas triangulum ab 1. & 4. modo tradito tertia tertij porismatur. Fient latera 3. 4. 5. Area vero 6. cubus scilicet 8. suo multatus latere. At si formes triangulum ab 1. & ¹². fiet triangulum ¹²⁴ ¹²⁴, cuius area. ¹²⁴ est latus multatum suo cubo ¹².

Proposito verò quolibet alio triangulo, cùm per quartam tertij porismatum. necesse sit illud formari à duobus planis similibus. hos planos sumens, & vtrunque per minorem ipsorum sigillatim diuidens, fient duo quotientes in eadem ratione, quorum minor erit vnitas, à quibus si formes triangulum, erit

V v iij

Fermat's descent (Chap., §X), from pp. 338–339 of the *Diophantus* of 1670.

From this, as Frenicle observes, and as Fermat must have known, a number of consequences follow. For instance, the fact that the equations $x^4 \pm y^4 = z^2$ have no non-trivial solutions is included in the above proofs (cf. *Fe*.I.327, Obs.XXXIII). Also, in a pythagorean triangle (a, b, c), a and b cannot both be squares, since then the area $ab/2$ would be twice a square; a and c cannot both be squares, since $a = r^2$, $c = s^2$ would give $s^4 - r^4 = b^2$. From this one can deduce Fermat's statement (*Fe*.I.341; cf. *Fe*.II.406) that no triangular number other than 1 is a fourth power. Triangular numbers are those of the form $\frac{1}{2}n(n + 1)$; if this is a fourth power, one of the integers n, $n + 1$ must be of the form x^4 and the other of the form $2y^4$, so that we have $x^4 - 2y^4 = \pm 1$. If $x > 1$, consider the pythagorean triangle $a = x^2$, $b = \frac{1}{2}(x^4 - 1)$, $c = \frac{1}{2}(x^4 + 1)$; in one case a and b would be squares, and in the other case a and c would be such, contradicting what has been proved before. Similarly, by considering the triangle $a = x^2z^2$, $b = \frac{1}{2}(x^4 - z^4)$, $c = \frac{1}{2}(x^4 + z^4)$, one sees that the equations $x^4 \pm z^4 = 2t^2$ have no non-trivial solutions; such, more or less, is the proof that Euler gave for this result in 1738 (*Eu*.I-2.47–49 in E98).

§XI.

Still following the rough outline supplied by Fermat's communication of 1659 to Huygens, we now turn back to quadratic residues and quadratic forms; it will appear later (cf. *infra*, §XVI) that the practice of the "method of descent" in the study of diophantine equations provided Fermat with a powerful incentive for proceeding in that direction.

To summarize at once what will be discussed presently, Fermat's results on that subject come under the following headings:

(i) The statement about every integer being a sum "of three triangular numbers, four squares, five pentagonal numbers, etc."; this is asserted as early as 1638 and frequently reiterated (cf. above, §V). As to triangular numbers, it amounts to saying that every integer of the form $8n + 3$ is a sum of (at most) three squares. Other statements about sums of three squares appear in the letter of 1658 to Digby (*Fe.*II.405) and in a marginal note to Diophantus (*Fe.*I.314–315, Obs.XXVII) which has all the appearances of belonging to an early period of Fermat's career. Cf. *infra*, §XIV.

(ii) As early as 1641, Fermat seems to have been aware of some of the main facts about the quadratic form $X^2 - 2Y^2$ (cf. *Fe.*II.221,224–226); so apparently was Frenicle (cf. *Fe.*II.231–241), who specified, however, in his *Traité des triangles* (*loc.cit.*) that his results had been obtained "by induction", i.e., experimentally. Frenicle's main motivation in such matters came from the study of pythagorean triangles; his interest in the form $X^2 - 2Y^2$ arose from the fact that the sum and the difference of the two minor sides of a so-called "primitive" triangle

$$(a, b, c) = (2pq, p^2 - q^2, p^2 + q^2)$$

(i.e., one for which a, b, c are mutually prime; cf. Chap.I, §V) are given by that form:

$$|b \pm a| = |p^2 - q^2 \pm 2pq| = |(p \pm q)^2 - 2q^2|.$$

As to Fermat, one gets the impression that very soon it was the quadratic form as such that attracted his attention.

(iii) Fermat's extant correspondence with Frenicle, including the communications which passed between them through Mersenne, is entirely concentrated within the years 1640 and 1641. From then on, and until 1657, Frenicle's name occurs but rarely in Fermat's letters. Indeed, as a glance at volume II of his *Œuvres* will show, his correspondence between 1642 and 1654 has little to offer in the way of mathematical content. It is only in 1654 that we learn for the first time of the progress Fermat has made with his number theory in all these years; after an exchange of letters with Pascal on probabilities (of great importance, but of no concern to us here), Fermat musters enough self-confidence to send to Pascal some of his discoveries on "numbers" (*Fe.*II.312–313). What is new there is that, for the first time, the quadratic forms $X^2 + 3Y^2$, $X^2 + 2Y^2$ appear, as well as the mention of a "general rule" for finding the two squares in the representation $p = a^2 + b^2$ of any prime of the form $4n + 1$. We find the form $X^2 \pm XY + Y^2$ mentioned in the notes on Diophantus (*Fe*.I.300–301, Obs.X–XII), but this is of course not to be separated from $X^2 + 3Y^2$, as shown by the identities

$$4(x^2 \pm xy + y^2) = (2x \pm y)^2 + 3y^2,$$

$$x^2 + 3y^2 = (x \mp y)^2 \pm 2y(x \mp y) + (2y)^2.$$

What Fermat says he has proved concerning these forms is as follows: every prime of the form $3n + 1$ can be written as $x^2 + 3y^2$ (*Fe*.II.313,403; cf. *Fe*.I.301, Obs.XII); every prime of the form $8n + 1$ or $8n + 3$ can be written as $x^2 + 2y^2$ (*Fe*.II.313,403); every prime of the form $8n \pm 1$ can be written in infinitely many ways as $|x^2 - 2y^2|$ (JEH.41). The latter fact had been known (experimentally, cf. above, (ii)) to Frenicle; it is stated in his letter of 1641 to Fermat (*Fe*.II.235).

(iv) In 1654, Fermat had signally failed to arouse in Pascal an interest for number theory. In 1656 the English adventurer and double agent Sir Kenelm DIGBY ("an arrant mountebank", as Evelyn dubbed him; "the very Pliny of our age for lying", according to another of his contemporaries) visited Toulouse; obviously he must have met Fermat; either on that occasion or later, Fermat received from him Wallis's *Arithmetica Infinitorum*, which had been printed in 1656 (cf. *Fe*.II.337, and above, §III). This episode revived in Fermat the hope of finding, at least across the seas, a sparring partner worthy of himself; his correspondence with Digby, and, through Digby, with the English mathematicians WALLIS and BROUNCKER occupies the next year and a half, from January 1657 to June 1658. It begins with a challenge to Wallis and Brouncker, but at the same time also to Frenicle, Schooten "and all others in Europe" to solve a few problems, with special emphasis upon what later became known (through a mistake of Euler's) as "Pell's equation". What would have been Fermat's astonishment if some missionary, just back from India, had told him that his problem had been successfully tackled there by native mathematicians almost six centuries earlier (cf. Chap.I, §IX)!

Various diophantine equations, both in integers and in fractions, are also included in that correspondence; some have already been considered above, e.g., the impossibility of a pythagorean triangle of square area; of others more anon. Here, where we are concerned with quadratic forms, we merely take note of a conjecture about the form $X^2 + 5Y^2$; Fermat had observed experimentally that, while primes of the form $20n + 3$, $20n + 7$ cannot be written as $x^2 + 5y^2$, any product of two such primes can be so written; he mentions (*Fe*.II.405) that he is convinced of this, but cannot prove it. As has been known since Lagrange (cf. *Lag*.III.775–6) this is just where the question of the class-number enters the picture. No wonder Fermat was puzzled!

(v) Finally, in his communication of 1659 to Huygens, Fermat, while repeating some of the above statements, refers to his work on Diophantus's "simple and double equations". Most of these lead to curves of genus 1 (cf. *infra*, §XV); but Fermat gives two examples which belong to the theory of quadratic forms. Take for instance, he says, the equation $2x^2 + 7967 = y^2$; "I have a general rule for solving this if possible, or else for finding that it is impossible, and I can do it, whatever the coefficients may be"; to this he adds a further example[7], the "double equation" $2x + 3 = y^2$, $3x + 5 = z^2$, which amounts to $2z^2 - 3y^2 = 1$. The reference to Diophantus indicates that here Fermat asks for solutions "in fractions", i.e., for rational solutions. In other words, he refers to the general problem, successfully treated by Lagrange in 1768, of solving any equation $x^2 = Ay^2 \pm Bz^2$ in integers.

---

[7] In *Fe*.II.435, line 2, $2N + 5$ is a misprint for $3N + 5$ (cf. *Fe*.IV.140, line 2, and *Huy*.II.461, line 10). As to the first equation, *Fe*.IV.139, line 2 from b., has 1967 instead of 7967; it matters very little which one is the correct reading; note that $7967 = 31 \times 257$ and $1967 = 7 \times 281$.

§XII.

Now we shall take up the topics listed in §XI; we begin with (iii) and (ii).

In Chap.I, §VIII, we have described the two formulas (5) and (6) as being of ancient origin; here it will be enough to write both as

(5) $(x^2 + Ay^2)(z^2 + At^2) = (xz \pm Ayt)^2 + A(xt \mp yz)^2,$

where A may be either a positive or a negative integer; the times when the transition from one case to the other could have caused some difficulty were long past. This point of view was familiar to Fermat and not a few of his contemporaries; they may have regarded (5) as a rather obvious generalization of the special case $A = 1$.

In §§VII, VIII and IX, we attempted, with Euler's help, to reconstruct Fermat's procedure concerning sums of two squares by means of lemmas 2 (§VIII) and 3 (§IX). Now, still following Euler, we note that lemma 2 subsists without any change, with its proof, if we replace the form $X^2 + Y^2$ by $X^2 + AY^2$. Of course composition has now to be understood as based on the identity (5). For (3) and (4), one has then to substitute

$$ax \pm Aby = qu, \quad ay \mp bx = qv,$$

$$a = ux + Avy, \quad b = \pm (uy - vx).$$

From this extension of lemma 2 one concludes, just as in §IX, that, if $A > 1$, a prime q cannot be written as $x^2 + Ay^2$ in more than one way; if it could, our result, applied to q and to $N = q$, would give $1 = u^2 + Av^2$. On the other hand, this same remark leads at once to Pell's equation if A is negative; this will be discussed more in detail in §XIII.

Again, just as in §VIII, we conclude from the lemma that, if a prime p divides a number $N = a^2 + Ab^2$, and if every one of the other prime divisors of N can be written as $x^2 + Ay^2$, then the same is true of p. We now reconsider in this light the proof given in §VIII for the form $X^2 + Y^2$.

Let us for a moment adopt Euler's terminology, calling an
odd prime p a *prime divisor* of the "formula" (or "form")
$X^2 + AY^2$ if it divides some integer $N = a^2 + Ab^2$ with a
and b prime to p; this is so if and only if $-A$ is a quadratic
residue modulo p (cf. §VII). Let p be such; take a, b prime
to p and such that p divides $a^2 + Ab^2$. Let r be the remainder
of the division of a by p, and put $a' = r$ or $p - r$, whichever
is smaller; b' being defined similarly from b, we can replace
a, b by a', b', and then divide a', b' by their g.c.d.; in other
words, we may assume that a and b are mutually prime and
$< p/2$. Put $N = a^2 + Ab^2$. If $A = 2$, we have $n < 3p^2/4$, so
that all prime divisors of N, other than p, are $< 3p/4$; then,
just as in §VIII, we may, using "complete induction", assume
that they can all be written in the form $x^2 + 2y^2$ (note that
this includes the prime 2 if N happens to be even), and,
applying lemma 2, we conclude that also p can be so written.
Take $A = 3$; then we get $N < p^2$; thus the same reasoning
is valid if N is odd. If N is even, a and b must be odd, and
we have $a \equiv \pm b \pmod 4$ if the sign is properly chosen.
Then $a' = \frac{1}{4}(a \pm 3b)$ and $b' = \frac{1}{4}(a \mp b)$ are integers, and so
is $N/4 = a'^2 + 3b'^2$; we can then replace N by $N/4$, and so
on if necessary, until we have replaced N by an odd integer
with the same properties. Thus every prime divisor of X^2
$+ 3Y^2$ can be written as $x^2 + 3y^2$. We can proceed similarly
for $A = -2$ since in that case $0 < a, b < p/2$ implies $|N| <$
$p^2/2$. As observed above, the representation $p =$
$x^2 + Ay^2$ must be unique if $A = 2$ or 3, but of course not if
$A = -2$, as appears from the identity (5) combined with
$1 = 3^2 - 2 \times 2^2$. Incidentally, as Euler was to observe, the
identity (5) shows that, for $A = 1, \pm 2, 3$, all odd divisors
(and not only all odd prime divisors) of any integer $N =$
$a^2 + Ab^2$ with a prime to b can be written as $x^2 + Ay^2$.

Obviously every integer $x^2 + 3y^2$, if not a multiple of 3,
must be of the form $3n + 1$; therefore the above results
imply that every prime divisor of $X^2 + 3Y^2$, i.e., every odd
prime for which -3 is a quadratic residue, must be of that
form; to complete the proof of Fermat's statement of 1654

about such primes, one has to prove the converse. This is the same kind of difficulty which was pointed out in §VII concerning the form $X^2 + Y^2$; following Euler again, we write $p = 3n + 1$ and use the identity

$$x^{3n} - 1 = (x^n - 1)(x^{2n} + x^n + 1).$$

For all x prime to p, p divides the left-hand side; in view of lemma 1, §VII, we conclude that, for some x prime to p, p must divide the second factor in the right-hand side, and therefore also

$$4(x^{2n} + x^n + 1) = (2x^n + 1)^2 + 3.$$

This shows that p is indeed a divisor of $X^2 + 3Y^2$.

As to the forms $X^2 \pm 2Y^2$, take first $p = 8n + 1$ and write

$$x^{8n} - 1 = (x^{4n} - 1)[(x^{2n} \mp 1)^2 \pm 2(x^n)^2].$$

As above, using lemma 1, we conclude that p is a divisor of $X^2 + 2Y^2$ and of $X^2 - 2Y^2$. Now let p be of the form $8n - 1$ or $8n + 3$. If $p = 8n - 1$, it cannot be a divisor of $X^2 + 2Y^2$, for we have seen that such a divisor would itself be of the form $x^2 + 2y^2$ and therefore of the form $8n + 1$ or $8n + 3$. For a similar reason, if $p = 8n + 3$, it cannot be a divisor of $X^2 - 2Y^2$. Now in both cases p is not a divisor of $X^2 + Y^2$, as has been shown in §VII; therefore, if we put $p = 2m + 1$, no two of the $2m$ integers

$$1^2, 2^2, ..., m^2, -1^2, -2^2, ..., -m^2$$

can be congruent modulo p, so that they exhaust the $2m$ congruence classes prime to p modulo p. This shows that, for every a prime to p, one of the integers a, $-a$ is a quadratic residue and the other is not. Take $p = 8n + 3$; as p is not a divisor of $X^2 - 2Y^2$, 2 is not a quadratic residue modulo p; therefore -2 is such a residue, which is the same as to say that p is a divisor of $X^2 + 2Y^2$; then it can itself be written as $x^2 + 2y^2$. Similarly, if $p = 8n - 1$, it is a divisor of

$X^2 - 2Y^2$ and can be written as $x^2 - 2y^2$. This completes the proof of Fermat's statements about the forms in question.

Concerning the number of representations of integers by the forms $X^2 + 2Y^2, X^2 + 3Y^2$, Fermat must have known that whatever method he had followed for sums of two squares could be applied with almost no change; lemma 3 of §IX remains valid for these forms, as well as its proof, provided one adds the assumption that $q > 3$, $N/q > 3$ if the form is $X^2 + 3Y^2$. Just as in §IX, one can conclude from this that all representations of an integer N can be derived by composition (in the sense of identity (5)) from those of the primes dividing N. In particular, every proper representation of an integer N^r can be derived by composition from one of N. For $r = 2$, this can be verified elementarily, just as in the case of sums of two squares (cf. above, §IX); in fact, the case $r = 2$, for the form $X^2 + 2Y^2$, has already played a role in Fermat's proof described in §X. At least the case $r = 3$ must also have been known to Fermat, since it is needed in some of the proofs to be discussed below in §XVI; we shall formulate it here as a lemma:

Lemma 5. Assume that $N^3 = a^2 + Ab^2$, with a, b mutually prime, and $A = 2$ or 3. Then one has $N = u^2 + Av^2$, with u, v such that $a = u^3 - 3Auv^2$, $b = 3u^2v - Av^3$.

The proof is similar to that of lemma 4 in §IX; of course it has now to be based on (5). Here, too, one would obtain an easier proof by making use of the ring $\mathbf{Z}[\sqrt{-2}]$ resp. $\mathbf{Z}[j]$ (cf. Appendix I).

The case of the form $X^2 - 2Y^2$ is more complicated, since here the solutions of $x^2 - 2y^2 = \pm 1$ have to be taken into account (cf. Chap.I, §§VIII and IX). Some essential facts about it were known to Fermat already in 1641 (cf. *loc.cit.* §XI(ii)), but his main statements are contained in his communication to Frenicle of which the latter got a few copies printed and added to his own pamphlet of 1657 (JEH.41–44). His motivation at that time was the study of the diophantine equation

(6) $(2X^2 - 1)^2 = 2Y^2 - 1.$

Fermat told Huygens in 1659 that this has no other solution in integers than $(1, 1)$ and $(2, 5)$ and that he has sent a proof of this to Frenicle (Fe.II.434 = Huy.II.461, and again Fe.II.441 = Huy.II.538). Genocchi proved it rather easily in 1883 ($Nouv.Ann.Math$.(III) 2, 306–310), using the fact (proved by Euler, and possibly known to Fermat; cf. above, §X) that the equations $X^4 \pm Y^4 = 2Z^2$ have no solution. For Fermat, equation (6) turned up in connection with a problem included in his challenge of 1657 "to the English mathematicians and all others", viz., to find a cube which, added to the sum of its divisors, makes up a square. As an example, he mentioned 7^3, which gives $1 + 7 + 7^2 + 7^3 = 20^2$ (Fe.II.332), but soon he added that he knew of other solutions (Fe.II.342).

As was also noticed by Frenicle, if the cube is p^3, with p a prime, Fermat's condition can be written

$$1 + p + p^2 + p^3 = (p+1)(p^2+1) = \text{a square.}$$

As 2 is not a solution, $p + 1$ and $p^2 + 1$ have the g.c.d. 2 and must be of the form $2x^2$, $2y^2$, where (x, y) is a solution of (6). Frenicle could not show that there is no such prime, but found out somehow that there is none between 7 and 10^{75} (cf. Huy.II.25 and JEH.23–24). Thus, in dealing with (6), Fermat's attention was drawn more particularly to the case where $p = 2x^2 - 1$ is a prime. This is how he proceeds.

From every representation $n = x^2 - 2y^2$ of a positive or negative integer n by the form $X^2 - 2Y^2$, one deduces "by composition" the representations

$$-n = (x \pm 2y)^2 - 2(x \pm y)^2$$

of $-n$; this is no more than the case $A = -2$, $z = t = 1$ of (5) (cf. Chap.I, §VIII). Thus, in dealing with that form, it is enough to consider only representations either of positive or of negative integers. In part because of the application to (6), Fermat restricts himself to the latter, i.e., to representations $n = 2y^2 - x^2$ of positive integers n, where of course he takes x, y to be positive integers, and we shall do the same; then $0 \leqslant x < y \sqrt{2}$. Also by composition, i.e., by

interuallum numerorum 2. minor autem 1 N. atque ideo maior 1 N. + 2. Oportet itaque 4 N. + 4. triplos esse ad 2. & adhuc superaddere 10. Ter igitur 2. adscitis vnitatibus 10. æquatur 4 N. + 4. & fit 1 N. 3. Erit ergo minor 3. maior 5. & sarisfaciunt quæstioni.

ς' ἐτός. ὁ ἄρα μείζων ἴσαι ς' ἱνὸς μ̅ β̅. δήσει ἄρα ἀςιθμὸς δ̅ μονάδες δ̅ τριπλασίονας ᾖ μ̅ β̅. ἓ ἔτι ὑπερέχειν μ̅ ἱ. τρὶς ἄρα μονάδες ς μ̅ μ̅ ἱ. ἴσαι εἰσὶν ς̅' δ̅ μονάσι δ̅. κỳ γίνεται ὁ ἀριθμὸς μ̅' γ̅. ἴσαιδ ἐλάσσαι μ̅ γ̅. ὁ δὲ μείζων μ̅ ἱ. κỳ ποιοῦσι τὸ πρόβλημα.

CONDITIONIS appositæ eadem ratio est quæ & appositæ præcedenti quæstioni, nil enim aliud requirit quàm vt quadratus interualli numerorum sit minor interuallo quadratorum, & Canones iidem hic etiam locum habebunt, vt manifestum est.

QVÆSTIO VIII.

PROPOSITVM quadratum diuidere in duos quadratos. Imperatum sit vt 16. diuidatur in duos quadratos. Ponatur primus 1 Q. Oportet igitur 16 — 1 Q. æquales esse quadrato. Fingo quadratum a numeris quotquot libuerit, cum defectu tot vnitatum quod continet latus ipsius 16. esto a 2 N. — 4. ipse igitur quadratus erit 4 Q. + 16. — 16 N. hæc æquabuntur vnitatibus 16 — 1 Q. Communis adiiciatur vtrimque defectus, & à similibus auferantur similia, fient 5 Q. æquales 16 N. & fit 1 N. ⅘ Erit igitur alter quadratorum ²⁵⁶⁄₂₅. alter verò ¹⁴⁴⁄₂₅ & vtriusque summa est ⁴⁰⁰⁄₂₅ seu 16. & vterque quadratus est.

ΤΟΝ ἐπιταχθέντα τετράγωνον διελεῖν εἰς δύο τετραγώνους. ἐπιτετάχθω δὴ τ̅ ιϛ̅ διελεῖν εἰς δύο τετραγώνους. κỳ τετάχθω ὁ πρῶτος δυνάμεως μιᾶς, δήσει ἄρα μονάδες ιϛ̅ λείξει δυνάμεως μιᾶς ἴσας ᾖ τετραγώνῳ. πλάσω τ̅ τετράγωνον ἀπὸ ς̅. ὅσων δὴ ποτε λείξει ποσῶν μ̅' ὅσων ἐςὶν ἡ τ̅ ιϛ̅ μ̅ πλδυ₅᷄. ἴσω ς̅ β̅ λείξει μ̅' δ̅. αὐτὸς ἄρα ὁ τετράγωνος ἴσαι δυνάμεων δ̅ μ̅' ιϛ̅ λείξει ς̅ ιϛ̅. ταῦτα ἴσα μονάσι ιϛ̅ λείξει δυνάμεως μιᾶς. κοινὴ προσκείσθω ἡ λείξις, κỳ ἀπὸ ὁμοίων ὅμοια. δυνάμεις ἄρα ε̅ ἴσαι ἀριθμοῖς ιϛ̅. κỳ γίνεται ὁ ἀριθμὸς ς̅. πέμπτων. ἴσαι ὁ μὲν ὅς εἰκοσιπεμπτῶν. ὁ δὲ ρμδ̅ εἰκοσιπεμπτῶν. ἓ οἱ δύο συντεθέντες ποιοῦσι

ῦ εἰκοσέπεμπτα, ἤτοι μονάδες ιϛ̅. καὶ ἴσιν ἑκάτερος τετράγωνο.

OBSERVATIO DOMINI PETRI DE FERMAT.

CVbum autem in duos cubos, aut quadratoquadratum in duos quadratoquadratos & generaliter nullam in infinitum vltra quadratum potestatem in duos eiusdem nominis fas est diuidere cuius rei demonstrationem mirabilem sane detexi. Hanc marginis exiguitas non caperet.

QVÆSTIO IX.

RVrsvs oporteat quadratum 16 diuidere in duos quadratos. Ponatur rursus primi latus 1 N. alterius verò quotcunque numerorum cum defectu tot vnitatum, quot constat latus diuidendi. Esto itaque 2 N. — 4. erunt quadrati, hic quidem 1 Q. ille verò 4 Q. + 16. — 16 N. Cæterum volo vtrumque simul æquari vnitatibus 16. Igitur 5 Q. + 16. — 16 N. æquatur vnitatibus 16. & fit 1 N. ⅘ erit

ΕΣΤΩ δὴ πάλιν τὸν ιϛ̅ τετράγωνον διελεῖν εἰς δύο τετραγώνους. τετάχθω πάλιν ἡ τ̅ πρῶτου πλδυᾶ ς̅' ἱνὸς, ἡ δὲ τ̅ ἑτέρα ς̅ ὁσωνδήποτε λείξει μ̅' ὅσων ἔςὶ ἡ τ̅ διαιρουμένου πλδρᾶ. ἴσω δὴ ς̅ β̅ λείξει μ̅ δ̅. ἴσονται οἱ τετράγωνοι ὃς μὲν δυνάμεως μιᾶς, ὃς δὲ δυνάμεων δ̅ μ̅' ιϛ̅ λείξει ς̅ ιϛ̅. βούλομαι τοὺς δύο λοιποῦ συντεθέντας ἴσους ᾖ μ̅ ιϛ̅. δυνάμεις ἄρα ε̅ μ̅' ιϛ̅ λείξει ς̅ ϝ̅ ἴσαι μ̅ ιϛ̅. κỳ γίνεται ὁ ἀριθμὸς ς̅ πεμπτῶν.

Fermat's "last theorem" (page 61 of the *Diophantus* of 1670).

applying (5) to the case $A = -2, z = 3, t = 2$, we see that either one of the substitutions

(7) $(x, y) \rightarrow (|3x - 4y|, 3y - 2x)$

(8) $(x, y) \rightarrow (3x + 4y, 3y + 2x)$

transforms any representation $n = 2y^2 - x^2$ of n into another one. Among these, as Fermat now observes, there is always one for which $0 \leqslant x \leqslant y$; in fact, if $x > y$, (7) transforms (x, y) into (x', y') with $x' < x, y' < y$, so that, by iterating (7) as often as may be needed, one comes to a representation (x_0, y_0) with $x_0 \leqslant y_0$; such a representation is called "*le plus petit couple*" by Fermat; we shall call it *minimal*. Applying (7) to it again, one gets a representation (x_0', y_0') with $x_0' \geqslant y_0'$, $x_0' \geqslant x_0, y_0' \geqslant y_0$; applying (8) repeatedly to (x_0, y_0) and to (x_0', y_0'), one gets two sequences $(x_i, y_i), (x_i', y_i')$ of representations of one and the same integer n, all distinct from one another unless $x_0 = 0$ or $x_0 = y_0$. From this Fermat concludes rightly that every solution of $n = 2y^2 - x^2$ can be derived by this process from a minimal one; "the proof of all this is easy", he writes, even though a critical reader may well feel that some further elaboration would be welcome. For the modern reader, the matter could best be clarified by working in the ring $\mathbf{Z}[\sqrt{2}]$ (cf. Appendix I); to write $n = 2y^2 - x^2$ is the same as to write $-n = N(\xi) = \xi\xi'$ with $\xi = x + y\sqrt{2}, \xi' = x - y\sqrt{2}$; the substitution (8) amounts to $\xi \rightarrow \varepsilon^2\xi$ with $\varepsilon = 1 + \sqrt{2}$, and (7) to $\xi \rightarrow \varepsilon^{-2}\xi$ or to $\xi \rightarrow -\varepsilon^2\xi'$, as the case may be; but it is hardly to be imagined that Fermat, even heuristically, would have proceeded in this manner, in spite of the precedent offered by Euclid in his tenth book (cf. Chap.I, §VIII). Anyway, the above treatment, applied to the case $n = 1$ and the "minimal representation" $1 = 2 \times 1^2 - 1^2$, supplies the fact that all solutions of $1 = 2y^2 - x^2$ can be derived from that one by the repeated application of (8), i.e. by repeated composition with the representation $1 = 3^2 - 2 \times 2^2$. Again, by composition with the minimal representation, this produces all solutions of $x^2 - 2y^2 = 1$, thus showing that Theon's "side and diagonal numbers" (cf. Chap.I, §VIII) give all the solutions of $x^2 - 2y^2 = \pm 1$.

Fermat's further deductions must have followed the pattern set by his theory of the form $X^2 + Y^2$, even though it cannot be said that this pattern emerges clearly from his communication to Frenicle. Just as he had proved that a prime $p = 4n + 1$ has one and only one representation as a sum of two squares (cf. above, §§VIII and IX), he must have found, using similar arguments, that any two representations of a prime by the form $X^2 - 2Y^2$ must be such that they can be derived from each other by composition with a representation of ± 1 (cf. the proof of lemma 2 in §VIII, and the above remarks about its extension to forms $X^2 + AY^2$); in view of what has been shown above, it amounts to the same to say that a prime p can have only one minimal representation $p = 2b^2 - a^2$ with $0 < a < b$. Then, by composition, one derives from this a representation $p^2 = 2B^2 - A^2$, with

(9) $A = |a^2 + 2b^2 - 4ab|, \qquad B = a^2 + 2b^2 - 2ab,$

of which it is easily verified that it is minimal. Conversely, if p is any odd prime, assume that p^2 has the representation $p^2 = 2B^2 - A^2$; this may be assumed to be minimal, so that $0 < A < B$; we have then $p^2 = A'^2 - 2B'^2$ with $A' = 2B + \delta A$, $B' = B + \delta A$, $\delta = \pm 1$; since A must be odd, δ may be chosen so that $A' \equiv p + 2 \pmod 4$. This gives

$$2B'^2 = (A' - p)(A' + p),$$

from which (just as in Fermat's proof in §X) one can conclude that A', B', p can be written as

$$A' = a^2 + 2b^2, \qquad B' = 2ab, \qquad p = 2b^2 - a^2$$

where a, b are positive integers; then A, B are given by (9). Moreover, the inequalities $p > 0$, $A < B$ if $\delta = +1$, and $p > 0$, $A > 0$ if $\delta = -1$, imply (when p, A, B are expressed in terms of a and b) $a < b$, so that $p = 2b^2 - a^2$ is the minimal representation of p.

Now assume, as in Fermat's problem, that $p = 2x^2 - 1$ and $p^2 = 2y^2 - 1$; these are minimal representations of p and of p^2, respectively. Applying to the latter what has been proved above, i.e. taking in it $A = 1$ and $B = y$, we find that A and B are given by (9), where $p = 2b^2 - a^2$ is a minimal representation of p. As we have just seen, however, a prime p can have only one such representation; therefore we have $a = 1$, $b = x$, and (9) gives now:

$$1 = |1 + 2x^2 - 4x|,$$

which is easily seen to have no other solutions than $x = 1$ or 2, giving $p = 1$ (which is not a prime) and $p = 7$. This completes the proof, so far as primes are concerned.

While Fermat is far from explicit on some of the points considered above, one may still plausibly argue that he could have filled in the missing links, had he cared to do so. Unluckily his communication to Frenicle, brilliant as it is, does not end there. "What has been proved for primes", he writes, "can be extended, as you know, to composite numbers; . . . I need not dwell at length on such easy matters". Therefore, he says, equation (6) has indeed no other solutions than (1, 1) and (2, 5).

His argument, however, fails for "composite" (i.e., non-prime) numbers, because such integers have in general more than one "minimal" representation, as he and Frenicle must have known for many years (cf. *loc.cit.* §XI(ii)). Apparently he had trapped himself by an imprudent use of the words "*le plus petit couple*", translated above by "minimal". Of course, for a given n, the smallest solution of $n = 2y^2 - x^2$ is "minimal", but the converse is not true.

There is no sign that Frenicle detected the fallacy in Fermat's final argument; nor did J. E. Hofmann when he published Fermat's piece in 1944. Indeed one may, without unfairness, suspect that the proof went somewhat over his head. Nevertheless, on this occasion at least, it seems clear that Fermat's quickness of mind had tripped him. The reputation for infallibility that some of his admirers have sought to build up for him is no more than a legend.

§XIII.

The study of the form $X^2 - 2Y^2$ must have convinced Fermat of the paramount importance of the equation $x^2 - Ny^2 = \pm 1$ to be solved in integers, or, in modern parlance, of the units of real quadratic fields. As we have seen in Chap.I, §§VIII–IX, Archimedes and the Indians had trodden this path long before him; of course he could not have known about the Indians, nor about Archimedes's "*problema bovinum*" (the epigram discussed in Chap.I, §VIII). He had read Theon of Smyrna (*Fe*.II.266), and therefore knew about the "side and diagonal numbers" (cf. Chap.I, §VIII), but this did not go beyond the case $x^2 - 2y^2 = \pm 1$.

When Fermat, in 1657, offered the equation $x^2 - Ny^2 = 1$ as a challenge problem to the English and all others, he was asking for solutions in integers; this had not been specified in the challenge that went to England, but "it goes without saying", as he wrote to Frenicle (*Fe*.II.334), since any novice ("*le moindre arithméticien*") can obtain a solution in fractions. It is just such a solution in fractions, however, that Wallis and Brouncker sent him at first (*Fe*.III.418). When Fermat pointed out to them that this was not what he meant, did he expect to make them confess their ignorance? Did he feel a touch of annoyance when, after only a few months' delay, he received from them a complete method of solution? In his answer, sent as usual through Digby, he says that he "willingly and joyfully acknowledges" the validity of their solution ("*libens agnosco, imo et gaudeo . . .*", *Fe*.II.402). Writing privately to Huygens the next year, however (*Fe*.II.433), he points out that the English had failed to give "a general proof". Huygens had already said as much to Wallis (*Huy*.II.211). A general proof, according to Fermat, can only be obtained by descent.

It may thus be surmised that Fermat's method of solution did not greatly differ from the one he got from Wallis and Brouncker[8], but that he had been able to extract from it a formal proof of the fact that it always leads to a solution. The method to be summarized now is the one which Wallis credits to Brouncker (*Wal*.II.797; for more details, cf. A. Weil, *Coll.Papers* III.413–420); it is the most satisfactory variant among several that are described in that correspondence, and is equivalent (in substance) to the Indian *cakravāla* method (Chap.I, §IX) as well as to the modern treatments based on continued fractions.

The starting point here is to assume the existence of a solution (x, y) of $x^2 - Ny^2 = 1$, and to transform the problem by successive steps into others, each of which has a solution in smaller numbers; as may be seen from Fermat's treatment of the pythagorean triangles of square area (above, §X), this is typical of Fermat's descent; the difference lies in the fact that in the latter case one seeks to prove that there is no solution, while in the former the purpose is to find one. At the same time, Bachet's method for the solution of the equation $ay - bx = \pm 1$ (cf. Chap.I, §IV) undoubtedly provided a model for Wallis, Brouncker and Fermat to follow, depending as it does on the following observation. For $a > b > 0$, assume that the equation $aY - bX = \pm 1$ has a solution (x, y) in positive integers (both not 0, since for Bachet and his contemporaries 0 was no more "a number" than 1 had been in the days of Euclid); of course a and b must be mutually prime. Write $a = bm + c$ with $0 < c < b$, and put $x = my + z$. Then we have $bz - cy = \mp 1$, and (y, z) is a solution in positive integers, respectively smaller than x and y, of the equation $bZ - cY = \mp 1$; this is similar to the original one, and incidentally has smaller coefficients. Repeating the process, one must eventually reach an equation

---

[8] Indeed J. Ozanam, describing a solution, identical with Brouncker's, on page 509 of his *Nouveaux éléments d'algèbre* of 1702, and illustrating it by treating the cases $N = 23$ and $N = 19$, attributes it to Fermat. Did he know more than we do, or had he merely misread the *Commercium Epistolicum*?

$rV - SU = \pm 1$ with $r > s > 0$, admitting a solution (u, v) with $u > v = 1$; this implies that $r = su \pm 1$; pushing the process either one or two steps further, as a modern arithmetician would do, one reaches a point where the solution is $(1, 0)$. Anyway, for the final equation, the solution is obvious, so that, running the process in reverse, one obtains a solution for the original problem. This is precisely the *kuṭṭaka* method of the Indian mathematicians (cf. Chap.I, §§IV and IX); as pointed out in Chap.I, it does not differ substantially from Euclid's algorithm for finding the g.c.d. of two integers, nor from the calculation of the continued fraction for a/b. The modern reader may also note that the first step in Bachet's solution can be written

$$\begin{pmatrix} a & b \\ x & y \end{pmatrix} = \begin{pmatrix} b & c \\ y & z \end{pmatrix} \cdot \begin{pmatrix} 0 & 1 \\ 1 & 0 \end{pmatrix} \cdot \begin{pmatrix} 1 & 0 \\ 1 & 1 \end{pmatrix}^m,$$

so that the whole process amounts to writing the matrix on the left-hand side as a product of the well-known generators of the group $GL(2, \mathbf{Z})$.

Brouncker's method for solving Fermat's problem follows the same pattern. Let N be a positive integer, not a square; assume that the equation $U^2 - NX^2 = \pm 1$ has a solution (u, x). Let n be the integer such that $n^2 < N < (n+1)^2$; this gives $u > nx$, so that we may put $u = nx + y$ with $y > 0$. Put at the same time $A = N-n, B = n, C = 1$. Then (x, y) is a solution of the equation

$$(10) \qquad\qquad AX^2 - 2BXY - CY^2 = \mp 1$$

where A, B, C are positive integers such that $B^2 + AC = N$. This will now be transformed by successive steps into equations of the same form with smaller and smaller solutions, leading (just as in Bachet's method) to the solution of the original problem.

Brouncker's procedure is to write $x = my + z$, where m is the largest integer less than the positive root of the equation $At^2 - 2Bt - C = 0$; he seems tacitly to assume that, for this choice of m, z must be >0 and $<y$. Then (y, z) is the solution of an equation similar to (10), and the procedure

may be continued until one finds an equation with an obvious solution $(u, 1)$; as this happens in all the numerical cases which Wallis and Brouncker used to test their method, including those which Fermat had specifically mentioned, they saw no need for digging deeper. Fermat knew better.

In order to fill the gaps in the above treatment, it is necessary to point out the special features of (10) which permit the successful application of Brouncker's process. For brevity's sake, we will adopt a modified Gaussian notation, writing (A, B, C) for the quadratic form in the left-hand side of (10), where it is understood that A, B, C are positive integers and not 0; moreover, also by a slight modification of Gaussian usage, we will say that (A, B, C) is *reduced* if the positive root of $At^2 - 2Bt - C$ is >1 and the negative root is > -1; this is so if and only if $|A - C| < 2B$, and then (C, B, A) is also reduced. For $A = N - n^2$, $B = n$, $C = 1$, (A, B, C) is obviously reduced. Now, for any reduced form $F = (A, B, C)$, let again m be the largest integer less than the positive root of $At^2 - 2Bt - C$. If we put

$$A' = -(Am^2 - 2Bm - C), \qquad B' = Am - B, \qquad C' = A,$$

m is the largest integer for which $A' > 0$; it is then easily seen that $B' > 0$, that $F' = (A', B', C')$ is again reduced, and that the substitution

$$X = mX' + Y', \qquad Y = X'$$

changes F into $-F'$. If $N = B^2 + AC$ is the "determinant" of F, F' has of course the same one. As there can be no more than finitely many forms (A, B, C) with a given determinant N, it follows that the successive quadratic forms F, F', etc. produced by Brouncker's process must eventually repeat themselves. This would surely have been noticed by Wallis and Brouncker if only they had pushed the treatment of their numerical examples a little further than they did.

Notations being as above, we will say that $F' = (A', B', C')$ is the form derived from $F = (A, B, C)$. As any example will show at once, the form (C, B, A) is then the one derived from (C', B', A'); indeed we have

$$C = -(C'm^2 - 2B'm - A'), \qquad B = C'm - B', \qquad A = C',$$

and m is the largest integer for which $C'm^2 - 2B'm - A' < 0$. Now, starting from any given reduced form $F_0 = (A_0, B_0, C_0)$, we get by Brouncker's process a sequence $F_1 = F_0', F_2 = F_1'$, etc., where for each i the form $F_i = (A_i, B_i, C_i)$ is the one derived from F_{i-1}; in an obvious sense, we may say that F_i is the i-th derived form of F_0. All of these are reduced forms with the determinant N; as we have seen, they must repeat themselves, so that we have $F_{i+p} = F_i$ for some $i \geqslant 0$ and some $p \geqslant 1$. For each i, put $G_i = (C_i, B_i, A_i)$; then G_0 is the i-th derived form of G_i, and G_p is the i-th derived form of G_{i+p}. As $F_{i+p} = F_i$, we have $G_{i+p} = G_i$, and therefore $G_p = G_0$, $F_p = F_0$; then, for all k, we have $F_{kp} = F_0$. We have thus proved that Brouncker's process leads eventually to a repetition of the initial form F_0, and to a periodic sequence of forms from then on; if the starting point was Brouncker's, i.e., the form $(N - n^2, n, 1)$ we must have $C_p = 1$ for some $p > 1$, and therefore $A_{p-1} = 1$. Consequently, the equation $F_{p-1} = 1$ has the trivial solution $(1, 0)$, from which, running the process in reverse, one deduces a solution of the original problem. As the sign of the right-hand side, in the successive equations $F_i(X, Y) = \pm 1$, alternates from one equation to the next, it is $\pm(-1)^{i+1}$ if one has started originally from $U^2 - NX^2 = \pm 1$. Therefore the solution $(1, 0)$ for $F_{p-1} = 1$ gives a solution (u, x) for the equation $U^2 - NX^2 = +1$ if p is even, and for $U^2 - NX^2 = -1$ if p is odd. In the latter case, in order to get a solution for $U^2 - NX^2 = +1$, one has only to run the process through to F_{2p}, or more simply (what in fact amounts to the same) apply the basic identity (5) to the case $A = -N$, $x = z$, $y = t$. In fact, we have seen in Chap.I, §IX, how the Indian mathematicians used the identity (5) (the so-called *bhāvanā*) for shortcuts in the *cakravāla* process whenever a solution for $U^2 - NX^2 = -1$, ± 2, ± 4 has been found. As Wallis and Brouncker also discovered, precisely the same shortcuts can be applied to Brouncker's method whenever, in running the process through, one finds a form F_i with $A_i = 2$ or 4. They also noticed that the whole process may sometimes be shortened by making (with the same notations as before) the substitution

$$X = (m+1)X' - Y', \qquad Y = X'$$

instead of the one used above; in modern terms, this amounts
to building up a so-called "half-regular" continued fraction
for $\sqrt{N}$ instead of the "regular" one; in the *cakravāla* method,
as described in Chap.I, §IX, it corresponds more or less to
the prescription that at each step the integer denoted there
by $N - x^2$ should be made small in absolute value, rather
than small and >0. As pointed out at the time, such shortcuts
can be numerically useful but make the theoretical discussion
much more cumbersome.

The facts underlying the above treatment can be observed
quite easily on each numerical case to which Brouncker's
process is applied. Wallis and Brouncker treated among
others the case $N = 13$, which is quite typical and later
became Euler's favorite example for his description of their
method, e.g., in his *Algebra* of 1770 (*Eu*.I-1.383). Fermat,
writing to Frenicle in 1657, proposes the cases $N = 61$ and
$N = 109$, adding (misleadingly, or rather mischievously)
that he has chosen quite small numbers "*pour ne vous donner
pas trop de peine*"; he must have known, of course, that the
smallest solutions in these two cases are respectively

$$(1766319049,\ 226153980),$$
$$(158070671986249,\ 15140424455100),$$

and selected the values 61 and 109 precisely for that reason.

A final step is still required in order to show that the
process leads to all solutions, or, in modern terms, that the
group of units of a real quadratic field is generated by the
one obtained by the above method. For this, one has only
to start again (as we did heuristically, and as Wallis and
Brouncker did without looking any further) from a given
solution (u, x) of $U^2 - NX^2 = \pm 1$. This was transformed
by the substitution $u = nx + y$ into a solution (x, y) of the
next equation $F(x, y) = \mp 1$, then into a solution (x', y') of
the next equation $F'(x', y') = \pm 1$ by the substitution
$x = mx' + y'$. Experience shows that this leads to smaller
and smaller pairs of integers, and ultimately to a pair $(x, 1)$
and then to $(1, 0)$, which is where the process terminates;

but, for this, one has to know that $0 < x' < x$, $0 < y' < y$, etc. As to the initial step $u = nx + y$, the assumption $u^2 - Nx^2 = \pm 1$ gives

$$n^2x^2 - 1 < u^2 < (n+1)^2x^2 + 1;$$

this implies $nx \leqslant u \leqslant (n+1)x$; it implies $nx < u < (n+1)x$ unless $x = 1$. As to the next step and all following ones, assume that (x, y) satisfies

$$F(x, y) = Ax^2 - 2Bxy - Cy^2 = \mp 1.$$

Put $f(t) = F(t, 1)$, $\varphi(t) = f(t) \pm y^{-2}$. As the roots of f lie in the intervals $[-1, 0]$ and $[m, m+1]$, we have

$$f(-1) > 0, \quad f(0) < 0, \quad f(m) < 0, \quad f(m+1) > 0.$$

As these four numbers are integers, this implies

$$\varphi(-1) > 0, \quad \varphi(0) < 0, \quad \varphi(m) < 0, \quad \varphi(m+1) > 0$$

unless $y = 1$. As our assumption on (x, y) implies that x/y is a root of φ, this proves that x/y lies in the interval $[m, m+1]$, as we wished to show.

One more point deserves mentioning. In 1654, writing to Pascal about his "inventions" on "numbers", Fermat cites a "general rule" for finding the representation of any prime q of the form $4m + 1$ as a sum of two squares (cf. above, §IX), but, as usual, gives no details. However, the application of Brouncker's process to $N = q$, and for instance to any one of the cases $N = 13, 61, 109$, shows at once that a form F_j with $A_j = C_j$ will occur in the sequence of reduced forms produced by that process; then we have $q = B_j^2 + A_j^2$. Thus, for $N = 13$, one finds $F_2 = (3, 2, 3)$; for $N = 61$, $F_5 = (5, 6, 5)$; for $N = 109$, $F_7 = (3, 10, 3)$. This could well be Fermat's "general rule".

A more careful observation would show even more: in those cases, the period $(F_0, F_1, ..., F_{p-1})$ in that sequence of forms has $p = 2j + 1$ terms, with j as above. A formal proof for this can be given as follows. Let (u, x) be the smallest solution for $u^2 - qx^2 = 1$; here u must be odd and x even, so that we may write

$$\frac{u+1}{2} \cdot \frac{u-1}{2} = q\left(\frac{x}{2}\right)^2.$$

The factors in the left-hand side are mutually prime since they differ by 1, and q must divide one of them; therefore one of them is of the form v^2, the other of the form qt^2, and we have $v^2 - qt^2 = \pm 1$. The sign cannot be $+$, since then (u, x) would not be the smallest solution of $u^2 - qx^2 = 1$. This gives $v^2 - qt^2 = -1$. As we have seen before, this implies that the number p of terms in the period $(F_0, F_1, ..., F_{p-1})$ is odd; put $p = 2j + 1$. Then $F_{2j+1} = F_0$, $C_{2j+1} = 1$ and consequently $A_{2j} = 1$. Now observe that, if a form $(1, B, C)$ with the determinant $N = B^2 + C$ is reduced, we must have $-1 < B - \sqrt{N} < 0$, $B = n$, $C = N - n^2$; in particular, we get $F_{2j} = (1, n, N-n^2)$. As before, put $G_i = (C_i, B_i, A_i)$ for all i; we get $G_{2j} = F_0$. As we have seen, G_j is the j-th derived form of G_{2j}, i.e., of F_0; it is therefore the same as F_j, and we have $A_j = C_j$, as was to be proved.

It does not seem unreasonable to assume that this may have been, in general outline, what Fermat had in mind when he spoke of his proofs in his letters to Pascal and Huygens; how much of it he could have made explicit must remain a moot question. Perhaps one of his most serious handicaps was the lack of the subscript notation; this was introduced later by Leibniz, in still imperfect form, and its use did not become general until rather late in the next century. Faced with a difficult or complicated proof, Fermat might well be content with a careful analysis of some typical numerical cases, convincing himself at the same time that the steps involved were of general validity. Actually he never succeeded in writing down his proofs. Were they such as Fermat himself, in his criticism of Wallis (cf. §III), has insisted that they should be in order to provide "the foundations for a new branch of science"? This is what we shall never know.

§XIV.

Now we come to item (v) of §XI. Just as Euler served us as a guide for looking at the forms $X^2 + AY^2$ for $A = 1, 2, 3$ and Wallis and Brouncker in discussing "Pell's equation", here we shall enlist the services of Lagrange (*Lag*.II.390–399).

We start from the equation $ax^2 + by^2 + cz^2 = 0$, to be solved in rational numbers; a, b, c are assumed not to have the same sign. Multiplying with c, and replacing z by z/c, we can rewrite this as $z^2 = Ax^2 + By^2$, where A and B are not both negative; they may be assumed to be squarefree, for otherwise, if e.g. A has a divisor $m^2 > 1$, we can replace x by x/m. We may take notations so that $|A| \leq |B|$. It is of course enough to look for solutions in integers; in fact, we may require x, y, z to be mutually prime; for, if e.g. x and z have the common divisor $d > 1$, d^2 will divide z^2 and Ax^2, hence By^2; as B is squarefree, d must then divide y, and we can replace x, y, z by $x/d, y/d, z/d$.

As any common divisor of x and B must now divide z, x must be prime to B, so that it has an inverse x' modulo $|B|$. As $Ax^2 \equiv z^2$ (mod $|B|$), we have $A \equiv (x'z)^2$ (mod $|B|$). This shows that the given equation can have no solution unless A is a quadratic residue (not necessarily prime to B) modulo $|B|$; whether this is so or not can be determined by trial and error if the more sophisticated methods based on the law of quadratic reciprocity are not available.

Now assume that A is such; put $A \equiv a^2$ (mod $|B|$). After replacing a either by the remainder r of the division of a by $|B|$ or by $|B| - r$, whichever is smaller, we may assume that $0 \leq a \leq \frac{1}{2}|B|$. As A is squarefree, we have $A \neq a^2$ unless $a = \pm 1$, $A = 1$. Discarding this case as trivial, we can write $a^2 - A = BB_1d^2$, where d is taken so that B_1 is squarefree. We have

$$|B_1| \leq \left| \frac{a^2 - A}{B} \right| \leq \frac{1}{4}|B| + 1$$

and therefore $|B_1| < |B|$ except in the trivial case $B = \pm 1$,

$A = \pm 1$. On the other hand, our basic identity (5) of §XII gives

$$B_1(Bdy)^2 = (a^2 - A)(z^2 - Ax^2) = (az \pm Ax)^2 - A(ax \pm z)^2.$$

If we put $X = ax \pm z$, $Y = Bdy$, $Z = az \pm Ax$, this shows that (X, Y, Z) is a solution of $Z^2 = AX^2 + B_1Y^2$.

As $A \equiv a^2$ (mod $|B_1|$), this process can now be repeated until we reach an equation of the form $Z^2 = AX^2 + B_nY^2$ with $|B_n| < |A|$. We can then put $A' = B_n$, $B' = A$; in order to proceed further, A' must be a quadratic residue modulo $|B'|$. Continuing the descent in this way, with steadily decreasing values of the coefficients A and B, we must either reach an equation of the form $Z^2 = MX^2 + NY^2$ where M is not a quadratic residue modulo $|N|$, in which case the original equation itself has no solution, or an equation $Z^2 = X^2 + NY^2$ with the obvious solution $(1, 0, 1)$; in the latter case, one obtains a solution of the original equation by reversing the process.

It will be apparent that this is a typical method "by descent", quite in conformity with Fermat's general principles; in fact, the underlying ideas are quite close to those described in §XIII for dealing with Pell's equation (and even more so, perhaps, to the Indian procedure described in Chap.I, §IX). Here again we have no evidence for believing that Fermat proceeded in this manner, except for his statement that he had solved the problem and for the fact that it fits in rather well with what we know of his methods.

We are in a far worse position if we attempt to deal with his famous statement about every number being a sum of "three triangles, four squares, five pentagons, etc." (cf. above, §§V and XI(i)). It occurs as early as 1638 (*Fe*.II.65), at a time when it seems most unlikely that he could have had any kind of proof for it. Writing to Pascal in 1654, he mentions it as being so far his most important result (*Fe*.II.313), adding merely that the proof rests upon his theorems on sums of two squares. In his *Diophantus* (*Fe*.I.305, Obs.XVIII), he writes that "it depends upon the deepest mysteries" in number theory, to whose exposition he plans to dedicate a whole volume. Writing for Huygens in 1659 (*Fe*.II.433), he repeats only the assertion about four squares, omitting the other parts of his earlier statement; should one attach some significance to this omission? This time he adds that the proof is by showing that "if some number were not of this kind" (i.e., not a sum of no more than four squares), "there would be a smaller one with the same property, etc." Indeed proofs of that kind are known (cf. Chap.III, §XI); they are more or less similar to the corresponding proofs for sums of two squares, but they presuppose a knowledge of Euler's identity on the product of sums of four squares, which is not an easy one to discover; certain it is that there is no trace of it in Fermat's writings.

As to "triangles" and sums of three squares, the mystery is even deeper. That every number is a sum of three "triangles" amounts to saying that every integer of the form $8n + 3$ is a sum of three squares; from this, as Legendre showed in his *Théorie des Nombres* (3e éd.,t.II, pp.331–356), one can derive (not quite easily, but at any rate elementarily) all of Fermat's further assertions about "four squares, five pentagons, etc." Sums of three squares are also mentioned in a marginal note to Diophantus (*Fe*.I.314–315, Obs.XXVII); here the question raised by Diophantus is whether an integer $N = 3n + 1$ is a sum of three squares (in integers, or perhaps in fractions); for this to be so, Fermat says, it is necessary (and by implication sufficient) that n should not satisfy any of the congruences

$$n \equiv (1 + 4 + 4^2 + \cdots + 4^{n-1}) + 2 \times 4^n \pmod{8 \times 4^n}.$$

This note has all the appearances of belonging to an early period of Fermat's career. His condition amounts to saying that the number $N = 3n + 1$ must not be of the form $4^n(8m + 7)$, which is indeed the condition for N to be a sum of three squares; Fermat's knowledge of this fact can only have been based "on induction", since as late as 1658 he was unable to prove that $2p$ is a sum of three squares if p is a prime of the form $8n - 1$; this latter fact is stated as "true in the sense of Conon", i.e., conjectured but still unproved, in his last communication to Wallis (*Fe*.II.405). To us, the chief interest of the above-mentioned observation on Diophantus is that it gives in substance the 2-adic expansion

$$\frac{-1}{3} = \frac{1}{1-4} \equiv 1 + 4 + 4^2 + \cdots + 4^{n-1} \pmod{4^n}.$$

But, as to "sums of three triangles", as long as no proof is produced which could with some plausibility be attributed to Fermat, the mystery must remain.

§XV.

If one takes literally the terms of Fermat's communication of 1659 to Huygens, one has to conclude that he took up such diophantine problems as $x^3 + y^3 = z^3$, $x^2 + 2 = y^3$, $x^2 + 4 = y^3$ at a comparatively late stage of his career. The first one of these, however, occurs already in a letter to Mersenne in 1638 (*Fe*.II.65), and again, along with a few more of the same kind, in 1640 (*Fe*.II.195), with the clear implication that these equations have no solution except the obvious ones. In 1659 his memory may have been at fault; more probably he wishes only to refer to the time when he had obtained for these questions a reasonably satisfactory treatment, rather than to what, even in 1640, may have been a mere conjecture based on plausibility arguments. Certainly a period of intensive work on diophantine equations of genus 1 began for him around 1641, when problems about pythagorean triangles were proposed to him by Frenicle ("for whose love", he once said, he worked "on numbers"; cf. *Fe*.II.241,244,265); of course both of them must have known their Diophantus in and out. Eventually, as we learn from Digby (*Fe*.II.362), Frenicle's interest in such matters slackened; fortunately Fermat's did not, or at least his Jesuit correspondent Billy, by dint of skilful questioning seasoned with effusions of lavish praise, succeeded in drawing him out long enough to obtain from him the material for what later became his *Inventum Novum* (cf. above, §I).

As we have observed in Chap.I, §X, the most significant problems in Diophantus are concerned with curves of genus 0 or 1. With Fermat this turns into an almost exclusive concentration on such curves. Only on one ill-fated occasion did Fermat ever mention a curve of higher genus, and there can hardly remain any doubt that this was due to some misapprehension on his part, even though, by a curious twist of fate, his reputation in the eyes of the ignorant came to rest chiefly upon it. By this we refer of course to the incautious words "*et generaliter nullam in infinitum potestatem*" in his statement of "Fermat's last theorem" as it came to be vulgarly called: "No cube can be split into two cubes, nor any biquadrate into two biquadrates, *nor generally any power beyond the second* into two of the same kind" is what he wrote into the margin of an early section of his *Diophantus* (*Fe*.I.291, Obs.II), adding that he had discovered a truly remarkable proof for this "which this margin is too narrow to hold". How could he have guessed that he was writing for eternity? We know his proof for biquadrates (cf. above, §X); he may well have constructed a proof for cubes, similar to the one which Euler discovered in 1753 (cf. *infra*, §XVI); he frequently repeated those two statements (e.g. *Fe*.II.65,376,433), but never the more general one. For a brief moment perhaps, and perhaps in his younger days (cf. above, §III), he must have deluded himself into thinking that he had the principle of a general proof; what he had in mind on that day can never be known.

On the other hand, what we possess of his methods for dealing with curves of genus 1 is remarkably coherent; it is still the foundation for the modern theory of such curves. It falls naturally into two parts; the first one, directly inspired by Diophantus, may conveniently be termed a method of ascent, in contrast with the descent which is rightly regarded as more peculiarly Fermat's own. Our information about the latter, while leaving no doubt about its general features, is quite scanty in comparison with Fermat's testimony about the former (e.g. *Fe*.I.291–292, Obs.III; *Fe*.I.297–299, Obs.VIII–IX; *Fe*.I.322–325, Obs.XXX; *Fe*.I.328–329, Obs.XXXIV; *Fe*.I.334–339, Obs.XLIII–XLIV) and with the abundant (and indeed superabundant) material collected by Billy in the *Inventum Novum* (= *Fe*.III.325–398).

In modern terms, the "ascent" is nothing else than a method for deriving new solutions, and in most cases infinitely many, from a few "visible" solutions for the equations of a curve of genus 1. What was new here was of course not the principle of the method; it had been applied quite systematically by Diophantus (cf. Chap.I, §X), and, as such, is referred to, by Fermat (*loc.cit.*) as well as by Billy, as "*methodus vulgaris*", i.e., as the traditional method (cf. also Bachet's prolix commentary to *Dioph*.VI.24$_b$). The novelty consisted in the vastly extended use which Fermat made of it, giving him at least a partial equivalent of what we would obtain by the systematic use of the group-theoretical properties of the rational points on a standard cubic (cf. *infra*, Appendix III). Obviously Fermat was quite proud of it; writing for himself in the margin of his *Diophantus* (*Fe*.I.334, Obs.XLIII), he calls it "*nostra inventio*", and again, writing to Billy: "it has astonished the greatest experts" ("*C'est une de mes inventions qui a quelquesfois estonné les plus grands maistres, et particulierement Monsieur Frenicle . . .*"; *Fe*.II.437). The record shows that Fermat treated in this manner the following types of problems, most of which occur already in Diophantus:

(I) "Double equations" of the form

$$Ax^2 + Bx + C = u^2, \qquad A'x^2 + B'x + C' = v^2,$$

provided either A and A', or C and C', are squares; it may be assumed that the left-hand sides have no common zero on the projective straight line, since otherwise this would define a curve of genus 0 (for these, cf. §XIV above). For our purposes, we will regard the curve as embedded in projective space, the homogeneous coordinates being $(1, x, u, v)$. If $A = a^2, A' = a'^2$, we have four rational points at infinity

$$P_{\pm,\pm} = (0, 1, \pm a, \pm a').$$

Similarly, if $C = c^2, C' = c'^2$, we have four "trivial" solutions

$$Q_{\pm,\pm} = (1, 0, \pm c, \pm c').$$

Of course, in Fermat's eyes, just as in Diophantus's, these are no "solutions". The traditional method, inherited from Diophantus (cf. Chap.I, §X), was to write, e.g. in the case $A = a^2, A' = a'^2$:

$$(a'u + av)(a'u - av) = B''x + C''$$

with $B'' = A'B - AB'$, $C'' = A'C - AC'$, and then

$$a'u + av = t, \qquad a'u - av = \frac{1}{t}(B''x + C''),$$

$$u = \frac{1}{2a'}\left(t + \frac{B''x + C''}{t}\right).$$

Substituting the latter value for u into the first one of the given equations, and taking $t = \pm t_0$, with $t_0 = B''/2aa'$, we obtain a linear equation for x, hence a solution. To clarify this in modern terms, consider, on the given curve, the function

$$f = a'u + av - t_0;$$

it is easily seen that this has simple poles at P_{++} and P_{--}, no pole at P_{-+}, and a zero at P_{+-}; in other words, one has constructed the point M given by the equivalence $M \sim P_{++} + P_{--} - P_{+-}$ (cf. Appendix III); similarly, for $t = -t_0$, one gets the point $M' \sim P_{++} + P_{--} - P_{-+}$. In the case $C = c^2, C' = c'^2$, the construction is of course quite similar. If all of A, A', C, C' are squares, Fermat has an additional construction, based (in modern terms) on the functions $a'(u \pm c) \pm a(v \pm c')$, which produces the points equivalent to $P_{++} + P_{--} - Q_{++}$, etc.

Now, after obtaining, by any of those means, a rational point $M = (x_0, u_0, v_0)$, Fermat substitutes $x + x_0$ for x, obtaining a new pair of equations where the constant terms are squares, after which the process can be started again, and so on indefinitely; this, he says, is "his invention". Of course it could not escape him that in some special cases the process may lead again and again to the same solutions; this corresponds to cases where one has started with points of finite order in the group of solutions. But usually he does not hesitate to assert that he is getting infinitely many solutions (cf. e.g. *Fe*.II.248–249,259–260,263), and invariably he is right.

(II) Next, consider the equation

$$Ax^4 + Bx^3 + Cx^2 + Dx + E = y^2;$$

in modern terms, we may regard it as defining a two-sheeted covering of the projective line; we assume of course that the left-hand side has no multiple root, and also that A or E is a square. If $A = a^2$, there are two rational points P_+, P_- at infinity, respectively given by the expansions

$$y = \pm(ax^2 + bx + c + \cdots)$$

with $b = B/2a, c = (4AC - B^2)/8Aa$. The *"methodus vulgaris"* is then to substitute for y, in the given equation, the value $y = ax^2 + bx + c$, which leads to a linear equation in x. In the language explained in Appendix III, this amounts to considering the function $y - ax^2 - bx - c$, which has a double pole at P_- and a zero at P_+; its remaining zero is then $M \sim 2P_- - P_+$. Similarly, if $E = e^2$, we have the two points $Q_\pm = (0, \pm e)$; putting $y = a'x^2 + b'x + e$ with suitable a' and b', we get a point $M' \sim 2Q_- - Q_+$. If we have both $A = a^2$ and $E = e^2$, then, following Fermat, we can use the substitution $y = \pm ax^2 + mx \pm e$, with m chosen so that the equation obtained in this way, after division by a power of x, becomes linear in x; this leads to points $\sim 2P_\pm - Q_\pm$, $2Q_\pm - P_\pm$. The process can then be iterated as above in (I), substituting $x + x_0$ for x after a solution (x_0, y_0) has been found.

(III) Similar methods can be applied to an equation

$$Ax^3 + Bx^2 + Cx + D = y^2,$$

where one assumes that $D = d^2$, or to an equation

$$Ax^3 + Bx^2 + Cx + D = y^3,$$

where either A or D is a cube. In these cases, the method amounts to the familiar one of intersecting a cubic with the tangent to the curve at a given rational point. Was Fermat

aware of this geometric interpretation? In his work on analytic geometry, he had dealt with the question of constructing the tangent to a plane curve given by its equation,
and had treated in particular the "semicubic parabola"
$ay^2 = x^3$, the "cissoid" $x(x^2+y^2) = ay^2$, and the "con-
choid" $(x^2+y^2)(y-a)^2 = b^2y^2$ (cf. *Fe*.I.159–161 and 218).
For lack of evidence, this intriguing question must remain
unanswered.

On the other hand, it is noteworthy that the simple idea
of intersecting a cubic with the straight line through two
previously known rational points does not seem to have
occurred to Fermat, either in geometric or in algebraic garb.
One finds it, apparently for the first time, and stated in full
generality, in some manuscript notes of Newton's
(*New*.IV.112–114), presumably written not long after the
appearance of Samuel Fermat's *Diophantus* of 1670 (the volume which contained Billy's *Inventum Novum*), and possibly
inspired by that publication.

(IV) The following problem is a mere variant of (I), into
which it can be transformed by a simple change of variable,
and it would hardly deserve a separate mention if Fermat
did not seem to have been rather proud of his way of handling
it (cf. *Fe*.I.334, Obs.XLIII, and *Inv.Nov*.II.1 = *Fe*.III.360).
It consists of the "triple equation"

$$Ax + B = u^2, \qquad A'x + B' = v^2, \qquad A''x + B'' = w^2$$

where B, B', B'' are squares; in the projective space with the
coordinates $(1, u, v, w)$, it defines a space quartic with eight
"visible" rational points. If $B = b^2$, one puts $x = 2bt + At^2$,
$u = b + At$, after which one has a "double equation" for
which the known rational points have to be used according
to the recipes explained above.

It will be noticed that in all this there is no number theory;
everything would be equally valid over any field, in particular
the field of real numbers. In Fermat's eyes such work has
therefore to be classified as "geometry"; we would describe
it as algebraic geometry, just as the work of Diophantus
and that of Viète of which it is an extension; as to this, cf.

Chap.I, §X, and Fermat's comments, quoted there, from his challenge of 1657 to the English mathematicians (*Fe*.II.335). In dealing with these questions, Fermat had a twofold motivation.

Firstly, while Diophantus had usually been content with one solution for any given problem, he had occasionally asked for more; in *Dioph*.V.8_b, for instance, he gave a recipe for finding three pythagorean triangles having equal areas, and Viète, in his *Zetetica* (*Zet*.IV.11 = *Op*.p.70–71), translated it into algebraic terms. If the triangle is taken to be, up to a proportionality factor, $(2pq, p^2 - q^2, p^2 + q^2)$, then, putting either $x = q/p$ or $x = p/q$, and calling A the area, the problem amounts to finding a curve $|x - x^3| = Ay^2$ with three rational points. Diophantus's procedure is to find a solution for $x - x^3 = x' - x'^3$, i.e., $x^2 + xx' + x'^2 = 1$, thus providing two triangles of area $A = x - x^3$, and then observing that $x + x'$ is a solution for $A = X^3 - X$; we would interpret this by saying that, since $(x, 1)$ and $(x', 1)$ are on the cubic $AY^2 = X - X^3$, the third intersection of this cubic with $Y = 1$, which is $(-x-x', 1)$, must also be rational; there is no indication, however, that Diophantus, Viète or even Fermat saw in this an application of an underlying general principle. What Fermat does, in contrast with Diophantus and Viète, is to start from just one pythagorean triangle, so as to get a curve $x - x^3 = Ay^2$ with one known rational point, then to intersect the cubic with the tangent at that point and iterate the process as often as desired; this, at least, is described by him as his basic method, but (not surprisingly) he has also various ways of reducing the same problem to a "double equation" and then applying to it the procedure described under (I) (cf. *Fe*.I.309–311, Obs.XXIII, and *Inv.Nov*.I.38 = *Fe*.III.348).

There was another reason why Fermat could not be content with the traditional method; it is that this "*methodus vulgaris*" often leads to solutions in negative numbers; in the traditional view, these are no solutions at all. Take for instance the problem (apparently raised and perhaps solved by Diophantus in his lost Porisms; cf. *Dioph*.V.16) of writing

the sum or the difference of two given cubes as the sum or
the difference of two other cubes; Viète dealt with it in his
Zetetica (*Zet*.IV.18–20 = *Op*.74–75); Bachet reproduced
Viète's treatment, including even his numerical examples
(and without any acknowledgment; cf. Chap.I, §XII), in his
commentary to *Dioph*.IV.2. The problem is of course to
derive a second rational point (x, y) from one known point
(a, b) on a curve $X^3 + Y^3 = A$, except that Viète and Bachet
split it up into various cases according to the signs of a, b,
x, y.

Here Viète, and Bachet after him, are content with the
"*methodus vulgaris*"; in effect, they intersect the curve with
its tangent at (a, b); this gives the solution

$$x = a \cdot \frac{a^3 + 2b^3}{a^3 - b^3}, \qquad y = b \cdot \frac{2a^3 + b^3}{b^3 - a^3}$$

which, from their point of view, is to be regarded as satisfactory only if x, y have the prescribed signs when a, b have
prescribed signs. For instance, if the problem is to find two
cubes whose sum is the difference of two given cubes p^3,
q^3, one would have to take, in our present notation, $a = p$,
$b = -q, p > q$, and require x, y to be positive, so that Viète's
formulas solve the problem if $p^3 > 2q^3$ but not otherwise.
Also, as Fermat points out, they can never solve the problem
of finding two cubes whose sum is the sum of two given
cubes. Fermat's prescription for this batch of problems
(*Fe*.I.291–292, Obs.III; I.297–299, Obs.VIII–IX) is to iterate
Viète's procedure as many times as may be needed in order
to obtain a solution as required, and then again as often as
necessary if further solutions are desired.[9] Here, too, he
gives no formal proof for the fact that this will always be

---

[9] Presumably Fermat did not know that the same idea had already
occurred to Albert Girard on the occasion of the latter's French translation
of Books V and VI of Diophantus; cf. *Les Œuvres Mathematiques de Simon
Stevin . . . reveu, corrigé et augmenté par Albert Girard*, Leyde, Elsevier 1634,
pp. 159–160.

successful, but seems content with his experience in such
matters.

As to the first discovery of this "method of ascent", we
have Fermat's testimony in his letter of 1644 to Carcavi
where he refers to the following problems about pythagorean
triangles (i.e., triples of integers a, b, c such that $a^2 + b^2 =
c^2$) proposed to him at one time ("*autrefois*") by Frenicle (cf.
Frenicle's letter of 1641 to Fermat, *Fe*.II.241) and which
had formerly eluded both of them; "*je ne voyois même pas de
voie pour y parvenir*", he writes to Carcavi ("I saw no way of
even tackling them"); but now, in 1644, he can solve them
(*Fe*.II.265):

(A) Find a triangle for which $a > b$, and $(a-b)^2 - 2b^2$ is
a square;

(B) Find a triangle such that c and $a + b$ are squares.

As Frenicle had indicated, they are closely related. To
see this, put $x = a/b$ in (A) and observe that the problem
amounts to the "double equation"

$$(11) \qquad x^2 + 1 = u^2, \qquad x^2 - 2x - 1 = v^2$$

to be solved in rational numbers subject to the condition
$x > 1$. As to (B), call d the g.c.d. of a, b, c, and write (cf. (2),
Chap.I, §V)

$$(12) \quad a' = \frac{a}{d} = 2pq, \quad b' = \frac{b}{d} = p^2 - q^2, \quad c' = \frac{c}{d} = p^2 + q^2,$$

where p is prime to q, $p > q$, and $p - q$ is odd. Call δ the
g.c.d. of c' and $a' + b'$; δ divides $a' + b' \pm c'$, i.e., $2p(p+q)$
and $2q(p-q)$, whose g.c.d. is 2; as c' is odd, δ is 1, so that d
is the g.c.d. of c and $a + b$. Now, if c and $a + b$ are squares,
d must be a square, and so are c' and $a' + b'$; moreover,
if p is prime to q, $p^2 + q^2$ cannot be a square unless
$p - q$ is odd. Putting $x = -p/q$, we find that (B) amounts
again to the double equation (11), but with the side-condition
$x < -1$.

These problems, and others of the same type, occur repeatedly in Fermat's correspondence of the years 1643 and
1644; obviously these are the years of his breakthrough in

such matters. To Carcavi, in 1644, he sends his solution (1517, 156, 1525) for (A). In 1643 he had sent to Mersenne his solution for (B):

$$a = 4565486027761, \quad b = 1061652293520,$$
$$c = 4687298610289$$

(*Fe*.II.261; cf. *Fe*.II.259–260,263); this corresponds to the values $p = 2150905$, $q = 246792$ in (12). Problems (A) and (B) are those which he indicates as examples for his iterative method in the margin of his *Diophantus* (*Fe*.I.336–338, Obs.XLIV), describing them as "very difficult questions" which could not have been solved otherwise; the details of his procedures for solving (B) can be found in the *Inventum Novum* (*Inv.Nov*.I.22,25,45,III.32 = *Fe*.III.339,340–341, 353–354,388–389); one of them is to observe that the problem is the same as to make $c(a+b)$ a square (since one may assume, as shown above, that $a + b$ is prime to c); then, putting $z = p/q$, one has to make $(z^2+1)(z^2+2z-1)$ a square, which can be done by the repeated application of the last one of Fermat's methods described above under (II) (*Inv.Nov*.III.32 = *Fe*.III.388–389). What is more, in the same marginal note (*Fe*.I.337) he makes bold to "assert confidently" that his solutions for (A) and (B) are the smallest possible ones; this was indeed verified by Lagrange in 1777 (cf. *infra*, §XVI, and Appendix V).

§XVI.

Experience soon shows that Fermat's iterative method leads rapidly to very large integers. An altogether different method is needed when one wishes, not to find more and more solutions for a given problem, but on the contrary to show that there are none, or at least that whatever solutions exist can all be derived from a few known ones by some regular process. A typical case for such a proof has already been described above in §X, where it was shown that the area of a pythagorean triangle cannot be a square; this amounts to proving that $x - x^3 = y^2$ has no other rational solution than $y = 0, x = 0$ or ± 1 (cf. above, §XV). For this, as we have seen, a "method of descent" is required, by which smaller and smaller solutions can be produced if it is assumed that one exists. Another example for this is given by the equation $x^2 - Ny^2 = 1$ ("Pell's equation") discussed above in §XIII; this is of genus 0, not of genus 1, but a method of descent was also required for its complete solution.

For such "negative" questions, Fermat quotes four examples in his communication of 1659 to Huygens (*Fe*.II.433–434). Strangely enough, the last of these is no other than his erroneous assertion about integers of the form $2^{2^n} + 1$ being primes (cf. above, §IV), which, little more than a year before, he had been careful to present as a mere conjecture (*Fe*.II.404). Huygens, who had been skeptical about this (*Huy*.II.212 = *Fe*.IV.122), took note of the discrepancy in the margin of the copy he took of that letter (*Huy*.II.462); one wonders whether ill health had momentarily distracted Fermat's attention at this point. As to the three other examples, all they have in common is that they apparently depended in part upon Fermat's theory of the quadratic forms $X^2 + AY^2$ for $A = 1, 2, 3$ and more specifically upon the results stated above as lemma 4 of §IX and lemma 5 of §XII. We shall first dispose quickly of two of these, which had already figured as challenge problems in Fermat's letter of 1657 to Digby (*Fe*.II.345) and which seem little more than straightforward consequences of those lemmas. In one case one has to show that $x^2 + 2 = y^3$ has no solution other than $(5, 3)$ in natural integers. In fact, if (x, y) is a solution, lemma 5 of §XII (or else the elementary theory of the ring $\mathbf{Z}[\sqrt{-2}]$, for which cf. Appendix I) shows that there are two integers u, v such that $y = u^2 + 2v^2, x = u^3 - 6uv^2$, $1 = 3u^2v - 2v^3$. The latter formula implies $v = \pm 1$, $3u^2 = 2 \pm 1$, hence $v = 1, u^2 = 1, y = 3, x = \pm 5$. In the other case one must solve $x^2 + 4 = y^3$. Here we apply lemma 4 of §IX. If x is odd, we must have $y = u^2 + v^2, x = u^3 - 3uv^2, 2 = v^3 - 3vu^2$; as v divides 2, it must be ± 1 or ± 2; if it were ± 1, x would be even; if $v = \pm 2$, then we have $2 = \pm(8 - 6u^2)$, hence $u^2 = 1, y = 5, x = 11$. On the other hand, if x is even, put $x = 2x', y = 2y'$; then $x'^2 + 1 = 2y'^3$, and x' must be odd. Put $z = (x' + 1)/2$, $t = (x' - 1)/2$; this gives

$$z - t = 1, y'^3 = z^2 + t^2.$$

Apply again lemma 4 of §IX; we get

$$z = u^3 - 3uv^2, \quad t = v^3 - 3vu^2, \quad 1 = z - t = (u-v)(u^2 + 4uv + v^2),$$

which obviously has no other solution than $u = 1, v = 0$ and $u = 0, v = -1$, giving $z = 1$ or $0, x' = \pm 1, x = \pm 2, y = 2$.

Quite different, and to us far more interesting, is the case of the equation $x^3 + y^3 = z^3$ to be solved in integers, or equivalently $X^3 + Y^3 = 1$ to be solved in rational numbers, this being the first one of the four sent by Fermat to Huygens in 1659. It may be regarded as a companion to the one discussed above in §X, both being cases for the "genuine" descent as it applies to curves of genus 1; such was also the problem stated, along with these two, in Fermat's communication of 1640 to Frenicle (*Fe*.II.195), viz., "find four squares in arithmetic progression". At that early date, Fermat had perhaps no more than plausibility arguments for the fact that these problems have no solution; but eventually he must have obtained a formal proof also for the third one, since we are told so by Billy in his *Inventum Novum* (*Inv. Nov*.II.11 = *Fe*.III.365).

How grateful we should be to the good Jesuit, had he shown some curiosity towards such "negative" statements, instead of tediously aligning one example after another of Fermat's iterative method, not a few of which must have been of his own devising! Some of them do not even make good sense. Thus we are left to reconstruct, as best we can, some of Fermat's most interesting proofs by descent, and again, as on former occasions, to seek Euler's and Lagrange's help. In doing so one should not forget that it is in the nature of such proofs to admit a number of possible variants, one as good as the other; examples for this abound in the writings of Euler; an explanation of this phenomenon is to be found only in the modern arithmetical theory of curves of genus 1. Without going into further details, it will be enough to mention that such proofs invariably proceed from the equations of one curve of genus 1 to those of another (e.g. in §X the curves which, in homogeneous form appear there as $pq(p+q)(p-q) = m^2$, $x^4 - y^4 = z^2$, $r^4 + 4s^4 = x^2$, i.e., in non-homogeneous form as $X^3 - X = Y^2$, $X^4 - 1 = Y^2$, $X^4 + 4 = Y^2$) and from there back to the original equation. The formulas by which one goes from one curve to the next define sometimes merely a birational correspondence, in which case there is no real progress; but mostly they define what used to be known as "transformation" in the classical theory of elliptic functions ("isogeny" in the modern terminology), and usually one of order 2 or 3; in special cases, such a transformation may be nothing else than complex multiplication, formulas for which are for instance implicit in the proofs of §X and in the one to be described below.

Because of these circumstances, any attempt at reconstruction can be no more than a hit or miss proposition; all it can do is to show one way in which Fermat could have obtained his results. As to the four squares in arithmetic progression, the problem was treated by Euler in 1780 in a paper, published posthumously (Eu.I-5.56–60 in E758); it is somewhat confusedly written, obviously by his assistants, at a time when he was totally blind (cf. Chap.III). A more direct and more elegantly presented proof can be found in J. Itard, *Arithmétique et Théorie des Nombres*, P.U.F., Paris 1973, p. 112. Another proof, seeking to exhibit the mechanism of the descent from the point of view of the modern theory of curves of genus 1, will be given in Appendix IV.

Now we turn to the equation $x^3 + y^3 = z^3$; this defines a curve of genus 1 in the projective plane. Fermat has left no indication about his treatment of that equation, except to say that it was by descent. It is mentioned by Euler for the first time in 1753, when he writes to Goldbach that he has found a proof, so different from the known one for $x^4 + y^4 = z^4$ that it leaves him no hope of extending them to a general proof for n-th powers (*Corr.*I.618). His proof for cubes is mentioned by him in 1760 (cf. Chap.III, §XIV) and then explained in full detail in the final chapter of his *Algebra* of 1770 (*Eu.*I-1.486), with the sole difference that in 1760 he indicates that it is based on the theory of the quadratic form $X^2 + 3Y^2$, while in 1770 he uses the field $\mathbf{Q}(\sqrt{-3})$. As we have repeatedly mentioned (cf. Appendix I), those two modes of exposition are essentially equivalent and can easily be translated into one another; Fermat would have used the former, of course, and we shall do the same here; otherwise we follow Euler closely.

One may assume that x, y, z are mutually prime; then one of them must be even, and the other two odd; changing their signs if necessary, we may rewrite the equation as $x^3 + y^3 = z^3$, where z is even and x, y are odd; put $x = p + q$, $y = p - q$. We have now $2p(p^2 + 3q^2) = z^3$. As $p \pm q$ is odd, $p^2 + 3q^2$ is odd; as z is even, p must then be a multiple of 4, and q must be odd. Writing our equation as

$$\frac{p}{4}(p^2 + 3q^2) = \left(\frac{z}{2}\right)^3,$$

we shall distinguish two cases, according as z is a multiple of 3 or not. If it is not, then the two factors in the above equation are mutually prime, so that each of them must be a cube:

$$p = 4r^3, \qquad p^2 + 3q^2 = s^3.$$

Apply lemma 5 of §XII; this gives

$$p = u(u + 3v)(u - 3v), \qquad q = 3v(u^2 - v^2),$$

where u, v must be mutually prime; as q is odd, v must be odd and u even. As $p = 4r^3$, this shows that $u/4$ is an integer and a cube, and that $u + 3v$ and $u - 3v$ are cubes. Putting now $u = 4e^3$, $u + 3v = f^3$, $u - 3v = g^3$, we get $f^3 + g^3 = (2e)^3$; this is a solution of the original equation, obviously in smaller numbers.

If z and therefore p are multiples of 3, q is not; we can now write

$$\frac{p}{4}\left[q^2 + 3\left(\frac{p}{3}\right)^2\right] = 9\left(\frac{z}{6}\right)^3.$$

Thus $p/4$ must be a multiple of 9, and the product of $p/36$ and $q^2 + 3(p/3)^2$ is a cube; as these two factors are mutually prime, they must both be cubes, so that we can apply lemma 5 of §XII and write

$$p = 36r^3, \qquad q = u(u^2 - 9v^2), \qquad \frac{p}{3} = 3v(u^2 - v^2),$$

and therefore

$$-4r^3 = v(v+u)(v-u).$$

Here again u, v must be mutually prime, u must be odd, and v even. Then $2v$, $v + u$ and $v - u$ must be cubes; as the first one is the sum of the other two, this is again a solution of the original equation. Getting smaller and smaller solutions, one comes to a contradiction, just as Fermat said.

Finally his "confident assertion", quoted above in §XV, about his numerical solutions for problems (A) and (B) deserves some comment. In view of the large numbers involved, particularly in the solution for (B), mere trial and error could of course not show that these are the smallest solutions. Lagrange cleared up the matter in 1777 (*Lag*.IV.377–398); after replacing the problem by the essentially equivalent one of solving $2x^4 - y^4 = z^2$ in integers (cf. *infra*, Appendix V), he dealt with this by applying to it Fermat's own method of descent; the details are substantially more complicated than in the case of Fermat's equation $x^4 - y^4 = z^2$, while the principles are the same. The conclusion, however, is different; instead of showing that there are no other solutions than the trivial ones, one finds now that all solutions can be derived from the obvious ones by the processes to be described in Appendices III and IV, or (what comes to the same, at least in this case) by Fermat's processes described above in §XV. In modern language, the space quartic defined by the "double equation" (11) of §XV, and the quartic curve

$2X^4 - 1 = Z^2$, are curves of rank 1, isogenous to one another. Taking into account the side-conditions implied by problems (A) and (B) as formulated by Frenicle and Fermat, one can then verify that Fermat's solutions are indeed the smallest ones (cf. Appendix V).

Had Fermat obtained such a proof? This is not inherently impossible, since Lagrange's method is no other than his own. One rather imagines, however, that he merely allowed himself to be guided by the analogy with Pell's equation, where the group of solutions is also, just as here, the product of an infinite cyclic group and a group of order 2; in that case he might have taken for granted the facts later proved by Lagrange. If so, his intuition had not deceived him.

§XVII.

This concludes our analysis of Fermat's writings "on numbers". Once properly assembled, they present a remarkably coherent picture.

His proofs have almost totally vanished. Writing them up, at a time when algebraic notation was still clumsy in the extreme, and models were altogether absent, would have cost a tremendous effort, and the complete lack of interest on the part of his contemporaries must have been depressing.

In his later years, Fermat must have feared that the fruit of his labors would get lost. Here is the melancholy conclusion of his final letter to Huygens:

"Such is in brief the tale of my musings on numbers. I have put it down only because I fear that I shall never find the leisure to write out and expand properly all these proofs and methods. Anyway this will serve as a pointer to men of science for finding by themselves what I am not writing out, particularly if Monsieur de Carcavi and Frenicle communicate to them a few proofs by descent that I have sent them on the subject of some negative propositions. Maybe posterity will be grateful to me for having shown that the ancients did not know everything, and this account may come to be regarded by my successors as the "handing on of the torch", in the words of the Great Chancellor of England [Bacon], following whose intention and motto I shall add: many will pass away, science will grow" ["*Voila sommairement le conte de mes resveries sur le suject des nombres. Je ne l'ay escrit, que parce que j'apprehende que le loisir d'estendre et de mettre au long toutes ces demonstrations et ces methodes me manquera. En tout cas cette indication servira aux scavants pour trouver d'eux mesmes ce que je n'estens point, principalement si Monsieur de Carcavi et Frenicle leur font part de quelques demonstrations par la descente que je leur ay envoyees sur le subject de quelques propositions negatives. Et peut estre la posterité me scaura gré de luy avoir fait connoistre que les anciens n'ont pas tout sceu, et cette relation pourra passer dans l'esprit de ceux qui viendront apres moy pour traditio lampadis ad filios, comme parle le grand Chancelier d'angleterre, suivant le sentiment et la devise du quel j'adjousteray, multi pertransibunt et augebitur scientia*": Fe.II.436 = Huy.II. 461–462|August 1659).

Less resignedly no doubt, but perhaps in a similar vein, Galois, on the eve of his death, was writing to his friend Chevalier: "*Après cela il y aura, j'espère, des gens qui trouveront leur profit à déchiffrer tout ce gâchis*" ("there will come those, I hope, who will find it worth their while to decipher all this mess").

It does not appear that Huygens reacted in any way to Fermat's letter: "There is no lack of better things for us to do", he had written to Wallis in 1658 (*Huy*.II.211 = *Fe*.IV.121), soon adding, however, a more balanced appraisal of Fermat's and Frenicle's arithmetical work: "Let them give up their problem-setting and follow your example by giving us what may bring credit to them and pleasure to us. There are indeed admirable properties of numbers whose demonstration would be well worth while in my opinion; such, surely, is the one in Fermat's letter at the bottom of page 185 . . ." The latter reference is to Fermat's last letter to Digby and his assertions on sums of squares and triangular numbers (*Fe*.II.403–404).

Nevertheless, if Huygens was undoubtedly more capable than most of his contemporaries of appreciating and of

criticizing arithmetical work, he was not prepared to take
up the torch proffered by Fermat. Only once did he come
at all close to number theory; this occurred in his *Descriptio
Automati Planetarii* (*Huy*.XXI.587–652), first printed posthumously in 1703 but probably composed between 1680
and 1687 (cf. *ibid*.p.585). Here, in connection with practical
problems about automata with dented wheels, one finds a
thoroughly original and masterly treatment of the best approximation of real numbers by fractions, based on the
continued fraction algorithm (*ibid*.pp.627–643). The topic,
which had been already touched upon by Wallis, was eventually taken up by Euler (cf. e.g. *Eu*.I-14.187–215 =
E71|1737; *Eu*.I-15.31–49 = E281|1757; etc.), then by La-
grange who gave full credit to Huygens, calling it *"une des
principales découvertes de ce grand géomètre"* (*Lag*.VII.43–44
= *Eu*.I-1.536); it is noteworthy, however, that Huygens
himself never bothered to get it published.

Fermat's torch had indeed been long extinguished when
Euler, in 1730, picked it up, kindled it anew, and kept it
burning brightly for the next half-century. That story will
be the subject of the next chapter. Here we merely note
that in 1742, after seeking for more than ten years with
little success to clear up the questions pertaining to sums
of two squares, Euler, at the height of his fame, wrote from
Berlin to Clairaut in Paris, asking him to institute a search
for Fermat's manuscripts (*Eu*.IV A-5.124; cf. Chap.III, §IV);
according to him, "many mysteries might still lie hidden
there". The answer was discouraging (*ibid*., p. 129); as he
reports to Goldbach (*Corr*.I.168) the search was hopeless,
since there was no interest any more for such matters (*"da
der goût für dergleichen Sachen bei den Meisten erloschen ist, so
ist auch die Hoffnung verschwunden"*). There was no remedy
but for Euler to do it by himself. And so he did, with Lagrange's valuable cooperation from 1768 onwards. Indeed,
by the time of Euler's death, not only had all the territory
explored by Fermat been regained, but further ground had
been conquered. If some mysteries remained, they concerned
questions which had lain beyond Fermat's horizon.

In one direction Fermat has cast his shadow well into the present century and perhaps the next one. Already Leibniz had announced it; here is what he wrote with prophetic insight in 1702:

". . . my hope is that those will come who will spread the seeds of the new science [i.e., Leibnizian analysis] further and reap a richer harvest, especially if they attend more diligently than heretofore to the advancement of Diophantine algebra, which Descartes' disciples have neglected because they failed to perceive any use for it in Geometry. I, on the other hand, remember having repeatedly suggested (what could seem strange to some) that the progress of our integral calculus depended in good part upon the development of that type of Arithmetic which, so far as we know, Diophantus has been the first one to treat systematically . . ." (*Acta Eruditorum* 1702, p. 219 = *Math.Schr.*ed. C.I.Gerhardt, (II) Bd. I, Halle 1858, p. 360).

In modern terms, Leibniz was trying to say that the study and classification of algebraic differentials and their integrals depended upon the methods of algebraic geometry for which, in his days, Diophantus, Viète and Fermat offered the only existing models. He mentions neither Viète nor Fermat; perhaps he had never studied them; but his intuition was not leading him astray.

When the same threads are later picked up, first by Euler and then by Lagrange, with brilliant success, the close connection between "Diophantine algebra" (as Leibniz called it) and elliptic integrals begins to appear below the surface. It may well be implicit in Euler's oft-repeated admonition to his fellow-analysts that number theory is, even from their point of view, no waste of time. Sometimes one can detect in his words an echo of Leibniz's far-sighted remark quoted above:

". . . the author has often mentioned that the Diophantine method, if further developed, will redound to the benefit of the whole of Analysis, so that he is far from feeling sorry for the prolonged efforts he has devoted to that branch of mathematics . . ." (*Eu.*I-2.428 in E255), and further on in the same paper: "this shows how much more there is left to do in Diophantine Analysis, which beyond any doubt could contribute greatly to the whole of Analysis, finite and infinitesimal. In fact, one main tool of integral calculus consists in transforming irrational differential expressions into rational ones, and this comes directly from Diophantine Analysis, so that one has every right to expect further progress from that same study . . ." (*ibid*.p.454).

In that paper, written in 1754, Euler proves by the "Diophantine method" the rationality of the cubic surface $x^3 + y^3 + z^3 = t^3$, after referring to Euclid's proof for the corresponding result for the equation $x^2 + y^2 = z^2$. As to curves of genus 1, Euler became interested in them quite early in his career (cf. *Corr*.I.31|1730). He deals extensively with them in a paper of 1738 where he rewrites in his own style Fermat's and Frenicle's treatment of the equations $x^4 \pm y^4 = z^2$ (*Eu*.I-2.38–58 = E98; cf. above, §X) and where he deals, also by descent, with a few equations of the same kind, all belonging to what we would now call the "lemniscatic" case. Did he immediately recognize a problem of the same nature when he raised, as early as 1730, the question of the rationality of the integral $\int dx/\sqrt{1-x^4}$ (*Corr*.I.47,51), or at the end of 1751 when he received Fagnano's *Produzioni Matematiche* containing a number of remarkable theorems about the rectification of the lemniscate and about "lemniscatic" integrals (cf. Chap.I, §I)? We have no way of telling; but Jacobi, upon getting from the St. Petersburg Academy, in 1834, the latest volume of Euler's posthumous papers, including several on Diophantine equations of the form

$$y^2 = A + Bx + Cx^2 + Dx^3 + Ex^4,$$

did not fail to observe (*Jac*.II.53) that Euler's procedure in such problems amounted to writing the multiplication formula for the elliptic integral

$$\int \frac{dx}{\sqrt{A+Bx+Cx^2+Dx^3+Ex^4}}$$

and added that this "memorable agreement" ("*consensus ille memorabilis*") between both theories could hardly have escaped the notice of "their common initiator". In this, as we know, he was not giving due credit to Fermat, whose work was far less familiar to him than Euler's; but, in matters pertaining solely to Euler, he was surely a good judge. He might have said the same about Lagrange, had a similar occasion arisen. Lagrange had taken up elliptic integrals in 1766 (*Lag*.II.5–33). In 1777 he wrote an important paper on the diophantine equations $2x^4 - y^4 = \pm z^2$ and related problems (*Lag*.IV.377–398; cf. above, §XVI). He came back to elliptic integrals in 1784 (*Lag*.II.253–312) with a paper where he is sometimes credited with having "rediscovered independently" the so-called Landen transformation, i.e., the transformation (or "isogeny") of order 2 for elliptic integrals.[10] Could he have been unaware that this same transformation had been a main tool in the infinite descent as practised by Fermat, by Euler, and again by himself in his own paper of 1777 (cf. Appendix IV)? Indeed Leibniz's prophecy was coming true.

In the nineteenth century, Fermat's legacy was once more threatened with extinction. Reduced to essentials, it consisted of three correlated principles. One was that algebraic geometry, and, to begin with, algebraic curves deserved to be studied, not just over the real or eventually the complex number-field, but over other fields as well, and signally over the field of rational numbers. Next came the principle that rational transformations were the proper tool in this study. Thirdly there was the infinite descent as it applies to diophantine problems. In the nineteenth century, however, the conspicuous success of the function-theoretic method in algebraic geometry led for a while to almost total victory for the complex numbers over their rivals, beginning with

---

[10] Cf. e.g. A. Enneper, *Elliptische Functionen* (Halle 1876), p. 307. Actually Lagrange knew Landen's paper of 1775 on this subject; he quotes it and comments briefly upon it in a letter of 1777 to Condorcet (*Lag*.XIV.41).

the theory of elliptic functions which swallowed up the study of curves of genus 1 (to the point of giving them the name by which they are usually known at present), and continuing with Riemann's theory of abelian functions. The development of so-called "pure" geometry, and particularly of the theory of projective spaces, initiated by Poncelet and developed by Chasles, Möbius and many others, could have restored the balance, had it not, from the beginning, operated almost invariably over the field of complex numbers; this tradition was continued in Italian geometry and beyond it. During the whole period, Fermat's name came to be associated almost exclusively with his "last theorem", i.e., with the equation $x^n + y^n = z^n$, which, however fruitful it turned out to be in the hands of Kummer, does not belong to our present subject.

A first faint reaction set in with the paper of 1890 of Hilbert and Hurwitz (*Acta Math.*14, pp. 217–224), where for the first time it was pointed out explicitly that the birational theory of algebraic curves provided the proper framework for the study of diophantine equations. Poincaré, in 1901, made a broader survey of the same ground, and, unwittingly perhaps, sought to make use of the trisection of elliptic functions as a tool for infinite descent. The true breakthrough came in 1922 with Mordell's celebrated paper (*Proc.Cambridge Phil.Soc.*21, pp. 179–192); here, if Fermat's name does not occur, the use of the words "infinite descent" shows that Mordell was well aware of his indebtedness to his remote predecessor. Since then the theory of elliptic curves, and its generalization to curves of higher genus and to abelian varieties, has been one of the main topics in modern number theory. Fermat's name, and his method of infinite descent, are indissolubly bound with it; they promise to remain so in the future.

Appendix I

Euclidean Quadratic Fields

As pointed out in §§VIII and XII, Fermat's results on the quadratic forms $X^2 + AY^2$ (where A is 1, ± 2 or 3) can best be understood by considering the rings $\mathbf{Z}[\theta]$ where θ is $i = \sqrt{-1}, \sqrt{\pm 2}$, or $j = (-1+\sqrt{-3})/2$. This will be explained now.

In each one of these cases we put $R = \mathbf{Z}[\theta]$; R is the ring of integers of the field $K = \mathbf{Q}(\sqrt{-A})$. We write $\xi \rightarrow \xi'$ for the automorphism of K which changes $\sqrt{-A}$ into $-\sqrt{-A}$, and we put $N(\xi) = \xi\xi'$; $N(\xi)$ is called the norm of ξ. We have

$$N(x+y\sqrt{-A}) = x^2 + Ay^2, \qquad N(x+yj) = x^2 - xy + y^2.$$

In all cases, we have $N(\xi\eta) = N(\xi)N(\eta)$; as pointed out in Chap. I, §VIII, this is equivalent to the basic identity for the form $X^2 + AY^2$, i.e., to (5) and (6) of Chap. I, §VIII, and to (5) of Chap. II, §XII.

The units of the ring R are the integers η of K which satisfy $N(\eta) = \pm 1$; these are ± 1, $\pm i$ for $A = 1$; ± 1 for $A = 2$; ± 1, $\pm j$, $\pm j^2$ for $A = 3$. For $A = -2$, there are infinitely many units, viz., all the integers of the form $\pm \varepsilon^n$ with $\varepsilon = 1 + \sqrt{2}$, $n \in \mathbf{Z}$ (cf. §XII). Two non-zero integers α, β are called *associates* if each is a multiple of the other in the ring R; this is so if and only if $\alpha^{-1}\beta$ is a unit. A non-zero integer $\alpha = a + b\theta$ will be called *normalized*, in the case

$A = 1$, if $a > 0$, $b \geqslant 0$; in the case $A = 2$, if $a > 0$ or $a = 0$, $b > 0$; in the case $A = 3$, if $a > b \geqslant 0$; in the case $A = -2$, if $-b < a \leqslant b$. Among the associates of a given integer, one and only one is normalized; this is easily seen for $A = 1, 2, 3$; for $A = -2$, the proof is essentially the one given by Fermat, and reproduced above in §XII, for the existence of a "minimal" solution (x, y) for an equation $n = 2y^2 - x^2$. In general, the product of two normalized integers is not normalized; this may be regarded as the source of Fermat's error discussed in §XII.

Let $\zeta = x + y\theta$ be any element of the field K; then there is at least one integer $\alpha = a + b\theta$ in K, such that $|N(\zeta - \alpha)| < 1$; indeed this will be so if we take for a, b the closest rational integers to x and to y, respectively. Consequently, if α, β are two integers, and $\beta \neq 0$, there are integers γ, ρ such that $\alpha = \beta\gamma + \rho$ and $|N(\rho)| < |N(\beta)|$; to obtain this, one has only to take γ so that $|N((\alpha/\beta) - \gamma)| < 1$. Then the "euclidean algorithm" shows, just as in Book VII of Euclid, the existence of a g.c.d. δ for any two integers α, β in the ring R; as in Euclid, this is defined as a common divisor of α and β such that every common divisor of α and β divides δ; clearly any two such integers must be associates, and there is one and only one normalized g.c.d. to α, β. Instead of the "euclidean algorithm", one may also take for δ anyone of the numbers $\gamma = \alpha\xi + \beta\eta$ for which $|N(\gamma)|$ has the smallest value when one takes for ξ and η all the elements of R; the process of "euclidean division" shows then that any such integer is a g.c.d. for α and β.

In the ring R, primes are defined as those integers, other than 0 and the units, which have no other divisors than their associates and the units. From the existence of the g.c.d. it follows that a prime that divides a product must divide one of its factors. If α is an integer, and $|N(\alpha)| > 1$, let π be one of its divisors for which $|N(\pi)|$ has its smallest value > 1; this must be a prime. From this it follows at once that every integer in R is a product of primes and a unit, and that its representation as such a product is "essentially" unique (i.e., unique up to unit factors); it is unique, up to the order of the factors, if one stipulates that its prime factors should be normalized.

To find the primes in the ring R, one needs to know the quadratic residue character of $-A$ modulo the rational primes; this has been determined above in §VII (for $A = -1$) and in §XII (for $A = \pm 2, A = 3$). Let p be a rational prime; if it is not a prime in R, it has in R a prime divisor π, which we may assume to be normalized, and $N(\pi)$ is then a divisor of $N(p) = p^2$, other than 1 and p^2 and therefore equal to $\pm p$. Now consider separately the four cases:

(a) if $A = 1$, put $\pi = x + yi$; then $p = x^2 + y^2$, and $p \equiv 1$ (mod 4) unless $p = 2$; in the latter case, $\pi = 1 + i$.

(b) if $A = 2$, put $\pi = x + y\sqrt{-2}$; then $p = x^2 + 2y^2$, and $p \equiv 1$ or 3 (mod 8) unless $p = 2$; in the latter case $\pi = \sqrt{-2}$.

(c) if $A = -2$, put $\pi = x + y\sqrt{2}$, where $-y < x \leqslant y$; then $p = 2y^2 - x^2$ and $p \equiv 1$ or 7 (mod 8) unless $p = 2$; in the latter case $\pi = \sqrt{2}$.

(d) if $A = 3$, put $\pi = x + yj$; then $p = x^2 - xy + y^2$. Here x, y cannot both be even, so that, among the associates of π, one (viz. π itself if y is even, πj^2 if x is even, and πj if x and y are odd) must be of the form $z + t\sqrt{-3}$ with integers z, t, so that $p = z^2 + 3t^2$; we have $p \equiv 1$ (mod 3) unless $p = 3$. In the latter case, p has the prime divisor $\sqrt{-3}$ (or in normalized form $2 + j = -j\sqrt{-3}$).

If, as above, p is not a prime, and π is a prime divisor of p, so is π'; it is easily verified that π' cannot be an associate of π unless $p = 2$ (in cases (a), (b), (c)) or $p = 3$ (in case (d)). Leaving the latter cases aside, p is in each case a "divisor of the form $X^2 + AY^2$" (as defined above in §XII) or, what amounts to the same, $-A$ is a quadratic residue modulo p. As to the converse, assume that p (other than 2 if $A = 1$ or ± 2, and other than 3 if $A = 3$) is a divisor of the form $X^2 + AY^2$; this means that p divides some integer $a^2 + Ab^2$, with a and b prime to p; as neither one of the integers $a \pm b\sqrt{-A}$ is a multiple of p in R, this implies that p cannot be a prime in R.

The problem of the number of representations of an integer N by one of the forms $X^2 + AY^2$ can now be better formulated as the problem of the number of elements of R with the norm N when two associates are not counted as distinct; in modern terms, this amounts to determining the zeta-function of the field $K = \mathbf{Q}(\sqrt{-A})$. As the various cases under consideration are quite similar, it will be enough to examine the case $A = 1$, $R = \mathbf{Z}[i]$. We have to solve $N = \alpha\alpha'$. If N has a prime divisor p of the form $4n - 1$, then, since p remains a prime in R, it must divide α or α' and must therefore divide both, so that N is a multiple of p^2; this gives $p^{-2}N = (\alpha/p)(\alpha'/p)$. This reduces the problem to the case when N has no such prime divisor. Then we can write

$$N = 2^s p_1^{r_1} \dots p_m^{r_m}$$

where each p_i is of the form $4n + 1$ and can therefore be written as $\pi_i\pi_i'$, where π_i is a prime in R. For any i, let λ_i, μ_i be the exponents of π_i, π_i' in α; they are also the exponents of π_i', π_i in α', and we have $\lambda_i + \mu_i = r_i$. Up to a unit factor i^ν, α must then be given by

$$\alpha = (1+i)^s \pi_1^{\lambda_1} \pi_1'^{\mu_1} \dots \pi_m^{\lambda_m} \pi_m'^{\mu_m}.$$

Clearly the number of possible choices for α is $\Pi(r_i + 1)$. This must be halved for the number of representations of N by $X^2 + Y^2$, since α and α' give the same representation. The representation is "proper" in the sense of §IX if and only if $\lambda_i = 0$ or $\mu_i = 0$ for all i, since otherwise α would be a multiple of p_i; thus the number of proper representations is 2^{m-1}. All this agrees of course with Fermat's results described in §IX.

We can now make use of the ring $\mathbf{Z}[\sqrt{2}]$ in order to reinterpret and clarify Fermat's communication of 1657 to Frenicle (JEH.41–44; cf. above, §XII). Let $\pi = a + b\sqrt{2}$ be a normalized integer, with $0 < a < b$; put $\rho = \varepsilon\pi'^2 = A + B\sqrt{2}$; this gives

$$A = a^2 + 2b^2 - 4ab, \qquad B = a^2 + 2b^2 - 2ab,$$

and it is easily seen that also ρ is normalized. Now assume that p is a rational prime of the form $p = 2x^2 - 1$, and that $p^2 = 2y^2 - 1$. Put $a = 1, b = x$, so that $p = -\pi\pi'$; then π, π' are primes in R. Put $\omega = 1 + y\sqrt{2}$, so that $\omega\omega' = -p^2 = -\pi^2\pi'^2$; as π, π' are primes, and as ω is not a multiple of p, this implies that ω is an associate of π^2 or of π'^2, or, what amounts to the same, of ρ or of $-\rho'$. But ω, ρ, $-\rho'$ are normalized; therefore ω must be either ρ or $-\rho'$, and we get $1 = \pm(1 + 2x^2 - 4x)$, just as in §XII, and $x = 1$ or 2. If p were not a prime, this proof could clearly not be carried out, despite Fermat's assertion to the contrary (cf. §XII).

Appendix II

Curves of Genus 1 in Projective Spaces

As we have seen, most of the diophantine problems studied by Fermat concern equations or sets of equations which, as we would say now, define elliptic curves, i.e., algebraic curves of genus 1; the problem in each case is to find rational (or sometimes integral) points on the curve. For a proper understanding of his methods it is convenient to translate them into the language now usual in algebraic geometry.

Much of what we have to say here, and indeed much of Fermat's work on this topic (cf. above, §XV) retains its validity over arbitrary ground-fields, including even those of characteristic $p > 1$; in order to simplify the language, we shall disregard the latter. So far as Fermat is concerned, only the field $\mathbf{R}$ of real numbers (the field of "geometry" as understood in his time), the field $\mathbf{Q}$ of rational numbers, and implicitly the field of all algebraic numbers, have any relevance.

Let Γ be an algebraic curve; we assume that it is irreducible and defined by equations with coefficients in a subfield K of the field $\mathbf{C}$ of complex numbers. By a point of Γ we will understand one with coordinates in $\mathbf{C}$; it will be called rational if its coordinates are in K, algebraic if they are so over K. A divisor means a formal sum $\mathfrak{a} = \Sigma a_i A_i$, where the A_i are points of Γ and the a_i are integers; Σa_i is the degree of $\mathfrak{a}$. If all the a_i are ≥ 0, $\mathfrak{a}$ is said to be positive, and we write $\mathfrak{a} \succ 0$. If P_1 is an algebraic point of Γ, and $P_1, P_2, \ldots, P_n$ are all its distinct conjugates over K, the divisor ΣP_i is said to be prime rational (over K); a divisor is called rational if it is a linear combination, with integral coefficients, of prime rational divisors. By a rational function $f(M)$ on Γ, one understands a rational function of the coordinates of the point M on Γ; it is called rational over K if its coefficients lie in K. One defines poles and zeros of functions on Γ, with their multiplicities, in the usual manner; every function has as many zeros as it has poles, each being counted as many times as its multiplicity indicates. If the zeros of f are A_1, A_2, ..., A_n, and its poles are B_1, B_2, ..., B_n, then $\Sigma A_i - \Sigma B_i$ is called the divisor of f and is written $\mathrm{div}(f)$; it is of degree 0; if f is rational over K, so is $\mathrm{div}(f)$. If a divisor $\mathfrak{a}$ is the divisor of some function f, it is said to be equivalent to 0, and one writes $\mathfrak{a} \sim 0$. Equivalences between divisors can then be written in the usual manner.

The curve Γ is said to be of genus 1 (or "elliptic") if, to every divisor $\mathfrak{a}$ of degree 1 on Γ, there is one and only one point $A \sim \mathfrak{a}$; we shall henceforth assume that this is so. Then, if $\mathfrak{m}$ is a rational divisor of degree $m > 0$ on Γ, the functions f on Γ, rational over K, such that $\mathrm{div}(f) \succ -\mathfrak{m}$ (and in particular the functions with the poles $M_1, \ldots M_m$ if $\mathfrak{m} = \Sigma M_i$) make up a vector-space of dimension m over K. In particular, if $m = 1$, the function f is unique up to a constant factor, and its divisor is $\mathrm{div}(f) = N - \mathfrak{m}$, where N is the point such that $N \sim \mathfrak{m}$; this point N, which is thus obtained as a zero of f, must then be rational.

Consequently, if there is on Γ a rational divisor of degree 1, there must also be on Γ at least one rational point A. In that case, the functions f on Γ, rational over K, such that $\mathrm{div}(f) \succ -2A$ (resp. $\succ -3A$) make up a vector-space V (resp. W) of dimension 2 (resp. 3) over K. Take any nonconstant $x \in V$; then V consists of the functions $\alpha x + \beta$ on Γ, with α, β in K; similarly, if we take y in $W - V$, W will consist of the functions $\lambda y + \mu x + \nu$ with λ, μ, ν in K. By suitable choices of x in V and of y in W, one may assume that they satisfy an equation of the form

$$y^2 = x^3 - ax - b$$

with a and b in K; here the right-hand side has no multiple root, since otherwise this would not define an elliptic curve. Then one can identify Γ with the plane cubic defined by this equation, or rather with the cubic defined in the projective plane P^2 by the homogeneous equation

$$Y^2Z = X^3 - aXZ^2 - bZ^3,$$

A being then the point at infinity $(0, 1, 0)$ on the cubic. If M, N, P are three points on the cubic, they lie on a straight line if and only if $M + N + P \sim 3A$; in particular, if, for every $P = (x, y)$ on the curve, we write $P' = (x, -y)$, we have $P + P' \sim 2A$. If, for every pair of points (M, N) on the curve, we define a point $M \overset{.}{+} N$ by the relation $M \overset{.}{+} N \sim M + N - A$, the law $(M, N) \to M \overset{.}{+} N$ defines the set of points on Γ as a commutative group of which the rational points make up a subgroup, A being the neutral element; when dealing with a given cubic, one may write $M + N$ instead of $M \overset{.}{+} N$ and 0 instead of A. We have $P = M \overset{.}{+} N$ if and only if M, N, P' are on a straight line, i.e., if P' is the third point of intersection with the cubic of the straight line through M and N. Similarly, $M \overset{.}{+} M$, for which one may write simply $2M$, is the point P such that P' is the intersection of the cubic with the tangent to the cubic at M.

Assume now merely that Γ has a rational divisor $\mathfrak{a} = A + A'$ of degree 2; we can again take a non-constant x on Γ, rational over K, such that $\mathfrak{a} + \mathrm{div}(x) \succ 0$. The space W of functions f, rational over K, such that $2\mathfrak{a} + \mathrm{div}(f) \succ 0$ is of dimension 4 and contains a subspace W' generated by $1, x, x^2$; take y in W and not in W'. After replacing y by $y + \lambda x^2 + \mu x + \nu$ with suitable values of λ, μ, ν, we have an equation

$$y^2 = a_0 x^4 + a_1 x^3 + a_2 x^2 + a_3 x + a_4$$

where the right-hand side has no multiple roots; this may be regarded as defining the curve Γ. In an obvious sense, the two points A, A' which make up the divisor $\mathfrak{a}$ are the points at infinity of the curve; they are rational if and only if a_0 is a square in K.

Similarly, if Γ has a rational divisor $\mathfrak{a}$ of degree $m \geqslant 3$, take a basis $(x_1, ..., x_m)$ of the space of functions f, rational over K, such that $\mathfrak{a} + \operatorname{div}(f) \succ 0$; map Γ into the projective space P^{m-1} of dimension $m - 1$ by taking, as the image of each point M of Γ, the point of P^{m-1} with the homogeneous coordinates

$$(x_1(M), ..., x_m(M)).$$

The image of Γ in P^{m-1} is then a curve Ω of degree m in P^{m-1}. Its intersection with the hyperplane

$$a_1 X_1 + \cdots + a_m X_m = 0$$

consists of the points $P_1 ..., P_m$ given by

$$\operatorname{div}\left(\sum_1^m a_i x_i\right) = \sum_1^m P_i - \mathfrak{a},$$

and m points $P_1, ..., P_m$ of Γ lie on a hyperplane if and only if $\Sigma_1^m P_i \sim \mathfrak{a}$. If $m = 3$, we thus get for Γ a model which is a plane cubic; the model Ω is a space quartic if $m = 4$. In the latter case, consider the 10 functions $x_i x_j$, for $1 \leqslant i \leqslant j \leqslant 4$; they are in the space of functions f satisfying $2\mathfrak{a} + \operatorname{div}(f) \succ 0$; as this space is of dimension 8, we have two linear relations between the $x_i x_j$; this amounts to saying that Ω is the intersection of two quadrics $\Phi(X) = 0$, $\Psi(X) = 0$, where Φ, Ψ are two quadratic forms in the X_i, X_j.

Finally, assume that we have on Γ two non-equivalent rational divisors $\mathfrak{a} = A + A'$, $\mathfrak{b} = B + B'$ of degree 2; take non-constant functions x, y, rational over K, such that $\mathfrak{a} + \operatorname{div}(x) \succ 0$, $\mathfrak{b} + \operatorname{div}(y) \succ 0$. Consider the 9 functions $x^i y^j$ with $i, j \leqslant 2$; they are in the space of functions f such that $2\mathfrak{a} + 2\mathfrak{b} + \operatorname{div}(f) \succ 0$; as this is of dimension 8, there must be a linear relation between the $x^i y^j$, i.e., a relation $F(x, y) = 0$, where F is a polynomial of degree $\leqslant 2$ in x and $\leqslant 2$ in y. Here y must occur in F, since otherwise x would be a constant; similarly x must occur in it. If F were of degree 1 in y, y would be of the form $R(x)$, R being a rational function. But it is easily seen that the divisor of poles of y would then be linearly equivalent to a multiple $m\mathfrak{a}$ of $\mathfrak{a}$. If $m > 1$, this would contradict the assumption $\mathfrak{b} + \mathrm{div}(y) \succ 0$; it would also contradict that assumption if $m = 1$, since $\mathfrak{b}$ is not equivalent to $\mathfrak{a}$. Consequently we have a relation

$$F(x, y) = 0$$

with F of degree 2 in x and of degree 2 in y. Diophantine equations of that type were first considered by Euler (cf. *infra*, Appendix V, and Chap.III, §XVI).

Appendix III

Fermat's "Double Equations" as Space Quartics

In our discussion of the "double equations" studied by Fermat and Euler and of the infinite descent as practised by them, we shall need some elementary facts about space quartics of genus 1. Let Ω be such a curve; as we have seen in Appendix II, it can be defined by a pair of equations $\Phi = \Psi = 0$, where $\Phi = 0$ and $\Psi = 0$ are the equations of two quadrics in P^3. For any ξ, call Q_ξ the quadric given by $\Phi - \xi\Psi = 0$; these quadrics make up the pencil of quadrics through Ω (including of course the quadric $\Psi = 0$, which may be written Q_∞).

For any quadratic form

$$\Phi = \sum_{i,j=1}^{4} a_{ij} X_i X_j$$

we shall write $\det(\Phi) = \det(a_{ij})$. A change of homogeneous coordinates with the determinant δ multiplies $\det(\Phi)$ with δ^2; if Φ is multiplied with a constant λ, $\det(\Phi)$ is multiplied with λ^4; therefore $\det(\Phi)$ is well defined up to a square factor when the quadric $\Phi = 0$ is given. For Φ, Ψ as above we shall write $F(\xi) = \det(\Phi - \xi\Psi)$. For $F(\xi) = 0$, Q_ξ is a cone; if Ω is a space quartic this cone cannot degenerate into a pair of planes. Clearly the "pencil" $\{Q_\xi\}$ must contain at least one cone; one may assume this to be Q_0; then 0 is a root of F; by taking coordinates so that $\Phi = X_3^2 - 2X_1X_2$, it is easily seen that 0 cannot be a double root of F unless the quadric $\Psi = 0$ goes through the vertex $(0, 0, 0, 1)$ of the cone $\Phi = 0$, in which case Ω is of genus 0 and not 1. Thus F is of degree 4 or 3 and has four distinct zeros on the projective straight line; in other words, the pencil of quadrics through Ω contains exactly four cones.

Now we will sketch a proof for the following result:

Proposition. There is a rational mapping ω from $\Omega \times \Omega$ to the curve $F(\xi) = \eta^2$ such that: (i) if M, M', N, N' are four points on Ω, then $\omega(M, M') = \omega(N, N')$ if and only if $M + M' \sim N + N'$; (ii) if A is a rational point on Ω, then the mapping $M \rightarrow \omega(A, M)$ is an isomorphism of Ω onto the curve $F(\xi) = \eta^2$.

For convenience we begin by recalling some well-known facts:

Lemma. A quadric $\Phi = 0$ (not a pair of planes) carries one family of straight lines or two according as $\det(\Phi)$ is 0 or not; if it contains a rational straight line, then $\det(\Phi)$ is a square.

If $\det(\Phi) = 0$, the quadric is a cone and the above assertions are obvious. If $\det(\Phi) \neq 0$, then, after suitably extending the groundfield, one can assume that the quadric contains a rational point A; take coordinates so that $A = (0, 0, 0, 1)$ and that the tangent plane at A is $X_3 = 0$; after possibly another rational change of coordinates, one can write, up to a constant factor:

$$\Phi = X_1^2 - dX_2^2 - 2X_3X_4;$$

here $\det(\Phi) = d \neq 0$. Then A lies on the two straight lines $X_1 = \pm X_2\sqrt{d}, X_3 = 0$. In particular, if we had assumed that the quadric contains a rational straight line, we could have chosen for A a rational point on it and proceeded as above; then $\sqrt{d}$ must be rational, and $\det(\Phi)$ must be a square. In any case, the two straight lines through A belong respectively to the two families

$$t(X_1 \pm X_2\sqrt{d}) = uX_3, \qquad u(X_1 \mp X_2\sqrt{d}) = 2tX_4.$$

Clearly each point of the quadric lies on one and only one

line in each one of these families; this implies that all straight lines lying on the quadric belong to one or the other of these families.

Now, going back to the quartic Ω defined by $\Phi = \Psi = 0$, take two points M, M' on Ω; call $\Delta_{MM'}$ the straight line through M and M' (or the tangent to Ω at M, if $M = M'$). If the quadric $\Phi - \xi\Psi = 0$ contains one point of $\Delta_{MM'}$ beside M and M', it must contain $\Delta_{MM'}$; thus Φ/Ψ has a constant value on $\Delta_{MM'}$, for which we write $\xi(M, M')$; clearly this is a rational function of the coordinates of M and M', with coefficients in the groundfield K. In view of the lemma, $F[\xi(M, M')]$ must be a square after the adjunction of the coordinates of M and M' to the groundfield; thus there is a rational function $\eta(M, M')$ of those coordinates, such that

$$F[\xi(M, M')] = \eta(M, M')^2.$$

Write Γ for the curve $F(\xi) = \eta^2$ and ω for the mapping

$$(M, M') \rightarrow (\xi(M, M'), \eta(M, M'))$$

of $\Omega \times \Omega$ into Γ; we will show that this has the properties described above. We begin with (ii).

As in Appendix II, take a groundfield K such that Φ and Ψ have their coefficients in it; assume that there is on Ω a point A with coordinates in K. For any ξ, take η such that $\eta^2 = F(\xi)$. Let Δ be one of the straight lines through A on Q_ξ; as shown in the proof of the lemma, it is rational over the field $K(\xi, \eta)$. So is therefore the point P, other than A, where Δ meets some quadric other than Q_ξ, say the quadric $\Phi = 0$, in the pencil $\{Q_\xi\}$. As P is then on Ω, we have $\Delta = \Delta_{AP}$, $\xi(A, P) = \xi$, and therefore $\eta(A, P) = \pm\eta$; replacing η by $-\eta$ if necessary, we may assume that $\eta(A, P) = \eta$, so that $\omega(A, P) = (\xi, \eta)$. Since at the same time P is rational over $K(\xi, \eta)$, this proves that $P \rightarrow \omega(A, P)$ is indeed a birational mapping, and therefore an isomorphism, between Ω and Γ. In particular, this shows that $\omega(A, P) = \omega(A, P')$ implies $P = P'$, a conclusion which remains valid even if A is not rational, since we can always enlarge K so as to make it such.

Now take M, M', P on Ω; put $\omega(M, M') = (\xi, \eta)$. Call P'

the fourth point of intersection of Ω with the plane Π through M, M', P; put $\mathfrak{a} = M + M' + P + P'$. As $\Delta_{MM'}$ and $\Delta_{PP'}$ lie in Π, they intersect at a point R. The quadric Q_ξ contains $\Delta_{MM'}$, hence R, hence $\Delta_{PP'}$; therefore we have $\xi(P, P') = \xi$.

Let N, N' be two more points on Ω; we have $N + N' \sim M + M'$ if and only if $N + N' + P + P' \sim \mathfrak{a}$; as has been shown in Appendix II, this is so if and only if N, N', P, P' lie in a plane; if so, the above proof shows that we have $\xi(N, N') = \xi(P, P') = \xi$, and therefore $\eta(N, N') = \pm\eta$. Now, M and M' being kept fixed, take N variable; as $\eta(N, N')$ is given rationally in terms of N, N', and as η is constant, $\eta(N, N')$ must be independent of N. Taking $N = M$, we get $N' = M'$. Therefore $\eta(N, N') = \eta$, and $\omega(N, N') = \omega(M, M')$, as asserted.

Conversely, assume that $\omega(N, N') = \omega(M, M')$, and take N'' such that $N + N'' \sim M + M'$. Then we have

$$\omega(N, N'') = \omega(M, M') = \omega(N, N').$$

As proved above, this implies $N'' = N'$, completing the proof of our proposition. Our result can be interpreted by saying that the equivalence classes of divisors of degree 2 on Ω are parametrized by Γ.

Our proposition can be applied for instance to the curves defined, in the language of Diophantus and of Fermat, by a "double equation"

$$ax^2 + 2bxy + cy^2 = z^2, \qquad a'x^2 + 2b'xy + c'y^2 = t^2.$$

Since they consider only curves of that form which have in projective space at least one "visible" rational point, we may, for our present purposes, assume that this is so. Then we see that the curve is isomorphic to the one given by

$$Y^2 = X[(b'X-b)^2 - (a'X-a)(c'X-c)].$$

Take for instance the problem

$$x^2 + 4y^2 = z^2, \qquad x^2 + y^2 = t^2$$

for which Euler proved (cf. above, §XVI) that there are no solutions other than the obvious ones. We see that these equations define a curve isomorphic to

$$Y^2 = -X(X-1)(X-4).$$

On the other hand, take the problem, considered by Fermat (cf. §XVI) of finding four squares in arithmetic progression. This amounts to

$$2y^2 = x^2 + z^2, \qquad 2z^2 = y^2 + t^2.$$

By the same rule as before, this curve is isomorphic to

$$Y^2 = X(X+2)(2X+1)$$

which is changed into the previous equation by substituting $-\frac{1}{2}X, \frac{1}{2}Y$ for X, Y. Thus, as Euler found out, the two problems are equivalent.

Appendix IV

The Descent and Mordell's Theorem

Our purpose is now to provide some background for Fermat's "method of descent" in his treatment of elliptic curves (cf. above, §XVI) in the light of modern work on the subject[11].

As has been shown in Appendix II, if an elliptic curve has at least one rational point, it can be exhibited as a cubic $y^2 = f(x)$, where f is a polynomial of degree 3. The rational points on the cubic make up a group, and Fermat's method of descent has been refined by Mordell in 1922 so as to show that this group is "finitely generated"; this means that there are on the cubic finitely many rational points such that all others can be derived from these by the group operation defined in Appendix II. The same is true if the groundfield, instead of being the field $\mathbf{Q}$ of rational numbers, is any algebraic number-field of finite degree over $\mathbf{Q}$. Let r be the largest integer such that there are on the cubic r rational points which do not satisfy any linear relation with integral coefficients not all zero; then r is called the rank of the curve. Even now no systematic method is known for determining the rank of a given cubic, let alone a set of generators for its group of rational points.

---

[11] Cf. e.g. J. W. S. Cassels, *J. London Math. Soc.* 41 (1966), pp. 193–291, and A. Weil, *Coll. Papers* I.47–57.

For simplicity we consider the case when the roots of f are rational integers α, β, γ, the cubic Γ being given by

(1) $$y^2 = f(x) = (x-\alpha)(x-\beta)(x-\gamma).$$

Put $x = X/Z$, where Z is a natural integer and X an integer prime to Z. This gives

$$(Z^2 y)^2 = Z(X-\alpha Z)(X-\beta Z)(X-\gamma Z).$$

As Z is prime to the other factors in the right-hand side it must be a square, $Z = T^2$. Traditionally one observes that the g.c.d. of any two of those factors divides a fixed integer, so that one may write $X - \alpha Z = AU^2$, where A is susceptible of only a finite number of values, therefore $x - \alpha = Au^2$ with $u = U/T$, and similarly $x - \beta = Bv^2$, $x - \gamma = Cw^2$, where u, v, w are rational and B, C can take only finitely many values. If one uses only the first one of these relations, one gets a rational point $(u, z=y/u)$ of the quartic

$$z^2 = A(Au^2+\alpha-\beta)(Au^2+\alpha-\gamma),$$

which is one of finitely many thus attached to Γ. On the other hand, after properly choosing the signs of u, v, w, one can write

(2) $$Au^2 + \alpha = Bv^2 + \beta = Cw^2 + \gamma$$

(3) $$x = Au^2 + \alpha, \qquad y = \sqrt{ABC} \cdot uvw.$$

Here ABC has to be a square; (2) defines a space quartic Ω in the (u, v, w)-space, and (3) defines a mapping of Ω into Γ. The original problem is thus reduced to finding rational points on finitely many such quartics. Either one of these methods is typical of the traditional descent.

An alternative procedure, equivalent to the one described above but more consonant with the modern theory would be as follows. Let (x, y) be a solution of (1); we can write, in one and only one way, $x - \alpha = Au^2$, $x - \beta = Bv^2$, $x - \gamma = Cw^2$, with rational u, v, w and squarefree integers A, B, C (positive or negative); equations (2), (3) are satisfied, and ABC has to be a square.

For any squarefree integers A, B, C (positive or negative) such that ABC is a square, call $\Omega(A, B, C)$ the space quartic defined by (2). The first step in the descent thus consists in reducing (1) to the finding of rational points on infinitely many quartics $\Omega(A, B, C)$. The next step will be to reject all but finitely many of these because they can have no rational points.

In homogeneous coordinates, $\Omega(A, B, C)$ may be regarded as defined by equations

(4) $AU^2 + \alpha T^2 = BV^2 + \beta T^2 = CW^2 + \gamma T^2;$

$\Omega(A, B, C)$ has a rational point if and only if (4) has a solution in integers U, V, W, T without a common divisor; a necessary condition for this is obviously that the same relations, regarded as congruences modulo some integer m, should have such a solution, whatever m may be. In the language of p-adic fields, this can also be expressed by saying that (2) should have a non-trivial solution in every p-adic field $\mathbf{Q}_p$; to this one can add the condition that it should have a non-trivial solution in the field $\mathbf{R} = \mathbf{Q}_\infty$ of real numbers.

In particular, take for p a common prime divisor of B and C; as ABC is a square, and A, B, C are squarefree, p does not divide A. Take $m = p^2$, and suppose that p does not divide $\beta - \gamma$; then the congruences

$$AU^2 + \alpha T^2 \equiv BV^2 + \beta T^2 \equiv CW^2 + \gamma T^2 \pmod{p^2}$$

imply firstly that p divides T, then that it divides U, then that it divides V and W; $\Omega(A, B, C)$ can then have no rational point (not even in $\mathbf{Q}_p$). Thus, if $\Omega(A, B, C)$ has rational points, and if we write a for the g.c.d. of B and C, a must divide $\beta - \gamma$. Write now $B = ab$, $C = ac$, with b prime to c; then we must have $A = bc$. As B and C are squarefree, a is prime to b and c, so that $|b|$ is the g.c.d. of A and C, and $|c|$ is the g.c.d. of A and B; therefore b divides $\gamma - \alpha$, and c divides $\alpha - \beta$.

This leaves us with finitely many curves $\Omega(A, B, C)$, out of which further arguments and the consideration of the real field may lead to reject some more; it is even possible to discard all except those which have points in all fields $\mathbf{Q}_p$ and in $\mathbf{R}$. In contrast, however, with classical results, going back to Legendre (cf. Chap.IV, §VI), on curves of genus 0, not all such quartics have rational points, nor is there any known procedure for finding one if such points exist. In order to proceed further, it will now be assumed that, among all quartics $\Omega(A, B, C)$, the quartics $\Omega_i = \Omega(A_i, B_i, C_i)$ $(1 \leq i \leq N)$ and no others have at least one rational point. Then, if $M = (x, y)$ is any rational point on Γ, there is one and only one i such that (2) and (3) are satisfied for $A = A_i, B = B_i, C = C_i$ and for some rational values for u, v, w; when that is so, we will say that M belongs to Ω_i and that it corresponds to the point $N = (u, v, w)$ on Ω_i.

Let $\Omega = \Omega(A, B, C)$ be one of the Ω_i; instead of defining it by (2), we write

$$\begin{aligned}
\Phi(U, V, W, T) &= \alpha(\beta-\gamma)(AU^2+\alpha T^2)+\beta(\gamma-\alpha)(BV^2+\beta T^2) \\
&\quad + \gamma(\alpha-\beta)(CW^2+\gamma T^2) \\
&= \alpha(\beta-\gamma)AU^2 + \beta(\gamma-\alpha)BV^2 \\
&\quad + \gamma(\alpha-\beta)CW^2 - \delta T^2
\end{aligned}$$

$$\Psi(U, V, W, T) = (\beta-\gamma)AU^2 + (\gamma-\alpha)BV^2 + (\alpha-\beta)CW^2$$

where we have put

$$\delta = (\beta-\gamma)(\gamma-\alpha)(\alpha-\beta);$$

then Ω may be regarded as defined by $\Phi = \Psi = 0$. With the notations of Appendix III, we have

$$\begin{aligned}
F(\xi) = \det(\Phi-\xi\Psi) &= \delta^2 ABC(\xi-\alpha)(\xi-\beta)(\xi-\gamma) \\
&= \delta^2 ABC \cdot f(\xi).
\end{aligned}$$

For any pair of points N, N' on Ω, let $\xi(N, N')$, $\eta(N, N')$ be defined as in Appendix III; put

$$\omega_0(N, N') = (\xi(N, N'), (\delta\sqrt{ABC})^{-1}\eta(N, N')),$$

so that ω_0 maps $\Omega \times \Omega$ into Γ.

Let $M = (x, y)$ be a point on Γ, belonging to Ω and corresponding to a point $N = (u, v, w)$ of Ω, so that the relations (2), (3) are satisfied. The tangent Δ_{NN} to Ω at N goes through the point with the homogeneous coordinates $(u', v', w', 0)$ given by $Auu' = Bvv' = Cww'$. Therefore $\xi(N, N)$ is given by

$$\xi(N, N) = \frac{\Phi(u', v', w', 0)}{\Psi(u', v', w', 0)}.$$

An easy calculation gives $\xi(N, N) = x$; then we have

$$\eta(N, N)^2 = F(x) = \delta^2 ABCf(x) = \delta^2 ABCy^2.$$

As η, according to its definition in Appendix III, is defined only up to a factor ± 1, we may, after changing its sign if necessary, assume that $\eta(N, N) = \delta\sqrt{ABC} \cdot y$, so that $\omega_0(N, N) = M$. If we write λ for the mapping $N \to \omega_0(N, N)$, this shows that λ is no other than the mapping of Ω into Γ defined by (3).

Now choose a rational point M_0 on Γ, belonging to Ω and corresponding to a point N_0 on Ω; by definition, this means that $\lambda(N_0) = M_0$. Write μ for the mapping $N \to \omega_0(N_0, N)$ of Ω into Γ; the proposition in Appendix III shows that this is an isomorphism of Ω onto Γ; write μ^{-1} for its inverse; we have $\mu(N_0) = \lambda(N_0) = M_0$. Take any rational point M on Γ; put $P = \mu^{-1}(M)$, so that $M = \mu(P)$. By definition, M belongs to Ω and corresponds to a point N on Ω if and only if (3) maps N onto M, i.e. if and only if $\lambda(N) = M = \mu(P)$, i.e., in view of the definition of λ and μ, if and only if $\omega_0(N, N) = \omega_0(N_0, P)$. Again by the proposition in Appendix III, this is so if and only if $2N \sim N_0 + P$ on Ω. Put $M' = \mu(N)$; as μ is an isomorphism and maps N, N_0, P onto M', M_0, M respectively, the relation $2N \sim N_0 + P$ on Ω is equivalent to $2M' \sim M_0 + M$. This proves that M belongs to Ω if and only if there is a rational point M' on Γ such that $2M' \sim M_0 + M$, in which case M corresponds to the point $\mu^{-1}(M')$ on Ω.

For the proof of Mordell's theorem, one also needs a quantitative estimate for the height of M' in terms of the height of M, these being defined as follows. Put again $M = (x, y)$ and $x = X/T^2$, with X prime to T; then we define the height $h(M)$ of M by $h(M) = \sup(|X|, T^2)$; similarly, if a point N of Ω has the homogeneous coordinates (U, V, W, T), these being rational integers without a common divisor, we put

$$h(N) = \sup(|U|, |V|, |W|, |T|).$$

If now the point M of Γ corresponds to the point N on Ω, we have

$$X = AU^2 + \alpha T^2 = BV^2 + \beta T^2 = CW^2 + \gamma T^2,$$

so that U^2, V^2, W^2 are bounded in terms of $|X|$ and T^2, i.e., in terms of $h(N)$. This gives:

$$h(N) \le c\sqrt{h(M)},$$

where c depends solely upon Γ. For an estimate of $h(M')$ in terms of $h(N)$, one could apply Siegel's theorem on heights (cf. C. L. Siegel, *Ges.Abh.*I.249–250, or A. Weil, *Coll. Papers* I.477); this would give

$$h(M') \le c'h(N)^\rho$$

where one can take for ρ any number $> \frac{1}{2}$, and where c' depends solely upon Ω, M_0 and ρ. For our present purpose, however, a cruder estimate will serve just as well. As before, we have $M' = \omega_0(N_0, N)$, and therefore $x' = \xi(N_0, N)$ if $M' = (x', y')$. Put $N_0 = (U_0, V_0, W_0, T_0)$; by definition, $x' = \xi(N_0, N)$ means that $\Phi - x'\Psi$ is 0 at some point $(U_0 + tU, V_0 + tV, W_0 + tW, T_0 + tT)$ with $t \ne 0$. As N_0 and N satisfy $\Phi = \Psi = 0$, this gives

$$(5) \quad x' = \xi(N_0, N) = \frac{\Phi(U_0, V_0, W_0, T_0; U, V, W, T)}{\Psi(U_0, V_0, W_0, T_0; U, V, W, T)}$$

where numerator and denominator are the bilinear forms respectively associated with the quadratic forms Φ, Ψ; these are linear forms in U, V, W, T, with coefficients which depend solely upon Ω and M_0. Of course, for any given M, the values of these forms need not be prime to each other;

nevertheless, if we write $x' = X'/T'^2$ with X' prime to T', those values give upper bounds for $|X'|$ and T'^2. Thus we get:

$$h(M') \leq c_1 h(N) \leq c_2 \sqrt{h(M)}$$

where c_1, and therefore $c_2 = cc_1$, depend solely upon Γ, Ω and M_0.

Now we come back to the finitely many quartics Ω_i defined above; we assume that on each we have somehow chosen a rational point N_i; let M_i be the corresponding point on Γ. For every rational point M, there must then be one and only one of the points M_i such that $M + M_i$ is equivalent to a divisor $2M'$, M' being a rational point of Γ. Here M' is not uniquely defined; it is so up to an element of order 2 in the group of rational points on the cubic Γ (in the sense explained in Appendix II); those elements are the point at infinity and the three points $(\alpha, 0)$, $(\beta, 0)$, $(\gamma, 0)$.

The process of the descent is then as follows. Starting with a rational point M of Γ, we can write it, in the sense of the group of such points on Γ, as $M \sim 2M' - M_i$, where M' is again rational; we have $h(M') \leq d_i \sqrt{h(M)}$, with d_i depending solely upon Γ, Ω_i and M_i; we can now repeat the process by writing $M' \sim 2M'' - M_j$ where we have $h(M'') \leq d_j \sqrt{h(M')}$, etc. Thus we get a sequence of points M, M', M'', etc. whose heights must ultimately become less than m if we take $m > \sup(d_i^2)$. Obviously there are only finitely many points $P_1, P_2, ..., P_r$ on Γ whose height does not exceed m. Thus the points M_i, P_j generate the group of rational points on Γ; this is Mordell's theorem. In principle, the same proof can be used for a cubic over any groundfield of finite degree over $\mathbf{Q}$.

For obvious reasons the above descent is called one "by bisection" (or by division by 2); division by any other integer would, in principle, be equally effective. While division, by 2 or by n, turns out to be most suitable for a general treatment of the problem, it is well-known from the classical theory of elliptic functions that division by n, which amounts to an extension of degree n^2 of the field of functions on the curve, may always be replaced by two "transformations" (now usually called isogenies) of order n. For instance, if one starts from the cubic Γ defined by (1), division by 2, leading to the quartic $\Omega(A, B, C)$, can be obtained by the successive adjunction of $u = \sqrt{(x-\alpha)/A}$ and of $v = \sqrt{(x-\beta)/B}$, each of these amounting to a transformation of order 2. Thus it is not surprising that the formulas used by Fermat, Euler and Lagrange in the descent define such transformations (except in the treatment of the curve $x^3 + y^3 = 1$, which depends upon transformations of order 3; cf. above, §XVI). As has been pointed out in §XVII, it may have been no mere coincidence that Lagrange re-discovered the "Landen transformation" some seven years after his major paper on equations of genus 1.

We will now illustrate our discussion of equation (1) by treating two examples, both originating from Fermat, which have been mentioned in §XVI. Take first the problem of the four squares in arithmetic progression; as Euler noted, and as has been shown in Appendix III, it is equivalent to the equation

(6) $$y^2 = x(x+1)(x+4)$$

which will be treated in the manner explained above. The quartics Ω_i are to be sought among the curves $\Omega(A, B, C)$ defined by equations

$$x = Au^2 = Bv^2 - 1 = Cw^2 - 4;$$

here we must have $A = bc$, $B = ca$, $C = ab$, with a, b, c dividing respectively 3, 4 and 1; thus A is ± 1 or ± 2, B is ± 1 or ± 3, and $C = AB$. For the curve to have real points, C must be >0. Consider the congruences

$$AU^2 \equiv BV^2 - T^2 \equiv CW^2 - 4T^2 \pmod{3^2};$$

if (A, B, C) is $(2, 1, 2)$, $(1, 3, 3)$, $(-1, -1, 1)$ or $(-2, -3, 6)$, they have no solution in integers not all divisible by 3; therefore these values should be rejected. This leaves the four curves $\Omega_0, \Omega_1, \Omega_2, \Omega_3$ respectively given by

$$(A, B, C) = (1, 1, 1), (2, 3, 6), (-1, -3, 3), (-2, -1, 2)$$

with the four points of Γ

$$M_0 = \infty, \quad M_1 = (2, 6), \quad M_2 = (-1, 0), \quad M_3 = (-2, -2)$$

belonging respectively to the quartics Ω_i. If we put $P_0 = (0, 0)$, $P_1 = (-4, 0)$, then the elements of order 2 in the group of rational points on Γ are M_2, P_0, P_1, which, together with the neutral element $M_0 = \infty$, make up a group of order 4; it is easily seen that $2M_1 = 2M_3 = P_0, M_1 + M_2 = M_3$, so that $\pm M_1, \pm M_3$ are elements of order 4.

Consider the real points on Γ; they make up two "branches", one B_0 consisting of the points for which $x \geq 0$ and extending to infinity, and an oval B_1 consisting of the points for which $-4 \leq x \leq -1$; any straight line in the plane must cut B_1 at two points or not cut it at all; in view of the definition of the addition operation on Γ, this shows that B_0 is a subgroup of index 2 of the group of real points, with $B_1 = B_0 + M_2$; every rational point on B_0 must belong either to Ω_0 or to Ω_1. If such a point M belongs to Ω_0 (resp. to Ω_1), we can write $M = 2M'$ (resp. $M = 2M' - M_1$); here, if M' lies on B_1, we can replace it by $M' - M_2$; thus, in carrying out the descent, it is enough to consider the rational points on B_0.

All that remains to be done, in order to carry out the descent in the manner explained above, is to obtain the estimate for $h(M')$ in terms of $h(M)$ if $M = 2M'$ or $M = 2M' - M_1$. Suppose first that M belongs to Ω_0; with the same notations as before, let $M = (x, y)$ correspond to $N = (U, V, W, T)$ on Ω_0; here M_0 corresponds to $N_0 = (1, 1, 1, 0)$. Writing $x = X/T^2$ as before, and $h = h(M) = \sup(X, T^2)$, we have

$$X = U^2 = V^2 - T^2 = W^2 - 4T^2,$$

and therefore

$$|U| \leq \sqrt{h}, \quad |V| \leq \sqrt{2h}, \quad |W| \leq \sqrt{5h}, \quad U \equiv W \pmod 2.$$

As to M', it is given by (5); writing $M' = (x', y'), x' = X'/T'^2$, we can write, in view of the fact that $U \equiv W \pmod 2$:

$$\frac{X'}{T'^2} = \frac{2(V - W)}{\frac{1}{2}(3U - 4V + W)},$$

where X', T'^2 are majorized by the numerator and denominator in the right-hand side. Putting $h' = h(M')$, we get now:

$$h' \leqslant 2(\sqrt{2} + \sqrt{5})\sqrt{h}.$$

Similarly, for M belonging to Ω_1, and for $M_1 = (2, 6)$, $N_1 = (1, 1, 1, 1)$, we get $h' < 4\sqrt{h}$. This shows that the group of rational points on the branch B_0 of Γ is generated by M_1 and points whose height is <54. As to the latter, let $M = (x, y)$ be such a point, with $x = X/T^2$; as $x > 0$, and as M belongs to Ω_0 or Ω_1, we have $X = U^2$ or $X = 2U^2$; X and T^2 are <54. It is easy now to verify that there are no such points other than P_1 and M_1. In conclusion, the only rational points on Γ are the eight points $M_0, M_2, P_0, P_1, \pm M_1, \pm M_3$. If now we write the equations for four squares in arithmetic progression:

$$2y^2 = x^2 + z^2, \qquad 2z^2 = y^2 + t^2,$$

we see that in the projective space P^3 they have no more than the solutions given by $x^2 = y^2 = z^2 = t^2$; this is what Father de Billy had been told by Fermat. Our other example will be treated in Appendix V.

Appendix V

The Equation $y^2 = x^3 - 2x$

The methods sketched above in Appendices III and IV will now be applied to Fermat's problems (A) and (B) (cf. above, §§XV and XVI).

As we have seen, the question amounts to the solution of the "double equation" (11) of §XV, i.e., of

(7) $$x^2 + 1 = u^2, \quad x^2 - 2x - 1 = v^2$$

in rational numbers, with the side-condition $x > 1$ in case (A) and $x < -1$ in case (B); (7) defines a space quartic Ω which, in homogeneous coordinates, may be regarded as given by $\Phi = \Psi = 0$ if one puts

(8) $\Phi = X^2 - 2XT - T^2 - V^2, \quad \Psi = U^2 - X^2 - T^2.$

To this we can apply the results of Appendix III; with the notations of that Appendix, we have $F(\xi) = \xi^3 - 2\xi$, and Ω is isomorphic to the cubic Γ given by

(9) $$\eta^2 = \xi(\xi^2 - 2).$$

We shall proceed with this as in Appendix IV, with the modifications required because the roots of the right-hand side are not all rational.

As in the case of (6) in Appendix IV, the group Γ_∞ of the real points of Γ consists of two branches B_0, B_1, defined respectively by $\xi > 0$ and by $\xi \leq 0$; B_0 is a subgroup of Γ_∞

of index 2, and we can write $B_1 = B_0 + M_2$ with $M_2 = (0, 0)$; M_2 is an element of Γ_∞ of order 2. As in the case of (6), Appendix IV, it will be enough to carry out the descent for the rational points on B_0.

Put $\xi = P/Q$, where P, Q are mutually prime positive integers; we have

$$(Q^2\eta)^2 = PQ(P^2 - 2Q^2)$$

and infer that Q must be a square, $Q = V^2$. The g.c.d. of P and of the last factor on the right is 2 or 1 according as P is even or odd. If P is even, we must have $P = 2U^2$, $P^2 - 2Q^2 = 2S^2$, $2U^4 - V^4 = S^2$; if P is odd, we can write $P = U^2, P^2 - 2Q^2 = S^2, U^4 - 2V^4 = S^2$. Thus, incidentally, the problem can be reduced to the consideration of the equations $2u^4 - 1 = \pm s^2$; this is how Lagrange had proceeded in his paper of 1777 (*Lag.*IV.377–398), where he begins by putting $S = X^2 + 2XT - T^2$ and observing that the equations $\Phi = \Psi = 0$ give $2U^4 - V^4 = S^2$; the correspondence thus defined between Ω (or Γ) and the quartic curve $2u^4 - 1 = s^2$ is a transformation (an "isogeny") of order 2, and the relationship between Ω (or Γ) and the curve $2u^4 - 1 = -s^2$ is similar.

Here, in analogy with Appendix IV, we shall use the bisection and the ring $\mathbf{Z}[\sqrt{2}]$, for which notations will be as in Appendix I. If P is odd, we put

$$\alpha = P + Q\sqrt{2} = U^2 + V^2\sqrt{2};$$

then α' is prime to α, and we have

$$\alpha\alpha' = N(\alpha) = P^2 - 2Q^2 = S^2.$$

If P is even, we put

$$\alpha = Q + \frac{P}{2}\sqrt{2} = V^2 + U^2\sqrt{2};$$

as Q is prime to P, it is odd; therefore α' is again prime to α, and we have

$$\alpha\alpha' = N(\alpha) = Q^2 - 2\left(\frac{P}{2}\right)^2 = -S^2.$$

In both cases, writing α as a product of primes and of a unit in the ring $\mathbf{Z}[\sqrt{2}]$, one finds that α, up to a unit, must be a square, so that it is of the form $\pm\beta^2$ or $\pm\varepsilon\beta^2$ according as $\alpha\alpha'$ is S^2 or $-S^2$; as before, we put $\varepsilon = 1 + \sqrt{2}$; β is an integer in $\mathbf{Z}[\sqrt{2}]$. In view of the above formulas we must have $\alpha > 0$, so that α is β^2 in one case, $\varepsilon\beta^2$ in the other. Put $\beta = Y - T\sqrt{2}$ and substitute β^2 resp. $\varepsilon\beta^2$ for α in the above formulas; one gets

$$U^2 = Y^2 + 2T^2, \qquad V^2 = -2YT$$

in one case, and

$$U^2 = Y^2 - 2YT + 2T^2, \qquad V^2 = Y^2 - 4YT + 2T^2$$

in the other. These pairs of equations define two space quartics Ω_0, Ω_1 in the projective space with the homogeneous coordinates (Y, T, U, V); in conformity with the notations of Appendix III, we put

$$\Phi_0 = 4YT + 2V^2, \qquad \Psi_0 = Y^2 + 2T^2 - U^2,$$

$$\Phi_1 = Y^2 - 4YT + 2T^2 - V^2, \qquad \Psi_1 = U^2 - (Y^2 - 2YT + 2T^2),$$

so that Ω_0 (resp. Ω_1) is defined by $\Phi_0 = \Psi_0 = 0$ (resp. $\Phi_1 = \Psi_1 = 0$); one may note that the substitution $Y = X + T$ changes Φ_1, Ψ_1 into Φ, Ψ as defined in (8), so that Ω_1 is projectively the same as Ω. Again with the notations of Appendix III, $F(\xi)$ is $4\xi(\xi^2 - 2)$ for Ω_0, $\xi(\xi^2 - 2)$ for Ω_1; the points $M_0 = \infty$ and $M_1 = (2, 2)$ of Γ belong respectively to Ω_0 and to Ω_1; M_0 corresponds to $N_0 = (1, 0, 1, 0)$ on Ω_0, and M_1 to $N_1 = (1, 0, 1, 1)$ on Ω_1. For $i = 0$ or 1, a rational point $M = (\xi, \eta)$ of Γ belongs to Ω_i if and only if there is a rational point $M' = (\xi', \eta')$ on Γ such that $M \sim 2M' - M_i$. If M corresponds to $N = (Y, T, U, V)$ on Ω_i, then we have $\xi = U^2/V^2$ if $i = 0$, $\xi = 2U^2/V^2$ if $i = 1$. Putting again $h(M) = \sup(P, Q)$ and $h(N) = \sup(|Y|, |T|, |U|, |V|)$, we have in both cases $h(N) \leqslant \sqrt{h(M)}$. As before, M' is given by $\xi' = \xi(N_i, N)$, where $\xi(N, N')$ is defined as in Appendix III for any pair of points N, N' on Ω_i. For $i = 0$, this gives, in view of (5), Appendix IV, and of the fact that $Y - U$ must be even:

$$\xi' = \xi(N_0, N) = \frac{T}{\frac{1}{2}(Y - U)}$$

and therefore $h(M') \leqslant h(N) \leqslant \sqrt{h(M)}$. For $i = 1$, Y, U and V must be odd, and T must be even; we get

$$\xi' = \xi(N_1, N) = \frac{\frac{1}{2}(Y - 2T - V)}{\frac{1}{2}(U - Y + T)}$$

and $h(M') \leqslant 2h(N) \leqslant 2\sqrt{h(M)}$. Consequently the sequence M, M', M'', ... produced by the infinite descent on the branch B_0 of Γ must lead to a point of height $\leqslant 4$. Since the only such point on B_0 is M_1, this shows that M_1 generates the group of rational points on B_0; in other words, those points are all given by

(10) $$A_n \sim M_0 + n(M_1 - M_0)$$

for some $n \in \mathbf{Z}$, while those on B_1 are given by

(11) $$A'_n \sim M_2 + n(M_1 - M_0);$$

we have $A_0 = M_0$, $A_1 = M_1$, $A_2 = (9/4, -21/8)$, etc. If $A_n = (\xi, \eta)$, then, as easily seen, we have $A_{-n} = (\xi, -\eta)$ and $A'_n = (-2/\xi, 2\eta/\xi^2)$. As shown above, we have $h(A_n) \leqslant \sqrt{h(A_{2n})}$, so that the heights of the points A_n are not bounded; the cubic Γ is "of rank 1", and there is no equivalence relation between the points M_0, M_1, M_2, other than the obvious one $2M_2 \sim 2M_0$.

Now we turn back to Fermat's original problems and to the curve Ω defined by (7). Let $N = (X, T, U, V)$ be a point of Ω; take a rational point P on Ω, e.g. $P = (1, 0, 1, 1)$; if we use it as in Appendix III in order to obtain an isomorphism between Ω and Γ, and write $M = (\xi, \eta)$ for the point of Γ corresponding to N, then we have $\xi = \xi(P, N)$; in view of (5), Appendix IV, this gives, after taking (8) into account:

(12) $$\xi = \frac{V - X + T}{X - U} = 2\frac{X + U}{V + X - T}.$$

From this one sees that the points of Γ corresponding to $N_0 = (1, 0, 1, -1)$ and to $N_2 = (1, 0, -1, 1)$ on Ω are respectively M_0 and M_2. As to the point corresponding to $P = (1, 0, 1, 1)$, it must be either $(2, 2)$ or $(2, -2)$, so that, after choosing properly the sign of η in the correspondence between Ω and Γ, we may assume that it is $M_1 = (2, 2)$; accordingly we write N_1 for P. Therefore all the rational points on Ω can be written in terms of N_0, N_1, N_2 by formulas similar to (10) and (11); ξ, regarded as a function on Ω, has the divisor $2N_2 - 2N_0$. Now consider the planes $T = 0$, $X + U = 0$, $V + X - T = 0$; if we put $N_3 = (1, 0, -1, -1)$, they determine on Ω the divisors

$$\operatorname{div}(T) = \sum_{i=0}^{3} N_i, \qquad \operatorname{div}(X+U) = 2N_2 + 2N_3,$$

$$\operatorname{div}(V+X-T) = 2N_0 + 2N_3.$$

Therefore the functions defined on Ω by

(13) $$\zeta = \frac{X + U}{T}, \qquad \theta = \frac{V + X - T}{T}$$

have divisors respectively given by

$$\operatorname{div}(\zeta) = N_2 + N_3 - N_0 - N_1,$$

$$\operatorname{div}(\theta) = N_0 + N_3 - N_1 - N_2.$$

If $N = (x, 1, u, v)$ is a point on Ω, the function $\zeta - (x+u)$, which has the two poles N_0, N_1, must have two zeros; obviously they are N and the point $N' = (x, 1, u, -v)$, so that we have, when that is so, $N' \sim N_0 + N_1 - N$. Similarly, using the function $\theta - (v+x-1)$, we see, for $N'' = (x, 1, -u, v)$, that we have $N'' \sim N_1 + N_2 - N$; for $N''' = (x, 1, -u, -v)$, this gives

$$N''' \sim N_2 - N_0 + M.$$

In particular, if we call $B_n = (x_n, 1, u_n, v_n)$ the point of Ω corresponding to the point A_n on Γ, this shows that the points corresponding to A_{1-n} and to A'_n are respectively

$$B_{1-n} = (x_n, 1, u_n, -v_n), \qquad B'_n = (x_n, 1, -u_n, -v_n).$$

Thus these are all the rational points on Ω.

We will now take this opportunity for illustrating a method introduced by Euler towards the end of his life for solving "by ascent" (in the sense given above to this word in §XV) diophantine equations of genus 1; this was based on what he called "canonical equations" (*Eu.*I-5.158 in E 778|1780; cf. *infra*, Chap.III, §XVI), i.e. equations $F(X, Y) = 0$ of degree 2 in X and 2 in Y; for slightly different applications of his method to a problem equivalent to the present one (given by the pair of equations $X + Y = Z^2$, $X^2 + Y^2 = T^4$), cf. *Eu.*I-5.77–81 = E 769|1780 and *Eu.*I-5.87–88 in E 772|1780.

As explained above in Appendix II, a "canonical equation" in the sense of Euler will hold between two functions on the given curve Ω if each of them has only two poles. As such functions we choose ξ and ζ, as defined by (12) and (13). Putting as before $x = X/T$, $u = U/T$, $v = V/T$, we have $x + u = \zeta$, and, in view of the relation $x^2 - u^2 = -1$, $x - u = -1/\zeta$; this gives

$$x = \frac{1}{2}\left(\zeta - \frac{1}{\zeta}\right), \qquad u = \frac{1}{2}\left(\zeta + \frac{1}{\zeta}\right).$$

At the same time formula (12) gives:

$$v - x + 1 = \xi(x-u) = -\frac{\xi}{\zeta}, \quad v + x - 1 = \frac{2}{\xi}(x+u) = \frac{2\zeta}{\xi}$$

and therefore:

$$v = \frac{\zeta}{\xi} - \frac{\xi}{2\zeta}, \qquad x - 1 = \frac{\zeta}{\xi} + \frac{\xi}{2\zeta}.$$

Comparing the values found above for x, we get a relation between ξ and ζ which can be written as a "canonical equation":

$$F(\xi, \zeta) = \xi\zeta^2 - \xi^2 - 2\xi\zeta - 2\zeta^2 - \xi = 0.$$

Moreover, the above formulas determine a birational correspondence, i.e. an isomorphism, between Ω (or, what amounts to the same, Γ) and the plane curve C defined by

$F = 0$; the points on C corresponding to N_0, N_1, N_2, N_3 are respectively

$$P_0 = (\infty, \infty), \quad P_1 = (2, \infty), \quad P_2 = (0, 0), \quad P_3 = (-1, 0).$$

Euler's "ascent", applied to the equation $F = 0$, is now as follows. Let $R_0 = (\xi_0, \zeta_0)$ be any rational point on C; as F is of degree 2 in ζ, the equation $F(\xi_0, z) = 0$ has, beside ζ_0, a second rational root ζ_1, given by

$$\zeta_1 = \frac{2\xi_0}{\xi_0 - 2} - \zeta_0 = -\frac{\xi_0(\xi_0 + 1)}{\zeta_0(\xi_0 - 2)};$$

this determines on C a point $S_0 = (\xi_0, \zeta_1)$; the divisor of the function $\xi - \xi_0$ is then $R_0 + S_0 - 2P_0$, so that we have

$$S_0 \sim 2P_0 - R_0.$$

Similarly, the equation $F(x, \zeta_1) = 0$ has a second solution ξ_1, given by

$$\xi_1 = \zeta_1^2 - 2\zeta_1 - 1 - \xi_0 = \frac{2\zeta_1^2}{\xi_0},$$

and determining on C a point $R_1 = (\xi_1, \zeta_1)$ for which we have

$$R_1 \sim P_0 + P_1 - S_0.$$

Repeating the same procedure, we get a sequence R_0, S_0, R_1, S_1, etc., and an easy induction gives:

$$R_n \sim R_0 + n(P_1 - P_0), \qquad S_n \sim 2P_0 - R_n.$$

If for instance we take as our starting point the point $R_0 = P_0$, the comparison with the formulas obtained above shows that the points R_n, S_n are those which correspond on C to the points A_n, A_{-n} on Γ, respectively. We have, in that case, $R_0 = S_0 = P_0, R_1 = P_1, S_1 = (2, -3/2)$, and then, using the above formulas:

$$R_2 = \left(\frac{9}{4}, -\frac{3}{2}\right), \qquad S_2 = \left(\frac{9}{4}, \frac{39}{2}\right),$$

$$R_3 = \left(338, \frac{39}{2}\right), \qquad S_3 = \left(338, -\frac{1469}{84}\right),$$

$$R_4 = \left(\frac{12769}{7056}, -\frac{1469}{84}\right), \quad S_4 = \left(\frac{12769}{7056}, -\frac{172325}{112812}\right), \text{ etc.}$$

Then the point $B_n = (x_n, 1, u_n, v_n)$ of Ω is the one corresponding to R_n by the above formulas, and the points $(x_n, 1, \pm u_n, \pm v_n)$, for $n = 0, 1, 2, \ldots$, are all the rational points on Ω. In particular, we have (apart from $x_0 = x_1 = \infty$, which correspond respectively to the points N_0, N_1 on Ω):

$$x_2 = -\frac{5}{12}, \qquad x_3 = \frac{1517}{156}, \qquad x_4 = -\frac{2150905}{246792},$$

$$x_5 = -\frac{16969358281}{38880655800}$$

Fermat's problem (A) required a solution with $x > 1$; problem (B) required $x < -1$; his solutions, which are indeed the smallest possible ones, correspond respectively to x_3 and to x_4.

Euler in old age.

Chapter Three
Euler

§I.

Until the latter part of the seventeenth century, mathematics had sometimes bestowed high reputation upon its adepts but had seldom provided them with the means to social advancement and honorable employment. VIÈTE had made his living as a lawyer, FERMAT as a magistrate; even in Fermat's days, endowed chairs for mathematics were few and far between. In Italy, the University of Bologna ("lo studio di Bologna", as it was commonly called), famous throughout Europe, had indeed counted Scipione del FERRO among its professors in the early sixteenth century; but CARDANO had been active as a physician; BOMBELLI was an engineer, and so was Simon STEVIN in the Netherlands. NAPIER, the inventor of logarithms, was a Scottish laird, living in his castle of Merchiston after coming back from the travels of his early youth. Neighboring disciplines did not fare better. COPERNICUS was an ecclesiastical dignitary. Kepler's teacher MAESTLIN had been a professor in Tübingen, but KEPLER plied his trade as an astrologer and maker of horoscopes. GALILEO's genius, coupled with his domineering personality, earned him, first a professorship in Padova, then an enviable position as a protégé of the Grand-Duke of Tuscany, which saved him from the worst consequences of his disastrous conflict with the Church of Rome; his pupil TORRICELLI succeeded him as "philosopher and mathematician" to the Grand-Duke, while CAVALIERI combined the Bologna chair with the priorate of the Gesuati convent in the same city.

Among Fermat's scientific correspondents, few held professorial rank. ROBERVAL, at the Collège de France (then styled Collège Royal), occupied the chair founded in 1572 from a legacy of the scientist and philosopher Pierre de la Ramée. The Savilian chair at Oxford, created for H. BRIGGS in 1620, was held by WALLIS from 1649 until his death in 1703; but his talented younger friend and collaborator William BROUNCKER, second Viscount, was a nobleman whose career as commissioner of the Navy, and whose amours, are abundantly documented in Pepys's diary. It was only in 1663 that Isaac BARROW became the first Lucasian professor in Cambridge, a position which he relinquished to NEWTON in 1669 to become preacher to Charles II and achieve high reputation as a divine. In the Netherlands, while Descartes's friend and commentator F. SCHOOTEN was a professor in Leiden, René de SLUSE, a mathematician in high esteem among his contemporaries and an attractive personality, was a canon in Liège. DESCARTES, as he tells us, felt himself, by the grace of God ("graces à Dieu": Discours de la Méthode, Desc.VI.9), above the need of gainful employment; so were his friends Constantin HUYGENS and Constantin's son, the great Christian HUYGENS. LEIBNIZ was in the employ of the Hanoverian court; all his life he preserved his love for mathematics, but his friends marveled sometimes that his occupations left him enough leisure to cultivate them.

Whatever their position, the attitude of such men towards mathematics was often what we can describe as a thoroughly professional one. Whether through the printed word or through their correspondence, they took pains to give proper diffusion to their ideas and results and to keep abreast of contemporary progress; for this they relied largely upon a private network of informants. When they traveled, they looked up foreign scientists. At home they were visited by scientifically inclined travelers, busy bees intent on disseminating the pollen picked up here and there. They eagerly sought correspondents with interests similar to their own; letters passed from hand to hand until they had reached whoever might feel concerned. A private library of reasonable size was almost a necessity. Booksellers had standing orders to supply customers with the latest publications within each one's chosen field. This system, or lack of system, worked fairly well; indeed it subsists down to the present day, supplementing more formalized modes of communication, and its value is undiminished. Nevertheless, even the seventeenth century must have found it increasingly inadequate.

By the time of Euler's birth in 1707, a radical change had taken place; its first signs had become apparent even before Fermat's death. The *Journal des Sçavans* was started in January 1665, just in time to carry an obituary on Fermat ("*ce grand homme*", as he is called there). Louis XIV's far-sighted minister Colbert had attracted Huygens to Paris in 1666, and the astronomer CASSINI in 1669, awarding to each a royal pension of the kind hitherto reserved to literati. In 1635 Richelieu had founded the *Académie françoise;* the more practical-minded Colbert, realizing the value of scientific research (pure no less than applied) for the prosperity of the realm, set up the *Académie des sciences* in 1666 around a nucleus consisting largely of Fermat's former correspondents; Fermat's great friend and former colleague Carcavi was entrusted with its administration and became its first secretary. In England, some degree of political stability had been restored in 1660 by the recall of Charles II; in 1662 the group of amateurs ("*virtuosi*") who had for some time held regular meetings in Gresham College received the charter which made of them the *Royal Society,* with Brouncker as its first president; in 1665 they started the publication of the *Philosophical Transactions,* which has been continued down to the present day. In 1698 the French academy followed suit with a series of yearly volumes, variously entitled *Histoire* and *Mémoires de l'Académie des Sciences.* In 1682 Leibniz was instrumental in creating, not yet an academy, but at least a major scientific journal, the *Acta Eruditorum* of Leipzig, to whose early issues he contributed the articles by which he was giving birth to the infinitesimal calculus.

Soon universities and academies were competing for scientific talent and sparing neither effort nor expense in order to attract it. Jacob BERNOULLI had become a professor in his native city of Basel in 1687; as long as he lived, this left little prospect to his younger brother and bitter rival Johann of finding academic employment there; at first he had to teach Leibniz's infinitesimal calculus to a French nobleman, the Marquis de l'HÔPITAL, even agreeing to a remarkable contract whereby the latter acquired an option upon all of BERNOULLI's mathematical discoveries. In 1695, however, Joh. Bernoulli became a professor in Groningen, eventually improving his position there by skilfully playing Utrecht against Groningen; finally he settled down in Basel in 1705 after his brother's death. No wonder, then, that in 1741 we find him congratulating Euler on the financial aspects of his Berlin appointment and suggesting at the same time that he would be willing (for a moderate yearly stipend, "*pro modico subsidio annuo*": *Corr*.II.62) to enrich the memoirs of the Berlin Academy with regular contributions of his own. In short, scientific life, by the turn of the century, had acquired a structure not too different from what we witness to-day.

§II.

Euler's father, Paul EULER, was a parish priest established in Riehen near Basel; he had studied theology at the university of Basel, while at the same time attending the lectures of Jacob Bernoulli; he had planned a similar career for his son, but placed no obstacle in the way of young Leonhard's inclinations when they became manifest. Clearly, by that time, a bright future was in store for any young man with exceptional scientific talent.

In 1707, when EULER was born, Jacob Bernoulli was

dead, and Johann had succeeded him; Johann's two sons Nicolas (born in 1695) and Daniel (born in 1700) were following the family tradition, except that, in contrast to their father and uncle, they loved each other dearly, as they took pains to make known (cf. e.g. *Corr.*II.291). Euler became their close friend and Johann's favorite disciple; in his old age, he liked to recall how he had visited his teacher every Saturday and laid before him the difficulties he had encountered during the week, and how hard he had worked so as not to bother him with unnecessary questions.

Three monarchs came to play a decisive role in Euler's career: Peter the Great, Frederic the Great, and the Great Catherine. Peter, a truly great czar perhaps, died in 1725; but he had had time to found Saint Petersburg, erect some of its most impressive buildings, and, most important of all for our story, to make plans for an Academy of Sciences modelled on what he had seen in the West; those plans were faithfully carried out by his widow. In 1725 the two younger BERNOULLIS, Nicolas and Daniel, were called there. Nicolas died the next year, apparently of appendicitis. About the same time an offer went to Euler to join the Petersburg Academy. He was not quite twenty years old; he had just won a prize for an essay on ship-building, never having seen a sea-going ship in his life. He had no early prospects at home. He accepted with alacrity.

From Basel he sailed down the Rhine to Mainz, then traveled to Lübeck, mostly on foot, visiting Christian Wolff on the way; this was a philosopher and follower of Leibniz, banished from Berlin (as he told Euler) by a king with little understanding for philosophy; his hobby-horse was Leibniz's theory of monads, and Euler was clearly not impressed. From Lübeck a ship took the young mathematician to Petersburg.

In those days academies were well-endowed research institutions, provided with ample funds and good libraries (cf. *Corr.*I.36). Their members enjoyed considerable freedom; their primary duty was to contribute substantially to the academy's publications and keep high its prestige in the international scientific world. At the same time they were the scientific advisers to the monarch and to state authorities, always on hand for such tasks, congenial or not, as the latter might find fit to assign to them; had it not been so, no state would have undergone the high expense of maintaining such institutions, as Euler once acknowledged to Catherine (cf. *Eu.*IV A-1, no. 1887|1766). In 1758, at the height of his fame, Euler (who had acquired in Petersburg a good command of the Russian language) did not find it beneath him, nor inconsistent with his continuing close relations with the Petersburg Academy, to translate for king Frederic some dispatches seized during military operations against the Russian army.

In 1727, however, the political situation had changed by the time Euler reached Petersburg. Under a new czar all academic appointments were in abeyance. On the strength of his prize essay, Euler was commissioned into the Russian navy, but not for long. Soon he was a salaried member of the academy, at first with the junior rank of "adjunct". When his friend Daniel left for Basel in 1733, he was appointed in his place; thus he could afford to marry, naturally into the local Swiss colony, and to buy for himself a comfortable house (cf. his letter of 1734 to his colleague G. F. Müller: *Eu.*IV A-1, no. 1683). His bride was the daughter of the painter GSELL; in due course she was to give birth to thirteen children, out of whom only three sons survived Euler; little is recorded of her otherwise. The eldest son Johann Albert, born in 1734, was to become one of his father's collaborators, and later a leading member of the academy.

Once Euler was thus well established in Petersburg, his productivity exceeded all expectations, in spite of the comparative isolation in which he had been left by Daniel Bernoulli's departure. It was hardly interrupted by a severe illness in 1735 and the subsequent loss of his right eye. He had beyond doubt become the most valuable member of the Academy, and his reputation had been growing by leaps and bounds, when two events in the higher spheres of European politics brought about a major change in his peaceful life. In Petersburg the death of the czarina in 1740, a regency, and the ensuing turmoils, seemed to threaten the very existence of the Academy. Just at this juncture Frederic II succeeded his father (the same king who had so cavalierly thrown Chr. Wolff out of Berlin) on the throne of Prussia; he immediately took steps directed towards the establishment, of an academy under his patronage, for which he sought out the most famous names in European science; naturally Euler was on the list. A munificent offer from Frederic, coupled with a fast deteriorating situation in Petersburg, brought Euler to Berlin in July 1741, after a sea-voyage of three weeks on the Baltic during which he alone among his family (or so he claimed) had been free from sea-sickness (cf. his letter to Goldbach of 1 August 1741: L. Euler and Chr. Goldbach, *Briefwechsel,* edd. A. P. Juškevič und E. Winter, Berlin 1965, p. 83). In the following year, to his great satisfaction, he was able to purchase an excellent house, well situated, and, by special royal order, exempt from requisition (*ibid.*p.130|15 Dec. 1742). There he lived for the next 24 years, with apparently the sole interruption of seasonal visits to the country estate he acquired in 1752 in Charlottenburg (cf. *Eu.*IV A-1, no.2782), and of a family trip to Frankfurt in 1750 to meet his widowed mother who was coming from Basel to live in Berlin with him; his father, who had been disappointed in his hope of getting Euler's visit in Basel (cf. *Corr.*II.57,451) had died in 1745.

With Euler's change of residence one might have expected that the steady flow of his publications would be diverted from Petersburg to Berlin; but far from it! He was not only allowed to keep his membership in the Petersburg Academy, but his pension from Petersburg was continued, and he was intent upon giving his former colleagues value for their money (cf. e.g. *Corr.*I.200|1743). Well might his great-grandson P.-H. Fuss describe Euler's Berlin period as "twenty-five years of prodigious activity". More than 100 memoirs sent to Petersburg, 127 published in Berlin on all possible topics in pure and applied mathematics

LEONARDI EULERI BASILIENSIS

imaginem

aeri incidendam curavit

grata Civitas

MDCCLI

Euler in middle age.

were the products of those years, side by side with major treatises on analysis, but also on artillery, ship-building, lunar theory; not to mention the prize-winning essays sent to the Paris academy (whose prizes brought substantial cash rewards in addition to high reputation; cf. Corr.I.497|1749); to which one has to add the *Letters to a German Princess* (one of the most successful popular books on science ever written: *Eu*.III-11 and III-12 = E 343, 344|1760–1761) and even a defense of christianity (*Rettung der göttlichen Offenbarung* ... , *Eu*.III-12.267–286 = E 92|1747) which did nothing to ingratiate its author with the would-be philosopher-king Frederic. At the same time Euler was conducting an increasingly heavy correspondence, scientific, personal and also official since the administrative burdens of the academy tended to fall more and more upon his shoulders.

As years went by, Euler and Frederic became disenchanted with each other. The king was not unaware of the lustre that Euler was throwing upon his academy, but French literati stood far higher in his favor. He was seeking to attract d'ALEMBERT to Berlin and was expected to put him above Euler as head of the Academy (cf. Corr.I.667,668,670,672|1763). Euler was spared this blow to his self-esteem; d'Alembert, forewarned perhaps by Voltaire's unpleasant experience with Frederic in 1753, enjoyed basking in the king's favor for the time of a short visit but valued his freedom far too highly to alienate it more durably. Nevertheless, as early as 1763, Euler's thoughts started turning again towards Russia.

Fortunately another political upheaval had just taken place there. In 1762 the czar's German wife had seized power as Catherine II after ridding herself and Russia of her husband. One of her first projects was to restore the Petersburg academy to its former glory. This was almost synonymous with bringing Euler back. Negotiations dragged on for three years. Finally, in 1766, the Russian ambassador in Berlin was instructed to request Euler to write his own contract. Frederic, realizing too late the magnitude of this loss, had tried to put obstacles in the way; he soon found that he could not afford to displease the imperial lady. In the same year Euler was back in Petersburg, after a triumphal journey through Poland where Catherine's former lover, king Stanislas, treated him almost like a fellow-sovereign.

By then Euler was losing his eyesight. He had lost the use of his right eye during his first stay in Petersburg. About the time when he left Berlin, or shortly thereafter, a cataract developed in his left eye. In 1770, in answer to a letter from Lagrange on number theory, he described his condition as follows: "*Je me suis fait lire toutes les opérations que vous avez faites sur la formule* $101 = pp - 13qq$ *et je suis entièrement convaincu de leur solidité; mais, étant hors d'état de lire ou d'écrire moi-même, je dois vous avouer que mon imagination n'a pas été capable de saisir le fondement de toutes les déductions que vous avez été obligé de faire et encore moins de fixer dans mon esprit la signification de toutes les lettres que vous y avez introduites. Il est bien vrai que de semblables recherches ont fait autrefois mes délices et m'ont coûté bien du tems; mais à présent je ne saurois plus entreprendre que celles que je suis capable de dévellopper dans ma tête et souvent je suis obligé de recourir à un ami pour exécuter les calculs que mon imagination projette*" ["I have had all your calculations read to me, concerning the equation $101 = p^2 - 13q^2$, and I am fully persuaded of their validity; but, as I am unable to read or write, I must confess that my imagination could not follow the reasons for all the steps you have had to take, nor keep in mind the meaning of all your symbols. It is true that such investigations have formerly been a delight to me and that I have spent much time on them; but now I can only undertake what I can carry out in my head, and often I have to depend upon some friend to do the calculations which I have planned": *Eu*.IV A-5.477|9 March 1770].

An operation was attempted in 1771 and was successful at first, but the eye soon became infected, and total or near-total blindness ensued. Except for this misfortune, and for a fire which destroyed his house in 1771 among many others in Petersburg, he lived on in comfort, greatly honored and respected; neither old age nor blindness could induce him to take a well-deserved rest. He had assistants, one of whom was his own son; others were sent to him from Basel, with the co-operation of his old friend Daniel Bernoulli; to one of them, N. FUSS, who had come to Petersburg in 1773 and later married a granddaughter of Euler's, we owe a vivid description of Euler's method of work in the last decade of his life (*Corr*.I.xliv-xlv). Hundreds of memoirs were written during that period; enough, as Euler had predicted, to fill up the academy publications for many years to come. He died suddenly on 18 September 1783, having preserved excellent general health and his full mental powers until that very day.

§III.

No mathematician ever attained such a position of undisputed leadership in all branches of mathematics, pure and applied, as Euler did for the best part of the eighteenth century. In 1745 his old teacher Johann Bernoulli, not a modest man as a rule, addressed him as "*mathematicorum princeps*" (*Corr*.II.88,92). "*Ce diable d'homme*" is how d'Alembert qualified him in comic disgust, writing to Lagrange in 1769, on finding himself anticipated by Euler in some results which he had felt rather proud of (*Lag*.XIII.136).

Here our concern will be with his work on number theory, which alone would have earned him a distinguished place in the history of mathematics, had he never done anything else; but it occupies only four volumes out of the more than seventy of his complete works. He seems to have worked on it chiefly by fits and starts. He was perhaps ill prepared for it by his apprenticeship under Johann Bernoulli, but took it up early in his career, and it remained a favorite subject with him until the end. This topic, as he well knew, was no more popular with his contemporaries than it had been in the seventeenth century (cf. Chap.II, §XVII; cf. also Daniel Bernoulli's disparaging remarks upon the subject of prime numbers, quoted below in §X). Euler, however, never tired of telling his readers that he felt no need to apologize for the time and efforts thus spent, that truth is one, and that no aspect of it may be neglected without damage to the whole (cf. e.g. *Eu*.I-2.62–64,295,428,461–462,519–522, etc.).

A substantial part of Euler's arithmetical work consisted in no more, and no less, than getting proofs for Fermat's statements as contained in his published works at the time (cf. *infra*, §IV). Our tentative reconstruction of Fermat's original proofs, given above in Chapter II, was largely based on the plausible but unprovable assumption that they could not have differed much from those later obtained by Euler. Thus these proofs need not be described again in detail; it will be enough to indicate the chronology of their discovery by Euler. Apart from the dates of presentation of his memoirs to the Petersburg academy, sometimes differing by several years from their dates of publication, one main source for this is his correspondence, chiefly the one with Goldbach, which was published by Euler's great-grandson P.-H. Fuss in 1843; it makes up the first volume of the *Correspondance Mathématique et Physique* (*Corr*.I); the second volume contains the letters from the Bernoullis to Euler, Goldbach and N. Fuss; Goldbach's conjecture on sums of primes occurs there in Goldbach's letter of 7 June 1742 (*Corr*.I.127). A fuller, abundantly annotated publication of the same correspondence is due to A. P. Juškevič and E. Winter (cf. Bibliography at the end of the present volume). Further correspondence between Euler and the Bernoullis is contained in Euler's *Opera Postuma* of 1862 (vol.I, pp. 519–588) and in G. Eneström's articles in *Bibl.Math.* from 1903 to 1907 (cf. Bibliography). Euler's correspondence with Clairaut, d'Alembert and Lagrange is published in full in *Eu*.IV A-5; further correspondence with various contemporaries (in the original text and in Russian translation) makes up the volume *Pis'ma k Učënym*, Moskva-Leningrad 1963 (*PkU*.); *Eu*.IV A-1 is a complete repertory, with brief summaries, of the whole of Euler's immense correspondence. So far as number theory is concerned, the most valuable information, at least for the years from 1730 to 1756, is to be found in his letters to Goldbach.

GOLDBACH had been born in Königsberg in 1690; a man of versatile tastes and talents, obviously well educated and well-to-do, he traveled extensively in his younger days (cf. *Corr*.II.183–184), seeking out learned men and scientists everywhere. In 1721 he met Nicolas Bernoulli in Venice; soon after that he started a correspondence with him, and then also with his younger brother Daniel. In 1725 he was largely instrumental in bringing them both to Petersburg, where he preceded them by a few months (cf. *Corr*.II.169); he seems to have been regarded by both brothers, and very soon also by Euler, as an influential friend and patron; Euler treated him all his life with a touching mixture of respect, esteem and affection. Perhaps he had started on his trip to Russia out of mere curiosity; but he stayed there until his death in 1764, at times in Petersburg and at times in Moscow. His chief hobbies were languages and mathematics, more particularly number theory, differential calculus and the theory of series; this made him an invaluable correspondent for Euler.

Goldbach left Petersburg for Moscow early in 1728; his correspondence with Euler starts in 1729; at first it is conducted entirely in Latin and continues so, less actively of course, even after Goldbach's return to Petersburg in 1732, at a time when they must have seen each other almost daily. Suddenly, on 21 August 1740, Euler drops into German with an urgent request for help (*"Die Geographie ist mir fatal"*: *Corr*.I.102); having already lost one eye, as he thinks, to cartographical work, he asks to be dispensed from a similar assignment which is not only alien to his regular duties but threatens to be fatal to his eyesight. Goldbach, justly alarmed, answers him on the very same day, also in German. In the following year Euler leaves for Berlin, and the correspondence continues steadily, basically in German, until close to the end of Goldbach's life. Their German, however, is not only sprinkled with French words, but it tends to drop into Latin, especially in the mathematical passages. Here is a specimen, taken from one of Euler's first letters after his arrival in Berlin (*Corr*.I.105–107|9 Sept.1741); after reporting on the gracious reception granted to him by the king and the royal family and on various other matters, Euler goes on to describe his latest ideas about number theory and "eulerian" integrals: *"Von den divisoribus quantitatis aa ± mbb, si a et b sint numeri inter se primi, habe ich auch curieuse proprietates entdeckt, welche etwas in recessu zu haben scheinen . . ."* ["about the divisors of $a^2 \pm mb^2$, if a and b are mutually prime, I have discovered curious properties, where some secrets seem hidden . . ."]. The "curious properties" are then described in Latin (cf. *infra*, §§V*i* and VIII).

§IV.

Number theory makes its appearance in Goldbach's reply to Euler's very first letter. On 13 October 1729, Euler, on Daniel Bernoulli's advice, had sent to Goldbach some early results on what, since Legendre, has been known as the gamma function (*Corr*.I.3–7); obviously he had been reading Wallis's *Arithmetica Infinitorum* (*Wal*.I.355–478). Goldbach's answer, on 1 December, is a cordial one, and competent enough, without bringing anything new; but it carries a fateful postscript: "Is Fermat's observation known to you, that all numbers $2^{2^n} + 1$ are primes? He said he could not prove it; nor has anyone else done so to my knowledge" (*Corr*.I.10).

Actually Goldbach had never read Fermat (as he told Euler a little later: *Corr*.I.26) and was quoting him from hearsay; the same may be said of other references to Fermat in his correspondence (e.g. *Corr*.II.168,238|1725). In those days Fermat's assertions seem to have circulated among "number-lovers" as a kind of folklore; Euler calls them "well-publicized" [*"in vulgus notas"*: *Eu*.I-2.38 in E 98|1738].

At first Euler responded somewhat coolly to Goldbach's question, by a mere expression of skepticism (*Corr*.I.18). But Goldbach insists (*Corr*.I.20), and suddenly, in June 1730, Euler catches fire. Not only is he now taking the matter seriously, but he has taken up Fermat's works and has started reading them (*Corr*.I.24).

One may ask which of Fermat's writings were accessible to him at the time. These included the *Varia Opera* of 1679 and the *Commercium Epistolicum* of 1658 (containing some important letters by Fermat) which Wallis had inserted into his monumental *Algebra* of 1693; Euler quotes both in 1732 (*Eu*.I-2.2 in E 26). Did he also have access to Fermat's observations in Samuel de Fermat's *Diophantus* of 1670 (cf. Chap.II, §I)? He makes no mention of them until 1748, but then quotes them thrice; writing to Goldbach (*Corr*.I.445–446|1748), he mentions for the first time Fermat's famous observation (*Fe*.I.291, Obs.II) about "Fermat's equation" $x^n + y^n = z^n$; then in his next letter (*Corr*.I.455|1748) he copies out in full the observation about sums of four squares, etc., and Fermat's intention to write a whole book on the subject (*Fe*.I.305, Obs.XVIII); in the same year he devotes a paper (*Eu*.I-2.223–240 = E 167|1748) to a diophantine problem stated by Fermat in the *Diophantus* (*Fe*.I.333, Obs.XL). Does it not look as if Euler had just discovered Fermat's *Diophantus* at that time?

At any rate, when Euler took up Fermat's writings in 1730, he found a statement there (apparently the one in Fermat's letter to Digby, no.XLVII of the *Commercium Epistolicum*; *Fe*.II.403) which impressed him even more than the conjecture about $2^{2^n} + 1$, just as it had impressed Huygens when it first appeared (cf. Chap.II, §XVII). As he tells Goldbach, Fermat had said that every number is a sum of four squares ("*non inelegans theorema*", Euler calls it, somewhat inadequately: *Corr*.I.24) and had further statements about sums of triangular and pentagonal numbers, of cubes, etc., "whose proofs would greatly enrich analysis". He had discovered a topic which was to haunt him all his life.

Soon another letter from Goldbach sends him back to Fermat and to Wallis's *Algebra*. Fermat had stated that a triangular number $\frac{1}{2}n(n+1)$ cannot be a fourth power (*Fe*.I.341, II.406; cf. above, Chap.II, §X). Goldbach thought he had proved, in the *Acta Eruditorum* of 1724, that such a number cannot even be a square, and he had communicated his "proof" to Nicolas and to Daniel Bernoulli in 1725 (*Corr*.II.170,237,239) and to Euler in 1730 (*Corr*.I.34). Euler pointed out the error at once; if one puts $x = 2n + 1$, the question amounts to $x^2 - 8y^2 = 1$ and is thus a special case of "Pell's equation". "Such problems", he writes (*Corr*.I.37), "have been agitated between Wallis and Fermat" (a clear reference to the *Commercium*) ". . . and the Englishman Pell devised for them a peculiar method described in Wallis's works". Pell's name occurs frequently in Wallis's *Algebra*, but never in connection with the equation $x^2 - Ny^2 = 1$ to which his name, because of Euler's mistaken attribution, has remained attached; since its traditional designation as "Pell's equation" is unambiguous and convenient, we will go on using it, even though it is historically wrong. Obviously the method referred to by Euler is the one Wallis credits to Brouncker (*Wal*.II.797; cf. *Wal*.II.418–429, and above, Chap.II, §XIII).

Euler must have been a careless reader in those days. Even so, another minor puzzle is hard to explain. Not only does Euler express the view that Fermat had not proved his assertion about sums of four squares ("*neque demonstrationem ejus habuisse videtur*": *Corr*.I.35|1730; cf. *Eu*.I-2.33 in E 54|1736), but at first he seems convinced that Fermat never said that he had proved it ("*neque ipse Fermatius demonstrare se posse affirmat*": *Corr*.I.30|1730). In Fermat's writings as published at the time, there are only two passages containing that statement. One, and presumably the one which Euler had noticed, is in Fermat's letter to Digby, no.XLVII of the *Commercium Epistolicum* (*Fe*.II.403 = no.XLVI in *Wal*.II.857–858) quoted by Euler in 1732 along with Fermat's *Varia Opera* of 1679 as a reference for the conjecture about $2^{2^n} + 1$ (*Eu*.I-2.2 in E 26|1732); the other is in Fermat's *Diophantus* (*Fe*.I.305, Obs.XVIII); on both occasions Fermat declares with great emphasis that he possesses complete proofs. In 1742, writing to CLAIRAUT, Euler not only describes the former passage correctly but expresses the belief that Fermat had indeed a proof, and deplores the loss of his manuscripts: *"Ce seroit un grand avantage . . . si l'on publioit ces démonstrations, peut être que les papiers de ce grand homme se trouvent encore quelque part"* ["It would be an excellent thing if these proofs were published; maybe the manuscripts of this great man can still be found": *Eu.*IV A-5.124]. When did Euler change his mind?

No less surprising is the story of his re-discovery of Fermat's theorem $a^{p-1} \equiv 1 \pmod{p}$, p being a prime and a prime to p. Euler was concerned with it as early as 1731. Later in life, coming back to it in the course of a systematic investigation of the multiplicative group modulo p, he calls it *"theorema eximium, a Fermatio quondam prolatum"* ["an outstanding theorem once brought forward by Fermat": *Eu.*I-2.510 in E 262|1755]. In his early days, however, we see him taking up the problem of the supposedly prime numbers $2^{2^n} + 1$ precisely as Fermat had done almost a century before, obviously in complete ignorance of what was readily available to him on pages 163, 177 of Fermat's *Varia Opera* (*Fe.*II.209,198). He begins by observing that $2^{p-1} - 1$ is a multiple of p when p is a prime, calling this *"theorema non inelegans"* on sending it to Goldbach in November 1731 (*Corr.*I.60); then, also experimentally (*"tentando"*: *Eu.*I-2.1 in E 26|1732), he finds that $a^{p-1} - b^{p-1}$ is a multiple of p when a and b are prime to p, discovering at the same time the counterexample $2^{32} + 1$ to Fermat's conjecture (cf. Chap.II, §IV). By June 1735, Euler has become vaguely aware at least of Fermat's letter of 1640 to Mersenne (*Fe.*II.195–199) when he writes to Ehler in Dantzig: *"Theorema . . . non est novum, sed ni fallor a Fermatio assertum, at sine demonstratione, ex sola inductione"* ["The theorem ($2^{p-1} \equiv 1$ mod p) is not new; if I am not mistaken it was stated by Fermat, but without proof, as being obtained merely by induction": *PkU.*295]; this, in the face of Fermat's reiterated statement that he had found a proof (with some difficulty, *"non sans peine"*: *Fe.*II.198). Writing to Ehler, Euler gives a somewhat clumsy proof based on the application of the binomial formula to $(1 + 1)^{p-1}$ and a suitable re-arrangement of the terms; he reproduces this in 1736 but adds to it the additive proof for Fermat's theorem based on the binomial formula for $(a+1)^p$ (*Eu*.I-2.33–37 = E 54|1736; cf. Chap.II, §IV). It is not until 1752, however, and perhaps not even then, that he appears clearly aware of Fermat's priority in this matter (*Eu*.II-5.141 in E 182). Obviously Euler, in his younger days, had been far too preoccupied with his own thoughts to pay close attention to the legacy of his predecessors.

§V.

Sometimes the text of a play is preceded by a short sketch of the characters "in the order of their appearance". Similarly, before embarking upon a detailed description of Euler's arithmetical work, and at the cost of some repetition, we shall list here its main themes in chronological perspective, in the order in which they appear in his writings and his correspondence. Some of these might have been classified by him as analysis rather than number theory; nevertheless he was well aware of the connections between those fields. As he once observed while proving an arithmetical theorem by the method of formal power-series: "from this, one may see how closely and wonderfully infinitesimal analysis is related not only to ordinary [i.e. algebraic] analysis but even to the theory of numbers, however repugnant the latter may seem to that higher kind of calculus" ["*ex hoc casu intelligere licet, quam arcto et mirifico nexu Analysis infinitorum non solum cum Analysi vulgari, sed etiam cum doctrina numerorum, quae ab hoc sublimi calculi genere abhorrere videtur, sit coniuncta*": *Eu*.I-2.376 in E 243|1751; cf. *infra*, §XXI].

(a) *Fermat's theorem, the multiplicative group of integers modulo N, and the beginning of group theory* (cf. §§VI,VII).

This begins in 1730 with the investigation of Fermat's conjecture about $2^{2^n} + 1$ (*Corr*.I.18|8 Jan.1730; cf. above, §IV) and the experimental discovery of Fermat's theorem in 1731 or 1732; the additive proof for it is obtained in 1735 (*PkU*.295). At about the same time Euler was re-discovering Bachet's solution of equations $ax - by = m$, or equivalently of congruences $ax \equiv m$ (mod b), or of double congruences $z \equiv p$ (mod a), $z \equiv q$(mod b) (Eu.I-2.18–32 = E 36|1735), of course by the euclidean algorithm, in obvious ignorance of the fact that this had been known to Bachet, Fermat and Wallis, not to mention Āryabhaṭa (cf. above, Chap.II, §XIII, and Chap.I, §IV). In our eyes its main feature is that it establishes the group property for congruence classes prime to an integer N modulo N by showing that, if a and b are prime to N, there is an x prime to N such that $ax \equiv b$ (mod N); indeed it is precisely thus that Euler came to use it in later years (cf. e.g. Eu.I-2.350 in E 242|1751, Eu.I-3.8 in E 283|1760, Eu.I-3.521 in E 554|1772, etc.)

For a more systematic investigation of the multiplicative group of integers modulo a prime p (the group mostly written nowadays as $\mathbf{F}_p^{\times}$) or more generally modulo an arbitrary integer N, one must then wait until Euler's Berlin period, after he had found an incentive for this in his experimental results on quadratic forms $mX^2 + nY^2$. From then on, Euler's progress can be traced through a steady stream of publications as well as in a posthumously published manuscript, the *Tractatus*, apparently composed around 1750 and then abandoned (Eu.I-5.182–283; cf. *infra*, §VI). Here we mark merely some of the main stages of that progress: from 1747 on, quadratic residues and m-th power residues; then the number of solutions of a congruence $x^m \equiv 1$ modulo a prime, at first (in 1749) by a difference argument; in the *Tractatus*, the multiplicative (i.e. group-theoretical) proof for Fermat's theorem and its extension to "Euler's theorem" $a^{\varphi(N)} \equiv 1$ (mod N) for any a prime to N; finally, in 1772, the existence of $\varphi(p-1)$ "primitive roots" modulo any prime p, with some of its main consequences.

(b) *Sums of squares and "elementary" quadratic forms* (cf. §§IX, XI).

As we have seen (cf. §IV), Euler was early fascinated by Fermat's assertions concerning sums of four squares, of three triangular numbers, etc. (*Corr*.I.24|25 June 1730). At first, just like Fermat (cf. Chap.II, §V), he merely made the rather trivial remark that no integers of the form $4m + 3$, or more generally $n^2(4m + 3)$, can be a sum of two squares, and no integer $n^2(8m + 7)$ can be a sum of three squares (*Corr*.I.44|17 Oct. 1730); this is of course easily proved by writing congruences modulo 4 or 8. In 1741 he mentions having known "for a long time" that $4mn - m - 1$ cannot be a square if m, n are positive integers (*Corr*.I.107), but it soon turns out (*Corr*.I.114) that in this he had relied upon Fermat's statement that $a^2 + b^2$, for a prime to b, has no prime divisor of the form $4n + 1$, from which indeed it follows easily. Only in 1742 is he able to send to Goldbach a proof for the latter fact (*Corr*.I.115), based of course on Fermat's theorem $a^{p-1} \equiv 1 \pmod{p}$.

This was the first stage in a seven years' campaign to prove all of Fermat's statements about sums of two squares (cf. *Corr*.I.134|1742, 313|1745, 415–419|1747; *Eu*.I-2.295–327|20 March 1749), at the end of which Euler could triumphantly report to Goldbach his final victory: "Now at last I have found a conclusive proof . . ." ("*Nunmehr habe ich endlich einen bündigen Beweis gefunden . . .*": *Corr*.I.493–497|12 April 1749; cf. *Eu*.I-2.328–337|1750). In the meanwhile, however, and from 1747 onwards, we find in Euler's letters and in Goldbach's replies more and more frequent references to sums of three and four squares, and, in connection with these, to orthogonal transformations in 3 and in 4 variables (cf. e.g. *Corr*.I.440|1747, 515–521|1750 and *Eu*.IV A-5.468, 478–480|1770). The decisive discovery of the identity for sums of four squares is communicated to Goldbach in 1748 (*Corr*.I.452; cf. *Eu*.I-2.368–369 in E 242|1751 and *Eu*.I-6.312 in E 407|1770). In the same letter of 1749 where Euler was sending to Goldbach the definitive proof for sums of two squares, he is also able to give a similar proof, based on his new identity, for the fact that every integer is at any rate a sum of four rational squares (*Corr*.I.495–497; cf. *Eu*.I-2.338–372 = E 242|1751). As to Fermat's original assertion about sums of four squares, it was left to Lagrange to take the final step in 1770 (*Lag*.III.189–201); but soon Euler followed this up with a beautiful memoir (Eu.I-3.218–239 = E 445|1772) giving a new variant of his own proof for two squares and then showing how it applied not only to the forms $X^2 + 2Y^2, X^2 + 3Y^2$ but also to sums of four squares. As to the "elementary" quadratic forms $X^2 + AY^2$ with $A = \pm 2, 3$, whose theory, at least in its main outlines, had already been known to Fermat (cf. Chap.II, §XII), Euler had been studying them since 1752 ($Corr$.I.597|1752; cf. Eu.I-2.467–485 in E 256|1753 and $Corr$.I.622|1755), finally achieving complete results for A = 3 in 1759 (Eu.I-2.556–575 = E 272|1759) and for A = ± 2 only in 1772 (Eu.I-3.274–275 in E 449|1772).

(c) *Diophantine equations of degree* 2 (cf. §XIII).

This topic, too, appears quite early in Euler's correspondence with Goldbach ($Corr$.I. 30–31,36–37|1730; cf. above, §IV), where in particular it is pointed out that infinitely many solutions of an equation $y^2 = ax^2 + bx + c$ in integers can be derived from the knowledge (however obtained) of one solution, combined with the solution of "Pell's equation" $X^2 - aY^2 = 1$. Those results are described in E 29|1733 (= Eu.I-2.6–17) and taken up again in fuller detail in E 279|1758 (= Eu.I-2.576–611; cf. $Corr$.I.629|1755) and in E 323|1759 (= Eu.I-3.73–111), where the connection between Pell's equation $X^2 - aY^2 = 1$ and the continued fraction for $\sqrt{a}$ turns up for the first time (cf. (f) *infra*).

Equations $y^2 = ax^2 + bx + c$, to be solved either in fractions or in integers, re-appear in Euler's *Algebra*, composed not later than 1767 (*Vollständige Anleitung zur Algebra*, 2 vol., St. Petersburg 1770: Eu.I-1.1–498 = E 387, 388) whose final section (*Von der unbestimmten Analytic: Eu*.I-1.326–498) was described by Lagrange, in a letter to d'Alembert, as "a treatise on diophantine equations, and in truth an excellent one" ["*un traité sur les questions de Diophante, qui est à la vérité excellent*": *Lag*.XIII.181|1770). Finally, in 1772, Euler extended his results to the most general equation of degree 2 in two unknowns, to be solved in integers (Eu.I-3.297–309 = E 452); curiously enough, in the introductory section of that paper which deals with rational solutions, there is no reference to Lagrange's important memoirs on that subject (*Lag.*II.377–535|1768, 655–726|1769; cf. above, Chap.II, §XIV, and Chap.IV, §III) which had been the subject of an exchange of letters between him and Euler in 1770 (*Eu.*IV A-5. 466–467,471–476,477; cf. above, §II).

(*d*) *Diophantine equations of genus* 1, *and others* (cf. §§XIV, XVI).

Such questions, including the traditional methods of solution which go back to Diophantus (cf. Chap.I, §X, and Chap.II, §XV) may well have been familiar to Euler from his student days. In 1724 we find his friend Daniel Bernoulli instructing Goldbach in the use of those methods and mentioning his own early interest in such matters[1] (*Corr.*II.202–203|1724; cf. *ibid.*190|1723). At that time Daniel did not know yet that the iteration of the traditional method yields infinitely   many   solutions   ("*quod   impossibile   censeo*": *Corr.*II.203); in 1730 he had found this out (*Corr.*II.356). By that time one may imagine him and Euler discussing such problems together; in 1730 Euler mentions a fruitless attempt to extend to equations $y^2 = P(x)$, with a polynomial P of degree 3, the idea which he had just applied successfully to the case when P is of degree 2 (*Corr.*I.37).

Some years later Euler devotes a whole memoir (*Eu.*I-2.38–58 = E 98|1738) to a detailed treatment of the equations $x^4 \pm y^4 = z^2, x^4 \pm y^4 = 2z^2$ (cf. Daniel Bernoulli's letter to him, *Corr.*II.451|1738); he refers to Frenicle's *Traité des Triangles* (cf. Chap.II, §X), reprinted in the *Mémoires* of the Paris Academy, vol. V (1729) and perhaps freshly arrived in Petersburg, but is clearly unacquainted with Fermat's

---

[1] In *Corr.*II.202, "*sextum vel septimum*" is surely a mistake, or misprint, for "*decimum sextum vel septimum*"; cf. *Corr.*II.190. Daniel could hardly have been solving difficult diophantine equations when he was "six or seven years old".

own proof (*Fe*.I.340–341, Obs.XLV; cf. Chap.II, §X). He
also adds the proof that no triangular number can be a
fourth power (cf. above, §IV, and Chap.II, §X) and a treatment of the equations $x^3 \pm 1 = y^2$.

Eventually he became aware of Fermat's assertion about
the equation $x^n + y^n = z^n$ (*Corr*.I.445–446|1748); "a very
beautiful theorem", he calls it in 1753 (*Corr*.I.618), adding
that he can prove it for $n = 3$ and $n = 4$, but not otherwise
(cf. *Corr*.I.623|1755). For $n = 4$ this was in fact included in
his results of 1738. For $n = 3$ his proof must have been the
one combining Fermat's descent with the theory of the
quadratic form $X^2 + 3Y^2$ (cf. *infra*, §XIV); he had already
made essential use of the "elementary" quadratic form
$X^2 - 2Y^2$ in dealing with a diophantine problem originating
from Fermat (*Eu*.I-2.223–240 = E 167|1748).

Also during the fifties, Euler studied Diophantus carefully,
hoping to bring some order into what had all the appearances
of a haphazard collection (*Eu*.I-2.399–458 = E 253,
255|1753–54). He did succeed in isolating a few of the Greek
geometer's favorite tricks, but otherwise the result of his
efforts was disappointing; only the later developments of
algebraic geometry were to throw light on a subject which
was not yet ready for even partial clarification. Perhaps this
is why Euler's interest in it seems to have flagged during
the latter part of his stay in Berlin. He came back to it only
by composing the last section of his *Algebra*, an excellent
treatise indeed, as Lagrange had called it, but containing
little new material if any.

From that time on, diophantine problems seem to have
become one of Euler's chief amusements. Noteworthy is a
series of papers, all written in 1780, and all published only
in 1830 (*Eu*.I-5.82–115,146–181 = E 772,773,777,778), in
which diophantine equations $\Phi(x, y) = 0$, where Φ is a polynomial of degree 2 in x and of degree 2 in y, are considered
for the first time; it is at this point that the relationship, in
Euler's work, between that subject and elliptic integrals appears most clearly (cf. §§XV,XVI).

(e) *Elliptic integrals* (cf. §XV).

As Euler knew, Leibniz and Joh. Bernoulli had already asked whether the differential

$$\omega = \frac{dx}{\sqrt{1 - x^4}}$$

can be integrated by means of logarithms or inverse trig-onometric functions and had guessed that it could not. When Euler asked the same question in 1730 (*Corr.*I.47,51) and then, in 1738, gave a proof for Fermat's theorem about the diophantine equation $X^4 - Y^4 = Z^2$, it must surely have occurred to him that any substitution transforming ω into a rational differential might well supply rational solutions for $z^2 = 1 - x^4$, and hence integral solutions for Fermat's equation. To his mind this gave perhaps some added flavor to integrals such as $\int \omega$ and the more general ones arising from the rectification of the ellipse, hyperbola and lemniscate (cf. e.g., *Eu.*I-20.8–20 = E 52|1735 and *Eu.*I-20.21–55 = E 154|1749); he was also meeting such integrals, in connection with problems in applied mathematics, in his investigations on the so-called "elastic curves" (cf. his letter to Joh. Bernoulli of 20 Dec.1738, *Bibl.Math.*(III) 5 (1904), p. 291, and the appendix *De curvis elasticis* to his treatise of 1744 on the calculus of variations, *Eu.*I-24.231–297 in E 65).

It is only in 1751, however, that FAGNANO's *Produzioni Matematiche*, reaching him in Berlin, opened his eyes to the fruitful field of investigation that lay unexplored there. Already a month after receiving those volumes, he was reading to the Berlin Academy the first memoir (*Eu.*I-20.80–107 = E 252|1752) of a long series (from *Eu.*I-20.58–79 = E 251 to *Eu.*I-21.207–226 = E 676|1777), at first only on the integrals considered by Fagnano, but leading up to the proof of the addition and multiplication theorems for integrals of the form

$$\int \frac{F(x)dx}{\sqrt{P(x)}}$$

where P is a polynomial of degree 4 and F an arbitrary polynomial or even a rational function. At the same time he was obtaining the general integral of the differential equation

$$\frac{dx}{\sqrt{P(x)}} = \pm \frac{dy}{\sqrt{P(y)}}$$

in the "canonical form" $\Phi(x, y) = 0$, with Φ of degree 2 in x and of degree 2 in y. As noted above in (d), this was later transported to the theory of diophantine equations.

(f) *Continued fractions, Pell's equation and recurrent sequences* (cf. §XII).

In Euler's letters to Goldbach, continued fractions make their first appearance in connection with Riccati's differential equation (*Corr.*I.58–59|1731, 63|1732), but soon also in an arithmetical context (*Corr.*I.68|1732); already at that time Euler was noticing the relation between periodic continued fractions, "recurrent" sequences (i.e. sequences (p_n) such that $p_n = ap_{n-1} + bp_{n-2}$ for all $n \geqslant 2$; cf. *Corr.*I.30|1730) and quadratic irrationals (cf. *Corr.*I.36|1730). Of course his use of the euclidean algorithm for solving equations $ax - by = \pm 1$ (*Eu.*I-2.20–21 in E 36|1735; cf. (a) above) also amounts to constructing the continued fraction for a/b (cf. *Corr.*I.243|1743, 299–301|1744), as may well have been obvious to him from the beginning and as he was to spell out in 1737 (*Eu.*I-14.194–196 in E 71) and again in Chapter XVIII (*De fractionibus continuis, Eu.*I-8.362–390) of his great *Introductio in Analysin Infinitorum* of 1748.

For many years he pursued this theme in its various aspects (cf. *Eu.*I-14.187–215 = E 71|1737, *Eu.*I-14.291–349 = E 123|1739, and *Eu.*I-15.31–49 = E 281|1757, supplementing the above-mentioned chapter XVIII of the *Introductio*); but the role played by the continued fraction algorithm in the solution of Pell's equation was only clarified by him in a memoir of 1759, first published in Petersburg in 1767 (*Eu.*I-3.73–111 = E 323) which may well have given Lagrange the final clue for his definitive treatment of that problem (cf. Chap.IV, §II).

(g) *Summation of $\zeta(2\nu)$ and related series* (cf. §§XVII,XVIII, XIX,XX).

One of Euler's most sensational early discoveries, perhaps the one which established his growing reputation most firmly, was his summation of the series $\Sigma_1^\infty n^{-2}$ and more generally of $\Sigma_1^\infty n^{-2\nu}$, i.e. in modern notation $\zeta(2\nu)$, for all positive even integers 2ν (*Eu*.I-14.73–86 = E 41|1735). This was a famous problem, first formulated by P. MENGOLI in 1650 (cf. G. Eneström, *Bibl.Math.* (III) 12 (1912), p. 138); it had resisted the efforts of all earlier analysts, including Leibniz and the Bernoullis (cf. e.g. Leibniz's letter to Joh. Bernoulli, *Math.Schr.*I-3.454|1697, and Joh. Bernoulli's letter to Euler, *Corr.*II.15|1737). Characteristically, before solving it, Euler had engaged in extensive numerical calculations in order to get good approximate values for these sums; it is largely with this intention, it seems, that he had developed the method traditionally known as "the Euler-MacLaurin summation formula", and in so doing had re-discovered the "Bernoulli numbers", whose true importance for number theory was not to emerge until the next century (*Eu*.I-14.42–72 = E 25|1732, *Eu*.I-14.108–123 = E 47|1735; cf. Part V, Chap.II, of the *Institutiones Calculi Differentialis* of 1755, *Eu*.I-10.309–336).

What Euler found in 1735 is that $\zeta(2) = \pi^2/6$, and more generally, for $\nu \geqslant 1$, $\zeta(2\nu) = r_\nu \pi^{2\nu}$, where the r_ν are rational numbers which eventually turned out to be closely related to the Bernoulli numbers. At first Euler obtained the value of $\zeta(2)$, and at least the next few values of $\zeta(2\nu)$, by a somewhat reckless application of Newton's algebraic results, on the sums of powers of the roots for an equation of finite degree, to transcendental equations of the type $1 - \sin(x/a) = 0$. With this procedure he was treading on thin ice, and of course he knew it.

Publication in Petersburg was slow; but soon Euler had sent the news to friends and colleagues all over Europe, and it became the topic for lively discussions between him and some of his correspondents. In the meanwhile he was sparing no efforts to consolidate his methods. Less than ten years after his first discovery, he was able to include in his *Introductio in Analysin Infinitorum* (published in Lausanne in 1748, but ready in manuscript in 1744; cf. *Corr*.I.292|1744) a full account of the matter, entirely satisfactory by his standards, and even, in substance, by our more demanding ones. This was based on a detailed treatment of trigonometric functions and of their expansions into infinite series and infinite products (*Eu*.I-8.133–212 = Chap.VIII–XI of the *Introductio*), to which he added a whole chapter on the "eulerian" products for $\zeta(s)$ and various L-series (*ibid*.284–312 = Chap.XV). As to his efforts to gain information about the numbers $\zeta(n)$ for odd $n > 1$, it is no wonder that they remained unsuccessful, since hardly any progress has been made with this problem down to the present day. Nevertheless they were not unfruitful; it was in fact while searching for such information that Euler hit upon the functional equation for $\zeta(s)$ and for the companion series

$$L(s) = \sum_{n=0}^{\infty} \frac{(-1)^n}{(2n + 1)^s}$$

(*Eu*.I-15.70–90 = E 352|1749). These remained as mere curiosities, and virtually unknown, until they received new life at the hands of Riemann in 1859.

(h) *"Partitio numerorum" and formal power-series* (cf. §XXI).

In 1740 a Berlin mathematician of French origin, Ph. NAUDÉ, wrote to Euler, asking (among other questions) in how many ways a given integer m can be written as a sum of μ distinct integers. Euler answered almost immediately (*PkU*.193) and followed this up within a few months with a memoir presented to the Petersburg Academy only shortly before taking his leave (*Eu*.I-2.163–193 = E 158|6 April 1741). He was to come back to this topic on several occasions, treating it in some detail in a chapter of his *Introductio* of 1748 (Chap.XVI = *Eu*.I-8.313–338) and for the last time in 1768 (*Eu*.I-3.131–147 = E 394).

He had seen at once that the key to Naudé's question lay in the consideration of suitable formal power-series such as $\Pi_{i=1}^{\infty}(1 + x^i z)$ and $\Pi_{i=1}^{\infty}(1 - x^i z)^{-1}$. In dealing with them he soon came across the remarkable infinite product

$$\prod_{i=1}^{\infty} (1 - x^i) = 1 - x - x^2 + x^5 + x^7 - \cdots + x^{51} + \text{etc.}$$

(*Eu.*I-2.191 in E 158|1741), which, expanded into a power-series as shown by the above formula, could not fail to strike him; not only are all the non-zero coefficients equal to ± 1, but the exponents are easily recognizable as "pentagonal numbers" $\frac{1}{2}n(3n + 1)$. It took Euler almost ten years before he was able to send a proof for this to Goldbach (*Corr.*I.522–524|1750).

As Jacobi found out, a full explanation of such matters has to be sought in the theory of theta-functions and of modular forms. This lay far beyond Euler's ken. Nevertheless one cannot but admire his power of divination when he tells Goldbach that the most natural method for proving Fermat's theorem on sums of four squares would be by considering the fourth power of the series $\Sigma_{n=1}^{\infty} x^{n^2}$ (not quite the theta-series $\Sigma_{n=-\infty}^{+\infty} x^{n^2}$, but close to it). This, too, had to wait for Jacobi.

(*i*) *Prime divisors of quadratic forms* (cf. §§VIII,IX).

The same letter to Goldbach, quoted in §III, and written by Euler almost immediately upon his arrival in Berlin (*Corr.*I.105–107|9 Sept. 1741), contains some momentous mathematical observations: firstly, as mentioned above in (*b*), the result about $4mn - m - 1$, i.e. in substance about the divisors of the quadratic form $X^2 + Y^2$; then a number of statements, obtained "by induction" (i.e. experimentally) on the prime divisors of the forms $X^2 - 2Y^2$, $X^2 - 3Y^2$, $X^2 - 5Y^2$; and finally his first results about the integrals

$$\int_0^1 \frac{dx}{(1 - x^n)^{p/q}}$$

(later known as "eulerian integrals of the first kind").

As he wrote to Clairaut a few months later, in response to a friendly word of enquiry about his occupations in Berlin, he was "enjoying perfect rest" there ["*je jouis . . . d'un parfait repos*": *Eu.*IV A-5.110|Jan.–Feb.1742] while waiting for the king to pay some attention to his academy; obviously he turned this period of leisure into one of intense creativity, particularly in the field of number theory. Within a year he had extended his observations to a large assortment of quadratic forms $X^2 + AY^2$, A being a positive or negative integer (*Corr.*I.146–150|28 Aug.1742, 161–162|27 Oct. 1742; cf. *ibid.*260|15 Oct.1743). Already in August 1742 he was quite close to the law of quadratic reciprocity and emphasizing the value of his discoveries: "I feel sure", he wrote, "that I am far from having exhausted this topic, but that innumerably many splendid properties of numbers remain still to be discovered" ["*Ich glaube aber fest, dass ich diese Materie bei weitem noch nicht erschöpfet habe, sondern, dass sich darin noch unzählig viele herrliche proprietates numerorum entdecken lassen . . .*": *Corr.*I.150). He rightly attached such importance to his conjectures that he never tired of repeating them to the end of his days, in more or less precise form, while making little progress towards proving them (cf. however *Eu.*I-3.280–281 in E 449|1772, and *Eu.*I-4.163–196 = E 598|1775, the latter memoir being based on Lagrange's *Recherches d'Arithmétique* of 1775). A clear formulation of the law of quadratic reciprocity appears in the conclusion (*Eu.*I-3.512 in E 552|1772) of a paper written in 1772 and published in Euler's *Opuscula Analytica* in 1783, the year of his death.

(*j*) *Large primes* (cf. §X).

Already in 1730, when taking up the Fermat numbers $2^{2^n} + 1$, Euler had met with the problem of deciding whether a given large number is a prime. As Fermat had known (cf. Chap.II, §IX), a non-square integer of the form $4n + 1$ is a prime if and only if it has one and only one representation as a sum of two squares and for this representation the two squares are mutually prime. The first mention of Euler's rediscovery of this criterion occurs in 1742 (*Corr.*I.134–135). In 1749, just before completing the proofs about sums of two squares (cf. (*b*) above), Euler applied this criterion to various primes and non-primes (*Eu.*I-2.321–327 in E 228). The same method is described more systematically in E 283|1760 (= *Eu.*I-3.1–45) and in E 369|1765 (= *Eu.*I-3,112–130); in the former paper Euler examines more specifically the primes of the form $a^2 + 1$ and also gives his method (based on the euclidean algorithm; cf. (*a*) and (*b*) above) for solving the congruence $x^2 \equiv -1$ modulo a given prime $p = 4n + 1$.

Finally, in his old age, Euler generalized Fermat's criterion to a large number of quadratic forms $mX^2 + nY^2$, communicating this new method to his friend and former Berlin colleague, the Swiss mathematician Béguelin (*Eu.*I-3.418–428 = E 498 and 708a|1778; cf. *Eu.*I-4.269–289,303–328,352–398 = E 708,715,718,719,725|1778). This was based on the concept of "*numerus idoneus*", i.e. "suitable number". According to the original, somewhat inexactly formulated definition, a number N is called "suitable" if (and only if) every odd integer which can be written in one and only one way as $a^2 + Nb^2$, with a prime to b, is prime; it is asserted that all the forms $mX^2 + nY^2$ with $mn = N$ have then the same property. Eventually this is modified by specifying that certain trivial cases have to be excluded. Euler also found at least a heuristic method for deciding whether an integer N is "suitable" in his sense (i.e., suitable for testing the primality of large numbers by a criterion generalizing Fermat's); he listed 65 such integers, from 1 to 1848, and used them to discover some large primes (larger then 10^7) which had not been known before.

In his *Disquisitiones* (Art.303), Gauss clarified the matter somewhat by indicating that Euler's *numeri idonei* are precisely those determinants of binary quadratic forms for which there is only one class in each genus. Euler had already encountered questions pertinent to the theory of genera when he had sought to investigate which integers can be written as $mx^2 + ny^2$ with rational x, y, when m and

n are given integers (Eu.I-4.1–24 = E 556|1772; cf. $Corr$.I.605|1753).

§VI.

We begin with Fermat's theorem (cf. above, §IV, and Chap.II, §IV). Fermat had discovered it while trying to factor such numbers as $2^n \pm 1$, $3^n \pm 1$, etc. As we have seen, it can be formulated and proved either additively, the proof being then based on the binomial theorem, or multiplicatively. Fermat's final formulation for it (Fe.II.209) had been of the latter type, and it may be surmised that he had proved it accordingly, even though he had himself discovered enough about binomial coefficients (cf. Chap.II, §II) to be capable of the other proof.

Euler's interest in this matter arose from Goldbach's insistence upon Fermat's conjecture about the numbers $2^{2^n} + 1$ (cf. above, §IV). Treading unwittingly in Fermat's footsteps, he began by discovering the theorem experimentally, at first in the case of $2^{p-1} - 1$, then in general (cf. §IV). In contrast with Fermat, however, he had absorbed the binomial theorem almost with his mother's milk; it is not surprising, therefore, that the first proof he obtained was the additive one (Eu.I-2.33–37 = E 54 | 1736; cf. PkU.295|1735).

Once in Berlin, when he took up in right earnest the investigation of sums of two squares (cf. *infra*, §IX), Fermat's theorem became one main ingredient in the proof of his first results ($Corr$.I.115–116|1742; cf. Chap.II, §VII, and *infra*, §IX); but this may not yet have provided an incentive for him to seek the proof of Fermat's theorem in the multiplicative properties of integers modulo a prime. One is tempted to look for such an incentive in the "*herrliche proprietates*" discovered empirically by him in that same year 1742, properties which he described to Goldbach with even more than his wonted enthusiasm ($Corr$.I.146–150|28 Aug.1742); there, having investigated quadratic forms

$X^2 + NY^2$ and examined, for various values of N, the dis-
tribution modulo $4N$ of the prime divisors of that form (i.e.,
in modern parlance, the rational primes which split in the
field $\mathbf{Q}(\sqrt{-N})$, this is what he found (*Corr.*I.148): "If a
prime of the form $4Nn + s$ is a divisor of $X^2 + NY^2$, then
so is every prime of the form $4Nn + s^k$; if two primes
$4Nn + s$, $4Nn + t$ are such, then so is every prime of the
form $4Nn + s^k t^i$ ". In fact (cf. *infra*, §VIII) he was discovering,
not only what is in substance the law of quadratic reciprocity,
but at the same time the concept of a subgroup of the multiplicative group modulo $4N$, and more precisely of a
subgroup of index 2 of that group.

Whether or not these observations gave Euler the primary
motive for exploring group-theoretical concepts further,
he obviously gave much thought to them, from 1745 to the
end of his life. Concurrently he became aware of Fermat's
authorship of his theorem (cf. above, §IV); perhaps he was
also impressed by Fermat's distinctly multiplicative formulation. However that may be, it is fascinating to watch
his progress towards a more and more conceptual approach
to number theory (belying the legend that he was an em-
piricist not much given to abstract thinking), a progress
evidenced by a vocabulary which became more and more
"modern" as years went by.

Already in 1747, while still sticking to the additive proof
for Fermat's theorem (*Eu.*I-2.65–67 in E 134), he finds it
worthwhile to point out that, in matters pertaining to divisibility by p (where p need not be a prime), integers a, b,
etc., may be replaced by $a \pm \alpha p$, $b \pm \beta p$, etc., i.e. by any
integers leaving the same residues in their division by p
(*ibid.*, p.77, Scholion); he discusses "residues arising from
the division of squares" (or of cubes, etc.) and begins to
consider the properties of such residues which arise from
Fermat's theorem. In particular he finds (*ibid.*, p.81, th.13)
that "if $a - f^n$ is divisible by the prime $p = mn + 1$, then
so is $a^m - 1$" (an easy consequence of Fermat's theorem),
and at once proposes the question of proving the converse,
which he has found invariably true (*ibid.*, p.82, Scholion).
Soon after that, in his first major paper on sums of two

squares (*Eu.*I-2.295–327 = E 228|1749), we find a more careful discussion of quadratic residues, still under the somewhat cumbersome designation as "the residues arising from the division of squares by the prime p" (*ibid.*, pp.312–313, Scholion); it is noted there that the number of such residues is exactly $\frac{1}{2}(p-1)$, and it is conjectured that, for $p = 4n + 1$, they are precisely the solutions of $x^{2n} \equiv 1 \pmod{p}$, this being a special case of the conjecture described above (with m, n replaced by $2n, 2$); it is the decisive case for the theory of sums of two squares.

Still in 1749, further progress is made (*Corr.*I.493–495; cf. *Eu.*I-2.328–337 = E 241|1750) by applying the difference operator to the congruence $x^{2n} \equiv 1 \pmod{p}$ in order to show that it cannot be satisfied for all x prime to the prime $p = 4n + 1$. As generalized by Euler a few years later (*Eu.*I-2.516 in E 262|1755), the argument is as follows. Call D the difference operator

$$(Df)(x) = f(x+1) - f(x).$$

By induction on m, one finds, for any $m \geqslant 1$:

$$(D^m f)(x) = f(x+m) - \binom{m}{1} f(x+m-1)$$
$$+ \binom{m}{2} f(x+m-2) - \cdots \pm f(x).$$

Therefore, if a prime p divides $f(x)$ for $x = 1, 2, ..., m + 1$, it must divide $(D^m f)(1)$. Also by induction on m, one finds that, if $f(x) = x^\mu$, $D^m f$ is $m!$ for $\mu = m$ and 0 for $\mu < m$; this gives $D^m f = m!$ for every monic polynomial

$$f(x) = x^m + a_1 x^{m-1} + \cdots + a_m.$$

Consequently, if $m < p - 1$, the congruence $f(x) \equiv 0 \pmod{p}$ cannot be satisfied for $x = 1, 2, ..., m+1$, nor still less for all x prime to p.

As will be seen presently, this ingenious argument was to be superseded later by a more powerful and more structural one. When it first appears in the above form, it is used for settling the question left open in 1747, by showing that, if $p = mn + 1$ is a prime and $a^m - 1$ is divisible by p, then a is an n-th power residue modulo p (*loc.cit.*pp.515–516, th.19). In fact, assume that $a^m \equiv 1 \pmod{p}$ and write the identity:

$$x^{mn} - a^m = (x^n - a)(x^{(m-1)n} + ax^{(m-2)n} + \cdots + a^{m-1}).$$

Fermat's theorem shows that the left-hand side is a multiple of p for all x prime to p, while Euler's argument shows that the second factor in the right-hand side cannot have that property; therefore one must have $x^n \equiv a \pmod{p}$ for some x. Put $a = -1$ and $n = 2$, and the proof reduces to the one reproduced above in Chap.II, §VII, which Euler had sent to Goldbach in 1749 (*Corr.*I.494–495; cf. above, §Vb, and *infra*, §IX); this had been the missing link in his earlier attempts to prove Fermat's theorems on sums of two squares.

Greatly encouraged, no doubt, by this success, and having seen through the press his *Introductio in Analysin Infinitorum* in 1748, Euler then[2] set to work on a treatise, perhaps intended as a similar "introduction" into number theory, for which he wrote sixteen chapters before giving up the attempt; these were published in 1849 under the title *Tractatus de numerorum doctrina* (*Eu.*I-5.182–283 = E 792); in part this almost looks like a first draft for sections I, II and III of Gauss's *Disquisitiones*. It begins with a lucid exposition of some elementary matters, chiefly (Chap.2–4) the calculation of the functions now usually denoted by $\sigma_0(n)$, $\sigma_1(n)$, $\varphi(n)$, i.e. respectively the number of divisors of n, their sum ($\int n$ in Euler's notation), and the number of integers prime to n and less than n (the "Euler function", for which Euler later used the notation πn; cf. e.g. *Eu.*I-4.105–115 = E 564|1775). Then it proceeds (Chap.5, nos.140–166) to

---

[2] Not earlier than 1749, since the use of the difference operator, discovered in that year, figures prominently there (*Eu.*I-5.222, nos.256–259). It should be observed that the marginal notes added to the first draft of the treatise, and inserted as *Additamenta* in *Eu.*I-5, are afterthoughts, presumably of later date; e.g. the *Additamentum* to no.565, p.278, rejects a wrong proof in the text.

an elementary discussion of what since Gauss has been known
as congruences with respect to a modulus; Euler's word for
the modulus, which he uses systematically in that sense, is
"the divisor". For a given divisor d, all integers $r + dx$ are
said to belong to the same "class" (no.14) and are to be
regarded as "equivalent" (no.154). Any representative for
the class of an integer a is called "a residue" of a, this term
being sometimes (but not always) confined to the residue r
of the division of a by d defined by $a = r + dx$, $0 \leqslant r < d$.
It is also specified that the mapping from the integers to
their "residues" (in the sense of the equivalence of those
residues) has the properties characterizing what we now
call a ring-homomorphism.

   The next chapter (Chap.6) gives a treatment of the residues
which arise, for a given "divisor" d, from the integers in an
arithmetic progression a, $a + b$, $a + 2b$, etc., or, in modern
language, of the cosets of the subgroup generated by b in
the additive group modulo d; of course this amounts to the
determination of the g.c.d. of b and d, and to the solution
of congruences $bx \equiv m$ (mod d), which had been the subject
of one of Euler's earliest papers[3] (Eu.I-2.18–32 = E 36|1735;
cf. above, §Va). This is followed by the corresponding investigation for the multiplicative group modulo d (Chap.7);
it is shown that if b is prime to d, the residues arising from
the progression 1, b, b^2, etc. make up a set closed under
multiplication and division modulo d, i.e. in effect a group;
from this, by considering its cosets in the multiplicative group
of integers prime to d modulo d, it is shown that its order,
i.e. the smallest integer $\nu > 0$ such that d divides $b^\nu - 1$,
must divide the order $\mu = \varphi(d)$ of the latter group. Therefore
d must divide $b^\mu - 1$.

   That is Euler's famous theorem; when d is a prime, it
reduces to Fermat's (cf. Chap.II, §IV). In later years, when

---

[3] Oddly enough, no justification is given in the text of Chapters 1 to
5 for the unique decomposition of integers into primes. The euclidean
algorithm is mentioned there only as an afterthought, in the *Additamenta*
to nos. 139 and 150.

publishing this proof (Eu.I-2.493–518 = E 262|1775 for
the case of a prime "divisor"; Eu.I-2.531–555 = E 271|1758
in the general case), Euler not only insisted that this was a
substantial generalization, justifying the introduction of the
"Euler function" $\varphi(n)$ "which otherwise might seem sterile"
(Eu.I-2.555, Scholion), but also pointed out that, even for
d prime, the above proof is better than the "additive" one;
it is "more natural", he says, since the latter depends upon
the binomial formula which seems alien to the subject ("*a
proposito non mediocriter abhorrere videtur*": Eu.I-2.510, Scholion). Incidentally, he also notes that $\varphi(N)$ is not in general
the smallest exponent n such that N divides $x^n - 1$ for all
x prime to N; e.g. if $N = p^\alpha q^\beta r^\gamma$, where p, q, r are distinct
primes, that property belongs, not to the product of
$p^{\alpha-1}(p-1)$, $q^{\beta-1}(q-1)$, $r^{\gamma-1}(r-1)$, which is $\varphi(N)$, but to their
l.c.m (Eu.I-2.532, Summarium); he had already made a
similar observation in his very first paper on number theory
(Eu.I-2.4 in E 26|1732) without being able to prove it.

Of course congruences are most interesting when the
modulus, or in Euler's language the divisor, is an odd prime
p, and we soon find him, in the *Tractatus*, concentrating on
that case (Chap.8). He begins by showing that, if a is prime
to p and p divides $a^n - 1$, then, since p also divides $a^{p-1} - 1$,
it divides $a^d - 1$ where d is the g.c.d. of n and $p - 1$; in
effect, he uses here the fact which would now be expressed
by saying that the group of residues of powers of a modulo
p is a homomorphic image of the additive group of integers
modulo $p - 1$. From this he deduces that, if n is itself an
odd prime and p divides $a^n - 1$ but not $a - 1$, then p must
be of the form $2nx + 1$; the case $a = 2$ is the relevant one
for the investigation of perfect numbers.

Still following Fermat, consciously or not (cf. Chap.II,
§IV), Euler observes next (no.255) that, for $p = 2m + 1$
and a prime to p, p divides

$$a^{2m} - 1 = (a^m - 1)(a^m + 1)$$

and must therefore divide either $a^m - 1$ or $a^m + 1$ but not
both; and he asks whether it could divide $a^m - 1$ for all a

prime to p. At this juncture he introduces his newly discovered method of finite differences to show that this can not happen; more generally, he uses it to show that, if $n < p - 1$, p cannot divide $a^n - 1$ for all a prime to p.

Here he narrowly missed an opportunity for showing the existence of "primitive roots" modulo an arbitrary prime, i.e. of an element generating the multiplicative group modulo p, with all the consequences which he was later to deduce from this fact. Indeed, with the means already at his disposal, it can be shown easily that, if a and b are elements of that group (or in general of any finite commutative group), there is always an element $a^\mu b^\nu$ whose order is the l.c.m of those of a and of b. Thus, if n is the greatest of the orders of the elements of that group, it must also be a multiple of the orders of all its elements, so that p must divide $a^n - 1$ for all a prime to p; thus n cannot be $< p - 1$.

Partly for lack of this result, most of the remainder of the *Tractatus* is tentative, obscure or just plain wrong. Its aim was clearly to give, in what would now be called group-theoretic terms, an account, firstly of quadratic residues, and then of m-th power residues modulo a prime p, at least for $m = 3, 4, 5$. Euler saw that, for each m, such residues make up a set closed with respect to multiplication and division: "a truly remarkable property", he calls it (*"insignem proprietatem"*: no.295), having already, as we have seen, encountered its analogue in connection with quadratic forms (cf. *infra*, §VIII). In our language, this is to say that the m-th power residues make up a subgroup G_m of the multiplicative group G modulo p; Euler even discusses, in terms of the cosets of G_m in G, what is now called the factor-group G/G_m, but, except for $m = 2$ when G_m is of index 2 in G, he fails to determine its structure or even its order; his attempt to show, for $m = 3, 4$ and 5, that, if m divides $p - 1$, G/G_m is cyclic of order m does not even make sense. Clearly his project was premature and had to be given up.

As usually happened with Euler, however, nothing was lost; whatever was most valuable in the *Tractatus* was to reappear in the publications of the next decade. The greater

part (pp.339–365 in *Eu*.I-2.338–372 = E 242) of his paper
of 1751 on sums of four squares is devoted to a careful
exposition of the elementary properties of quadratic residues
and non-residues, at first without assuming that "the divisor"
(i.e. the modulus) is a prime. When it is an odd prime p,
Euler shows that the residues make up a subgroup of index
2 of the multiplicative group modulo p, with all that this
implies, which to us seems so elementary as to appear trivial
but in his days was new and needed spelling out. In particular
he can now point out that, if -1 is a quadratic residue
modulo $p = 2n + 1$, then quadratic residues must occur
in pairs $\{r, -r\}$, so that n must be even; as to the converse
he merely notes that he has looked in vain for a "direct"
proof (i.e., apparently, one not depending upon the difference operator: p.365, Scholion).

His ostensible purpose in that paper had been to deal
with sums of four squares (cf. *infra*, §XI). Then, for a few
years, the *Tractatus* seems to have been set aside; the next
paper to be based on it, significantly entitled *Theoremata
circa residua ex divisione postestatum relicta* (i.e. theorems on
m-th power residues modulo a prime: *Eu*.I-2.493–518 =
E 262) was read to the Berlin academy in 1755; it describes
Euler's latest version (the one reproduced above) of his
method of finite differences, with his criterion $a^n \equiv 1 \pmod{p}$
for an integer a to be an m-th power residue modulo a prime
$p = mn + 1$. Next in line comes the paper entitled *Theoremata
arithmetica nova methodo demonstrata* (*Eu*.I-2.531–555 =
E 271|1758), based this time on Chapters 4 to 7 of the *Trac-
tatus;* the "new method" is here the group-theoretic investigation of the additive and multiplicative groups modulo
an arbitrary integer N, with the calculation of the order
$\varphi(N)$ of the latter group, all this leading up to Euler's theorem
that N divides $a^{\varphi(N)} - 1$ for all a prime to N.

Perhaps nothing more could then have been extracted
from the *Tractatus*, apart from some conjectures related to
the quadratic reciprocity law and a few bold inductions about
cubic and biquadratic residues, a proof for which lay well
beyond Euler's powers then and later. For some time his

interest in such matters seems to have slackened; to revive it, some outside impulse was needed. Lagrange was the one to supply it.

For many years Goldbach had been alone in taking an interest (an amateurish one, but real nonetheless) in Euler's arithmetical work. Their correspondence was almost completely interrupted between 1756 and 1762 by the Seven Years' war; in the meanwhile Goldbach was getting old; he died in 1764. Lagrange's correspondence with Euler begins in 1754 (cf. Chap.IV, §I) when he was only 18 years old; his first letter seems to betray a raw beginner, scarcely worthy of Euler's attention; characteristically Euler preserved it, although he did not answer it. Lagrange's next letter, however, written a year later, and outlining an altogether new and original treatment of variational problems, made a deep and lasting impression on Euler, who from then on, in spite of the difference in age, treated Lagrange as an equal. Lagrange had been born in Turin in 1736; he lived there, except for a trip to Paris in 1763–1764, until 1766, when he moved to Berlin as Euler's successor. In 1787 he finally settled down in Paris; by that time he had no active interest in number theory any more. He died in 1813.

Much if not all of Lagrange's work is based on that of Euler; this applies in particular to his contributions to number theory, about which he once wrote to Euler:

"*Je suis très charmé que mes recherches sur les problemes indeterminés aient pu meriter votre attention; le suffrage d'un savant de votre rang est extrêmement flatteur pour moi, surtout dans une matiere, dont vous êtes le seul juge compétent que je connoisse. Il me semble qu'il n'y a encore que Fermat et vous qui se soient occupés avec succés de ces sortes de recherches, et si j'ai été assés heureux pour ajouter quelque chose à vos decouvertes je ne le dois qu'à l'étude que j'ai faite de vos excellens ouvrages.*" ["I am most happy that my work on indeterminate problems should have been found worthy of your attention; the favorable opinion of a scientist of your rank is most flattering for me, particularly in a field where, to my knowledge, you are the one and only competent judge. It seems to me that only

Fermat and you have been successful so far in dealing with such matters, and, if I have been so lucky as to add to your discoveries, I owe it to nothing else than the study I have made of your excellent works": *Eu*.IV A-5.471|12 Feb.1770].

This was no mere politeness or flattery; in his correspondence with d'Alembert and others, Lagrange expressed himself about Euler more freely and more critically, no doubt, but always with the same sense of deep indebtedness and admiration.

This quotation, and the one in §II above, are part of an exchange of letters between Euler and Lagrange concerning the latter's great memoir on indeterminate equations (*Lag*.II.377–535), published in the *Nouveaux Mémoires* of Berlin in 1769; they had reached Euler in 1769 or early in 1770. Partly, no doubt, because of this exchange, Lagrange composed in 1770 a continuation of that memoir (*Lag*.II.655–726) which Euler received in 1770 or early in 1771 (cf. *Eu*.IV A-5.488|20 May 1771); it contained a proof for the fact that a congruence of degree n modulo a prime p has at most n solutions modulo p (*loc.cit*.pp.667–669).

This was precisely the point which Euler had missed when writing the *Tractatus;* one may surmise that he was quick to realize it and revise his thinking accordingly about the multiplicative group modulo a prime. At any rate, scarcely more than a year later, we find him, on 18 May 1772, presenting three papers[4] on m-th power residues to the Petersburg academy (*Eu*.I-3.240–281 = E 449,497–512 = E 552,513–543 = E 554).

Strangely enough, Lagrange's proof (which, incidentally, is made explicit only for congruences of degree 3, but is rightly stated to be general) had been based on a modification of Euler's argument by finite differences. It may well have

---

[4] Only the first one appeared in the *Novi Commentarii* of the Academy, in 1774. The other two were included by Euler in his *Opuscula Analytica* of 1783; perhaps he had waited with them, hoping to have more to say on the matter, but then, feeling his time to be running out, he wished to see them printed quickly.

been inserted by Lagrange into his paper as an afterthought; at any rate no use is made of it there, which is perhaps why Euler did not quote Lagrange in this connection. Euler's proof for it is the modern one. Nowadays one expresses it concisely by saying that the integers modulo a prime make up a field (as was well known to Euler, in substance if not in words) and that the classical argument, already known to Descartes (*loc.cit.* Chap.II, §VII), for proving that an equation $f(x) = 0$ of degree n has at most n roots is valid over any field; in fact, if a is a root, one can divide $f(x)$ by $x - a$ and write

$$f(x) = (x-a)f_1(x);$$

this cannot be 0 unless $x = a$ or $f_1(x) = 0$; as f_1 is of degree $n - 1$, induction on n proves the assertion. In order to express the same idea, Euler observes (*Eu.*I-3.249) that, if a is a solution of the congruence $f(x) \equiv 0 \pmod{p}$, then one can write $f(a) = mp$, where m is an integer, so that a is a root of $f(x) - mp$ in the usual sense. Writing now

$$f(x) - mp = (x-a)f_1(x)$$

and applying induction as above, or alternatively (as Euler does) proceeding by successive divisions until all the roots have been exhausted, one gets the desired result.

This theorem (made explicit, incidentally, only for the case $f(x) = x^n - 1$, although the proof is quite generally valid) is then used to show the existence of primitive roots modulo any prime p; the general design of the proof is apparent, even though some clumsiness (whether due to Euler or to his assistants) tends to obscure the details. On the one hand Euler shows in substance that there is for each n a polynomial F_n of degree $\varphi(n)$ (which later came to be called "the cyclotomic polynomial" of that degree) such that $x^n - 1$ is the product of the polynomials $F_d(x)$ taken over all the divisors d of n (including 1 and n); the roots of F_n, whether "real" or "imaginary", are then precisely the primitive n-th roots of unity. The proof is by induction on the number of prime factors of n, distinct or not, even though it is carried out only up to three factors (giving occasion to

Gauss, in a fit of surliness, to claim that Euler's result had been proved "only by induction", i.e., in his use of the word, by experimentation); actually Euler obtains $F_n(x)$ by dividing out $x^n - 1$ by all the "cyclotomic polynomials" belonging to the divisors of n other than n. Thus, for each n, we may write

$$x^n - 1 = F_n(x)G_n(x)$$

where G_n is the product of the polynomials F_d for d dividing n and $<n$. For the validity of Euler's argument the fact would still be needed that the coefficients of F_n and G_n are integers; this is easily seen, since they are obtained by successive divisions by monic polynomials (i.e., polynomials with highest coefficient 1) with integral coefficients, but Euler fails to mention that point. The final step is then to take $n = p - 1$ and to note that the polynomial $x^{p-1} - 1$, taken modulo p, has $p - 1$ distinct "real" roots 1, 2, ... , $p-1$, so that its divisor $F_{p-1}(x)$ must itself have as many roots modulo p, from among the values 1, 2, ... , $p - 1$, as its degree $\varphi(p - 1)$ indicates; these are the "primitive roots" modulo p. This argument amounts to saying that, over the field of integers modulo p (or indeed over any field), every divisor of a polynomial which splits into a product of linear factors must itself split into such factors; translated into Euler's language, this is easily seen to hold, even though Euler does not make the point quite explicitly.

As to the consequences which Euler draws from the existence of primitive roots, in that same memoir and in the two companion ones presented on the same day to the Academy, some seem obvious to us and can indeed not have cost much effort to Euler; others concern theorems and conjectures related to the quadratic reciprocity law and will be discussed later. Noteworthy is Euler's increasing awareness that in such matters he is dealing, not with individual integers, but with what we would call congruence-classes, i.e., in modern notation, with elements of the prime field $\mathbf{F}_p = \mathbf{Z}/p\mathbf{Z}$; for this concept he introduces at first the word *ordo* (*Eu.*I-3.242), then the word *species* (*Eu.*I-3.519); for the group of quadratic residues modulo p, and for the set of non-residues, each consisting of $\frac{1}{2}(p-1)$ "*species*", he has the word *classis* (*residuorum* resp. *non-residuorum*). Even more boldy, he introduces the notation α/β for the congruence-class γ defined by $\beta\gamma = \alpha$, carefully explaining that no fraction is meant here, but the congruence-class with the representative $(\alpha + np)/\beta$ when α, β are representatives for their respective congruence-classes, and n is so chosen that $(\alpha + np)/\beta$ is an integer, as is always possible if β is prime to p (*Eu*.I-3.521). For the pair α, $1/\alpha$, he has the word *sociata* (*Eu*.I-3.524), or *socia* (*Eu*.I-3.248), or *reciproca* (*Eu*.I-3.503), and applies this concept to an elegant proof for the quadratic character of -1 modulo a prime $p = 2n + 1$, as follows: observing that, if α is a quadratic residue, then so is $1/\alpha$, he concludes that the n quadratic residues can be arranged into pairs $\{\alpha, \alpha'\}$ with $\alpha' = 1/\alpha$, where $\alpha \neq \alpha'$ unless $\alpha^2 = 1$, i.e. unless $\alpha = \pm 1$ (*Eu*.I-3.507,525); thus n is even or odd according as -1 occurs among the quadratic residues or not. Of course it did not escape him that this can also be derived directly from the existence of primitive roots; indeed, if r is such a root, then, since p divides

$$r^{2n} - 1 = (r^n - 1)(r^n + 1)$$

but not $r^n - 1$, r^n has the residue -1 modulo p; on the other hand, for any m, r^m is or is not a quadratic residue according as m is even or odd; hence the desired result (*Eu*.I-3.255,260,262). In the same vein, when Euler received from Lagrange a paper containing two proofs for "Wilson's theorem" (*Lag*.II.425–438|1771; cf. Chap.II, §VII), he sent him at once one of his own, based on the observation that, if r is a primitive root modulo the prime $p = 2n + 1$, then

$$(p-1)! \equiv r^{1+2+\cdots+(p-2)} = r^{n(2n-1)} \equiv (-1)^{2n-1} = -1 \pmod{p}$$

(*Eu*.IV A-5.496|1773; cf. *Eu*.I-4.91–93 in E 560|1773).

§VII.

The use of the words "real" and "imaginary" (or "impossible") in connection with solutions of congruences, in

Euler's treatment of primitive roots modulo a prime (*Eu.*I-3.249, Scholion, and 252, Coroll.1, **E** 449|1772) should not pass here without comment. In the usage of the eighteenth century, and more specifically in Euler's, the word "imaginary" appears with two different meanings; it denotes either what we would call a complex number, i.e. an element of the field $\mathbf{R}(\sqrt{-1})$, or else, more vaguely because the concept could not yet be made precise, an element of any algebraic extension of some groundfield when this extension could not be embedded in the field $\mathbf{R}$ of real numbers. The first concept, motivated by the work of the Italian algebraists of the sixteenth century on equations of degree 3 and 4, had been introduced by Bombelli in Book I of his *Algebra* of 1572, in close imitation of Euclid's theory of irrationals of the form $a + \sqrt{b}$ in his Book X ("*binomia*" in the terminology of Campanus's Latin translation of Euclid; accordingly, *binomio* is Bombelli's word for a complex number $a + \sqrt{-b}$). Further developments belong to the history of algebra and analysis rather than to number theory; suffice it to say that Euler played a decisive role in extending to complex numbers the main operations of analysis.

On the other hand, in a series of letters addressed in 1735 to the Dantzig burgomaster C. L. G. EHLER for perusal by his fellow-citizen, the scientist H. KÜHN (*PkU.*282–329, notably pp. 297–301 and 324–328), Euler stated his views about the successive extensions of the concept of number, first from ordinary to positive and negative integers, then to rational and to various kinds of (real) irrational numbers, and beyond them to algebraic extensions; at each step, in Euler's view, all that is needed is that the usual algebraic rules should continue to hold and no contradiction should ensue; as to the last point, Euler is clearly satisfied with his own extensive experience and that of all competent mathematicians. In this view, every polynomial must split into as many linear factors as its degree indicates; for instance, "I decide", he says ("*iam statuo*": *loc.cit.*p.325) that $x^3 - ax^2 - bx - c$ consists of three factors $x - p, x - q, x - r$. Having worked out to his satisfaction the example

$x^3 - 8 = 0$, where two complex roots appear, Euler rests his case, and indeed Kühn finds this so convincing that he withdraws all his objections.

In the language of Euler and his contemporaries, the so-called "fundamental theorem of algebra" (no more and no less, in modern terms, than the fact that $\mathbf{R}(\sqrt{-1})$ is algebraically closed) could be expressed by saying that all "imaginary" roots of algebraic equations can be written as $a + b\sqrt{-1}$; an equivalent statement, as Euler knew, is that every polynomial (with real coefficients) can be written as a product of factors of degree 1 or 2. The question acquired special importance in integral calculus because the integration of rational functions by logarithms and inverse trigonometric functions depends upon it (cf. *infra*, §XV). In 1742 Euler convinced himself that all rational functions can be so integrated; in 1703 Leibniz had expressed the opposite view (*Math.Schr.*, ed. C. I. Gerhardt (II) Bd.I.360); even in 1742 Nicolas BERNOULLI, an older cousin of Euler's friend Daniel, thought he could offer a counterexample and did succeed in shaking Euler's confidence for a brief moment (*Corr.*I.170–171; cf. *Corr.*II.695 and L. Euler, *Opera Postuma* . . . Petropoli 1862 = Kraus Reprint 1969, pp. 525–539) before finally accepting Euler's arguments. In spite of contributions by d'Alembert, Euler, Lagrange, Gauss, the matter cannot be said to have been fully understood until much later (until Cauchy and his successors, from the point of view of analysis; until Artin and Schreier, from the purely algebraic point of view); this story need not concern us here.

In Euler's memoir E 449|1772 on m-th power residues, however, the groundfield is implicitly the prime field $\mathbf{F}_p = \mathbf{Z}/p\mathbf{Z}$ of integers modulo a prime p. Euler's word for a solution of a congruence $f(x) \equiv 0 \pmod{p}$ is *casus* ("a case") or also *solutio*, i.e. a solution for the problem of making $f(x)$ divisible by p; if $x = a$ is such, then he says that all the values $a + mp$ are to be counted as one and the same "case" ("*pro unico casu*": *Eu.*I-3.249, Scholion). All such "cases" are "real" if $f(x) = x^{p-1} - 1$; therefore the same is true if $f(x)$ is any polynomial dividing $x^{p-1} - 1$ (*Eu.*I-3.252, Coroll. 1); on the other hand some of them will be "impossible" and "so to say imaginary" ("*quasi imaginarii*": *Eu.*I-3.249, Scholion) if $f(x) = x^n - 1$ and n is not a divisor of $p - 1$. Undoutedly what Euler has here in mind, in still undeveloped and obscure form, is the concept, introduced by Galois in 1830, of "Galois imaginaries" as they came to be called at one time, i.e. of the finite algebraic extensions of the field of integers modulo p, or in other words the finite fields.

§VIII.

All his life Euler was deeply concerned with the representation of integers, at first as sums of two squares, then by quadratic forms (he said "formulas") $X^2 + NY^2$ or more generally $\mu X^2 + \nu Y^2$ where N, μ, ν are given integers, positive or negative. As to forms $aX^2 + bXY + cY^2$, it is only in his old age, and then only under the influence of Lagrange, that he paid any attention to them (cf. *infra*, §IX), except that the form $X^2 + XY + Y^2$ had, unavoidably but furtively, made an appearance in connection with his work on cubic residues and the form $X^2 + 3Y^2$ (*Eu.*I-2.572–574 in E 272|1759).

The multiplicative nature of such questions is evidenced by "Brahmagupta's identity" (cf. above, Chap.I, §VIII, and Chap.II, §XII):

$$(x^2 + Ny^2)(z^2 + Nt^2) = (xz \pm Nyt)^2 + N(xt \mp yz)^2.$$

At least in special cases, Euler must have been aware of this identity quite early, perhaps in connection with his first attempts to deal with Pell's equation and related problems (cf. *Corr.*I.36–37|1730, *Eu.*I-2.6–17 = E 29|1735, and *infra*, §XIII) long before he found it necessary to spell it out in the case $N = -2$ (*Eu.*I-2.237 in E 167|1748); for the simplest case $N = 1$, cf. *Corr.*I.134|1742, and Euler's first memoir on sums of two squares where it is still mentioned as "noteworthy" ("*notatu dignum*": *Eu.*I-2.300 in E 228|1749). The general case, implicit in a letter from Goldbach

(*Corr.*I.612|1753) and surely well known to Euler by that time, is stated in the latter's letter of August 1755 (*Corr.*I.629) and then again, as a theorem of capital importance ("*theorema eximium*": *Eu.*I-2.6G0 in E 279|1758) in a memoir on indeterminate equations, where it leads at once to the conclusion that the representation of integers by a form $X^2 + NY^2$ is a multiplicative problem and should first of all be studied in the case of a prime (*ibid.*,p.601).

As Euler knew, Fermat had treated the cases $N = 1, 2, 3$ and had announced his results in letter XLVII of the *Commercium* of 1658 (*loc.cit.* above, §IV; cf. Chap.II, §§VIII and XII). But, when Euler started taking up those problems seriously, he had long been aware of Fermat's earlier statement that a sum $a^2 + b^2$, with a prime to b, has no prime divisor $p = 4n - 1$ (*Fe.*II.204 = *Varia Opera*, p.161; cf. Chap.II, §VII), nor consequently any divisor $4n - 1$, be it prime or not. Obviously this had made a deep impression on his mind; quite early ("*vor langer Zeit:* *Corr.*I.107|1741; cf. *Corr.*I.114|1742) he had transformed it into the statement that neither $4mn - m - 1$ nor $4mn - m - n$ can be a square if m, n are positive integers; indeed, if $4mn - m - 1$ (resp. $4mn - m - n$) were a square a^2, this would imply

$$m(4n-1) = 1 + a^2 \text{ resp. } (4m-1)(4n-1) = 1 + 4a^2,$$

so that $4n - 1$ would divide a sum of two mutually prime squares.

Thus Euler was confronted not with one but with three types of questions, which may be formulated thus, for any given integer N (positive or negative, but other than 0 and -1):

(i) *Which are the primes that can be written as* $a^2 + Nb^2$, a and b being integers?

(ii) *What are the "prime divisors" of the form* $X^2 + NY^2$, if by this one understands with Euler the odd prime divisors (necessarily prime to N) of some integer $a^2 + Nb^2$ with a prime to Nb?

(iii) *Which odd integers are divisors of some integer* $a^2 + Nb^2$ *with a prime to* Nb?

As Euler must have realized rather soon, the integers (prime to $2N$) that satisfy (iii) are precisely those modulo which $-N$ is a quadratic residue (cf. *Eu.*I-2.209, *Annot.*9, and 219, *Annot.*18, in E 164|1744; cf. also no.300 of the *Tractatus, Eu.*I-5.229–230). In fact, if $a^2 + Nb^2 = mn$ with a prime to Nb, a and Nb must be prime to m; then there is b' such that $bb' \equiv 1 \pmod{m}$ (cf. above, §V*a*) and we have $-N \equiv (ab')^2 \pmod{m}$. The converse is immediate. If A = 1, for instance, this shows that an answer to (ii) must amount to the determination of the quadratic residue character of -1 modulo any odd prime (cf. above, §VI, and Chap.II, §VII). If for convenience we introduce the Legendre symbol (n/p), defined whenever p is an odd prime and n an integer prime to p, and equal to $+1$ or -1 according as n is a quadratic residue modulo p or not (cf. Chap.IV, §VI), then (ii) amounts to asking for the determination of $(-N/p)$ for all primes p, not dividing $2N$. On the other hand, from the point of view of the theory of quadratic fields, (i) and (ii) amount to finding the behavior of a rational prime p in the field $\mathbf{Q}(\sqrt{-N})$: if it divides $a^2 + Nb^2$ with a prime to Nb, then, in that field, it splits into the ideal prime factors $(p, a \pm b\sqrt{-N})$; it will satisfy (i) if those factors are the principal ideals $(a \pm b\sqrt{-N})$. Thus (i) lies deeper than (ii); to use Euler's words for this, it is *"altioris indaginis"* (*Eu.*I-2.603 in E 279|1758); it depends upon the theory of ideal-classes in quadratic fields, or equivalently of classes of binary quadratic forms, while (ii), as will be seen below (cf. Appendix I), depends upon no more than the law of quadratic reciprocity.

Throughout the years Euler was tireless in exploring all aspects of the above questions and bringing together whatever bits of information he could gather about them. Already in 1741 we find him reporting to Goldbach the *"curieuse proprietates"* he has discovered concerning the prime divisors of $X^2 - 2Y^2, X^2 - 3Y^2, X^2 - 5Y^2$ (*Corr.*I.107|9 Sept.1741); in the latter case this already went beyond anything that Fermat had stated. The next year the *"curieuse proprietates"* had blossomed forth into "a very pretty arrangement" (*"eine sehr artige Ordnung"*: *Corr.*I.146|28 Aug.1742) which he at once started to describe on the basis of his experiments, while predicting that this was only the prelude to further "splendid properties" ("*herrliche proprietates*": ibid., p.150) yet to be discovered. Not later than 1744 ("*vor einiger Zeit*": *Corr.*I.279|4 July 1744) he put together all his empirical observations on the prime divisors of $X^2 + NY^2$ (covering 16 positive values and 18 negative values for N), added theoretic conclusions in the form of a series of *Annotationes*, and sent the whole to Petersburg. This is the memoir E 164 (= *Eu.*I-2.194–222) which the Academy published in 1751.

From now on we shall assume N to be square-free; this does not restrict the generality, so far as (ii) and (iii) are concerned. In modern terminology and notation, the answer to (ii) is best described by means of the "Dirichlet character" attached to the field $\mathbf{Q}(\sqrt{-N})$. This is a certain function ω of integers, taking the values ± 1 and 0, periodic of period $D = |N|$ if $N \equiv -1 \pmod 4$ and $D = 4|N|$ otherwise; it is such that a prime p satisfies (ii) if and only if $\omega(p) = 1$; in other words, one has, for all primes p not dividing $2N$:

$$\omega(p) = \left(\frac{-N}{p}\right).$$

A construction and full characterization of ω will be given in Appendix I.

Let K_N be the set of congruence classes prime to $4|N|$ modulo $4|N|$ on which ω takes the value 1; thus a prime satisfies (ii) if and only if its congruence class modulo $4|N|$ belongs to K_N. As will be shown in Appendix I, K_N can be characterized by the following properties:

(A) K_N *is a subgroup of index 2 of the multiplicative group modulo* $4|N|$. This implies that it contains all quadratic residues prime to $4|N|$ modulo $4|N|$.

(B) -1 *belongs to K_N or not, according as $N < 0$ or $N > 0$.*

(C) K_N *admits the period D and no period $D' < D$.* By this is meant that, if r and s are prime to $4|N|$ and $r \equiv s \pmod D$, then both belong to K_N whenever one of them does, and that no divisor D' of D, smaller than D, has the same property.

This characterization of K_N may thus be regarded as providing a complete answer to question (ii).

Euler's first major discovery, on the basis of his experimental data, was to observe that the prime divisors of $X^2 + NY^2$ are precisely those primes which belong to a certain number of arithmetic progressions

$$P(r) = \{r, 4|N| + r, 8|N| + r, \text{etc.}\}$$

where r is prime to $4|N|$ and $0 < r < 4|N|$ (*loc.cit.*, *Annot.*3, 13; cf. *Corr.*I.147|1742).

Here it is well to keep in mind Dirichlet's theorem according to which every progression such as $P(r)$ contains infinitely many primes; were it not so, Euler's experimentation could hardly have been carried out so successfully. Because of this, the set of congruence classes modulo $4|N|$ determined by the integers r for which $P(r)$ has the property in question is no other than the set K_N introduced above.

At first, writes Euler to Goldbach, there seems to be no regularity ("*keine Ordnung*": *Corr.*I.147) in the numbers making up the set K_N; nevertheless they proceed according to "a beautiful law" ("*nach einer schönen lege*", as he puts it in his mixed German and Latin: *ibid.*). In the first place, all powers and products of powers of elements of K_N belong to K_N (*Corr.*I.148,150; *Annot.*8,16); in our language, this means that K_N is a subgroup of the multiplicative group modulo $4|N|$. Since obviously every integer satisfying (iii) is a product of primes or powers of primes with the same property, this implies the important fact that also all such integers must belong to some progression $P(r)$ with r in K_N (cf. *loc.cit.* pp.220–222, Schol.2,3); perhaps it was this that led Euler to the discovery of the group property for K_N. The same remark shows in particular that all odd integers $a^2 + Nb^2$ with a prime to Nb must belong to some $P(r)$ with r in K_N; for $b = 2$ and a prime to $2N$, it shows that all quadratic residues modulo $4|N|$ belong to K_N (*Annot.*7, 16). Furthermore, if r belongs to K_N, $-r$ belongs to it if $N < 0$ but not if $N > 0$ (*Annot.* 3, 13); and the number of elements of K_N is half the number of congruence classes prime to

$4|N|$ modulo $4|N|$ (*Annot.*5, 14) i.e. $\frac{1}{2}\varphi(4|N|)$ if φ is the "Euler function". Thus Euler had found that K_N has the properties (A) and (B) listed above.

As to (C), Euler tells us that, if r belongs to K_N, so does $r + 2|N|$ for $N \equiv -1$ (mod 4) (*Annot.*11, 19), while this is not so otherwise (*Annot.*12, 20). This means, in the former case, that K_N admits the period $|N|$; in fact, if r and s are prime to $4|N|$, they are both odd, so that, if $r \equiv s$ (mod $|N|$), s must be $\equiv r$ or $\equiv r + 2|N|$ (mod $4|N|$). In view of this, and of Euler's specific mention of the fact that this is not so for $N \equiv 1$ or 2 (mod 4), we may surmise that he had indeed looked for other periods; if we credit him with having perceived that there are none, we may thus conclude that he was in possession (conjecturally of course) of the full characterization of the set K_N and thus of a complete answer to question (ii).

With this Euler had wandered far ahead of the paths trodden by Fermat. Hidden in his guesses, as perhaps the most valuable nugget of his treasure-trove, lay the law of quadratic reciprocity (cf. Appendix I, and above, §Vi), which he was not to recognize as such until quite late in life. On one occasion he did push his experiments even further; while composing for the *Tractatus* two highly tentative chapters on cubic and biquadratic residues, he looked for conditions for 2, 3, 5, 6, 7, 10 to be cubic residues and for 2, 3, 5 to be biquadratic residues modulo a prime p (*Eu.*I-5.250–251, nos.407–410, and 258–259, nos.456–457). As to the latter question, it arises only for $p \equiv 1$ (mod 4), since otherwise, as Euler notes (*ibid.*p.252, no.419), every quadratic residue is a biquadratic residue; thus Euler can write $p = a^2 + 4b^2$ and find that the biquadratic character of 2, 3 resp. 5 modulo p depends only upon the values of a and b modulo 4, modulo 6 and modulo 5, respectively. Similarly, for cubic residues, it is enough to consider the case when p can be written as $p = a^2 + 3b^2$, and then the cubic residue character of 2, 3, 5, 7 resp. 10 depends only upon the values of a and b modulo 3, modulo 9, modulo 15, modulo 9, modulo 21 and modulo 5, respectively, in a manner correctly described by Euler. These brilliant but wholly isolated observations were lost to the mathematical world; in 1849, when the *Tractatus* was first published, Gauss, and then Jacobi and Eisenstein, had already proved all that Euler had guessed, and of course much more (cf. *Eu*.I-5, pp.XXI–XXIV).

§IX.

In the years 1741 to 1744, while assembling experimental data on the prime divisors of the forms $X^2 + NY^2$, Euler sought to use his conclusions so as to generalize his former remark about integers $4mn - m - n$ not being squares. For instance, as he writes to Goldbach (*Corr*.I.162|27 Oct. 1742), since his observations imply that the form $X^2 + NY^2$, for $N > 0$, has no divisor $r = 4Nx - 1$, an integer $4Nmn - m - n$, for positive m and n, can never be a square; in fact, if it were equal to a^2, this would give

$$(4Nm-1)(4Nn-1) = 1 + 4Na^2.$$

Perhaps, he adds, the latter statement will be easier to prove than the former one (*ibid.*; cf. *Corr*.I.179|5 Jan.1743). Almost no letter passed between the two friends, from 1741 to 1744, without this topic being somehow touched upon; even in his memoir of 1744 he included at the end a list of "formulae" $4Nmn - A(m \pm n)$ whose value, in view of his observations, can never be a square for m and n positive and prime to A (*Eu*.I-2.220–222; cf. *Corr*.I.260,278). Eventually he must have given up his hopes of finding direct proofs for such statements. It is amusing, however, to observe Goldbach's obstinate efforts to prove the simplest one of them, the one relating to $4mn - m - n$, even though this was unnecessary in view of Euler's proof (sent to Goldbach on 6 March 1742, and repeated on 19 January 1743: *Corr*.I.115–117,191–192) that $X^2 + Y^2$ has no prime divisor $p = 4n - 1$. With some prodding and some help from Euler, Goldbach was at last successful (*Corr*.I.255–256|28 Sept. 1743; cf. Euler's answer, pp. 258–259|15 Oct.1743); his proof, perhaps his only contribution of value in the whole correspondence, is by descent; had he and Euler but known it, it contains the first germ of the Lagrangian reduction theory for binary quadratic forms (cf. below).

Except for this episode which had led into a blind alley, Euler, for many years, concentrated his efforts upon the proof of Fermat's statements about the "elementary forms" $X^2 + NY^2$ for $N = 1, \pm 2, 3$, corresponding to fields $\mathbf{Q}(\sqrt{-N})$ with a euclidean algorithm (cf. Chap.II, §§VIII, XI and Appendix I); in those fields integers can be decomposed into primes, just as in ordinary arithmetic, so that every prime divisor of such a form $X^2 + NY^2$ can be represented by it; in other words, the answer to question (i) of §VIII is the same as for (ii). Those facts had been stated and presumably proved by Fermat; Euler had to reconstruct the proofs, and it took him many years.

His first step in this direction had been to deal with the easier part of (ii) for the form $X^2 + Y^2$, by proving that no prime $P = 4n - 1$ can be a prime divisor of that form, or in other words that $(-1/p) = -1$ for such a prime; this had been an immediate consequence of Fermat's theorem $a^{p-1} \equiv 1 \pmod{p}$ (*Corr.*I.115–117|1742; cf. Chap.II, §VI). Decisive progress was made with question (i), also for $N = 1$, in 1747; in a letter to Goldbach (*Corr.*I.416–417) Euler describes his newly found proof; it is the one which has been reproduced in Chap.II, §VIII; the crucial lemma[5] (lemma 2 of Chap.II, §VIII) appears in that letter as Theorem 2. Thus it was proved that every prime divisor of $X^2 + Y^2$ can be written as $a^2 + b^2$ and is consequently of the form $p = 4n + 1$.

It still remained to be shown that every prime $p = 4n + 1$ is a divisor of $X^2 + Y^2$, i.e. that $(-1/p) = 1$ for such a prime; as we have seen, this was accomplished in 1749 (cf. above, §§V*b* and VI). As to the letter of 1747, its contents

---

[5] Undoubtedly the more general statement (*Corr.*I.134|1742) that "if ab and a are sums of two squares, so is b" had been premature, in spite of Euler's assertion that he could prove it quite rigorously ("*rigidissime*").

went, in part almost word for word, into Euler's first memoir
on sums of two squares $(Eu.I-2.295-327 = E\ 228|1749)$,
including unfortunately a rather negligent and imperfect
presentation of the final part of the argument. This was
followed in 1753 by a very lucid treatment, by the same
method, of the form $X^2 + 2Y^2$ $(Eu.I-2.467-485$ in $E\ 256|1753)$,
and in 1759 by its extension to $X^2 + 3Y^2$ $(Eu.I-2.556-575$
$= E\ 272|1759$; cf. $Eu.I-2.481-482$ in $E\ 256|1753)$; the proofs
for these cases are in substance those described in Chap.II,
§XII.

It had thus been proved that every prime divisor of
$X^2 + NY^2$, for $N = 2$ or 3, can be written as $a^2 + Nb^2$ and
thus must be $\equiv 1$ or $3 \pmod 8$ for $N = 2$, and $\equiv 1 \pmod 3$
for $N = 3$; Euler had still to prove the converse. For $N = 3$,
this was done in 1759 $(Eu.I-2.573$ in $E\ 272|1759$, Prop.9);
the proof, which has been reproduced in Chap.II, §XII, is
based on the identity

$$x^{3n} - 1 = (x^n - 1)(x^{2n} + x^n + 1).$$

In the *Scholion* to that proof $(ibid.,p.575)$, Euler mentions
that he knows also how to prove that a prime $p = 8n + 1$
is a divisor of $X^2 + 2Y^2$ (but not how to prove the similar
result for $p = 8n + 3$); undoubtedly he alludes there to
the proof published in 1772, and reproduced above in
Chap.II, §XII, this being based on the identity

$$x^{8n} - 1 = (x^{4n} - 1)[(x^{2n} - 1)^2 + 2x^{2n}]$$

$(Eu.I-3.274-275$ in $E\ 449|1772$, no.85). In this last paper,
however, he can go further, thanks to "a communication
from a friend" (perhaps Lexell); this friend, in fact, merely
pointed out that Euler's argument for proving that every
prime divisor of $X^2 + 2Y^2$ can be written as $a^2 + 2b^2$ is
equally valid for $X^2 - 2Y^2$ $(ibid., p.275$; cf. $Eu.I-3.183$
in $E\ 427|1772)$. Therefore a prime $p = 8n + 3$ cannot
be a divisor of $X^2 - 2Y^2$, so that, for such a prime, we have
$(2/p) = -1$; as we have $p \equiv 3 \pmod 4$, which implies
$(-1/p) = -1$, we get $(2/p) = 1$, as was to be proved (cf.
Chap.II, §XII). This completes the determination of $(\pm 2/p)$

for all odd primes, since it shows that $(-2/p)$ is 1 if and only if $p \equiv 1$ or 3 (mod 8) and since $(-1/p)$ is known. Incidentally it is curious to note that in a paper presented to the Petersburg Academy on the very same day as the one just mentioned $(Eu.\text{I-3.511}$ in E 552|1772, Scholion 3; cf. also $Eu.\text{I-4.38}$ in E 557|1773, nos.24–25) Euler, or whoever was writing for him, appears to be unaware that the result $(-2/p) = 1$, for $p = 8n + 3$, has just been proved. By that time, perhaps, Euler's papers were sometimes little more than transcripts of his conversations with his assistants.

Thus, in 1772, Euler had fully caught up with Fermat in the matter of the prime divisors of the forms $X^2 + Y^2$, $X^2 \pm 2Y^2, X^2 + 3Y^2$, and of the representation of integers by those forms. At one point he went beyond Fermat by making an observation which attracted Lagrange's attention and then was widely generalized by Gauss (cf. $infra$, Chap.IV, §VI). Gauss proved that, if p is an odd prime, $\sqrt{\pm p}$ is contained in the field of p-th roots of unity, the sign being determined so that $\pm p \equiv 1$ (mod 4). This implies that the "cyclotomic polynomial" $X^{p-1} + X^{p-2} + \cdots + 1$ splits over the field $\mathbf{Q}(\sqrt{\pm p})$, so that there are two homogeneous polynomials $F(X, Y)$, $G(X, Y)$ of degree $\frac{1}{2}(p-1)$, with integral coefficients, satisfying an identity

$$4(X^{p-1}+X^{p-2}Y+\cdots+Y^{p-1}) = F(X, Y)^2 - (\pm p)G(X, Y)^2.$$

For $p = 3$, for instance, one can take $F = 2X + Y, G = Y$, and this was a main ingredient in Euler's treatment of $X^2 + 3Y^2$. Similarly, if also $q = pn + 1$ is a prime, the above identity, for $X = a^n, Y = b^n$, gives:

$$4(a^{q-1}-b^{q-1}) = (a^n - b^n)[F(a^n, b^n)^2 \mp pG(a^n, b^n)^2],$$

which, by the same argument as in the case $p = 3$, shows that $\pm p$ is a quadratic residue modulo q. Euler carried this out for $p = 5$ and $p = 7$ $(Eu.\text{I-3.280–281}$ in E 449|1772, nos.96–97); so did Lagrange a little later, but he failed in his attempt to extend it to $p = 11$ $(Lag.\text{III.789–792}|1775)$.

As to the representation of primes by forms $X^2 + NY^2$

(question (i) of §VIII) for other values of N than those considered by Fermat, Euler had for a long time to remain content with the experimental results he had inserted in his memoir of 1744. Already in 1742, after communicating to Goldbach the "*herrliche proprietates*" described above in §VIII, he had touched upon that question, announcing rashly that every prime divisor of $X^2 + NY^2$ can be represented by that form and giving as examples the cases $N = 1, 2, 3, 7$ (for which it is true) and $N = 5, 6$, for which it is false (*Corr*.I.146–148). In 1743 he still thought that if a prime p is a quadratic residue modulo $4N$, for $N > 0$, then it can be written as $a^2 + Nb^2$ (*Corr*.I.264). In his memoir of 1744 he included a list of experimental findings concerning 12 positive values of N, and the values $N = -2, -3$; as typical examples we quote:

1° If p is a prime divisor of $X^2 + 5Y^2$, it can be written as $a^2 + 5b^2$ for $p \equiv 1$ or 9 (mod 20); $2p$ can be so written for $p = 3$ or 7 (mod 20) (*Eu*.I-2.196, Theor.11).

2° All prime divisors of $X^2 + 7Y^2$ can be written as $a^2 + 7b^2$ (*ibid*.,p.197, Theor.14).

3° If p is a prime divisor of $X^2 + 17Y^2$, either p or $9p$ can be written as $a^2 + 17b^2$ when p is a quadratic residue modulo $4 \times 17 = 68$; otherwise $3p$ can be so written (*ibid*.,p.199, Theor.23).

4° If p is a prime divisor of $X^2 + 14Y^2$, it can be written either as $a^2 + 14b^2$ or as $2a^2 + 7b^2$ when it is $\equiv \pm 1$ (mod 8); otherwise $3p$ can be so written (*ibid*.,p.203, Theor. 35).

He must have found such phenomena puzzling; even ten years later, in a memoir significantly entitled "*Specimen de usu observationum in pura mathematica*" ("On the use of observations in pure mathematics": *Eu*.I-2.459–492 = E 256|1753) and dedicated largely to the form $X^2 + 2Y^2$, he repeated some of his earlier statements, commenting that "they seem certain but will surely be quite hard to prove" (*loc.cit*.p.486). In that same year he was pointing out to Goldbach that a prime $p = 4Nx + 1$ is not necessarily representable by the form $X^2 + NY^2$ "*in integris*", i.e. as $a^2 + Nb^2$, a and b being integers; nevertheless he still felt sure (and rightly so; cf. Chap.IV, Appendix I) that it can always be so represented rationally (*"in fractis"*: *Corr*.I.605–606|1753); in other words, the equation $p = x^2 + Ny^2$ must have a rational solution (x, y), even if it has none in integers. In 1772 he had a memoir presented to the Academy (*Eu*.I-4.1–24 = E 556) on the integers which can be represented rationally by a quadratic form $\mu X^2 + \nu Y^2$; little is new there, apart perhaps from some formulas containing special cases of the composition of quadratic forms (p.2, Theor.1, and p.4, Theor.2; cf. Chap.IV, §VI), not to mention a confused and confusing variant (pp.18–24) of Lagrange's method of solution for equations $ax^2 + by^2 + cz^2 = 0$ (cf. Chap.II, §XIV, and Chap.IV, §II); amazingly Lagrange is not mentioned there, although his paper (*Lag*.II.377–535|1768) had for some time been in the hands of Euler, who had commented upon it in a letter to Lagrange (*Eu*.IV A-5.466|16 Jan.1770).

Except for this puzzling item, there is no sign of Euler's having made any progress with these matters until he received from Lagrange in 1775, along with a volume of the Turin *Miscellanea*, the volume of the Berlin *Mémoires* containing Lagrange's *Recherches d'Arithmétique* (= *Lag*.III.695–758|1775). On sending this to Euler, Lagrange, with his usual lack of self-assurance, has expressed how eagerly he was awaiting Euler's verdict on his latest work:

"Si vos occupations et l'etat de votre santé vous ont permis de jetter les yeux sur le peu que j'ai donné dans les Memoires de Berlin et de Turin, je vous supplie de vouloir bien m'en dire votre avis; ma plus grande ambition a toujours été d'obtenir votre suffrage . . ." ["If your occupations and your state of health have let you glance at the little I have contributed to these volumes, I entreat you to give me your opinion on them; my highest ambition has always been to win your approval . . .": *Eu*.IV A-5.503–504|10 Feb.1775].

It would be wrong to tax him with insincerity, or even to suspect him of fishing for compliments (cf. Chap.IV, §II). Anyway, Euler greeted that work with even more than his customary enthusiasm:

"Il est bien glorieux pour moï", he wrote, *"d'avoir pour successeur à Berlin le plus sublime geomètre de ce siecle . . . J'ai parcouru*

avec la plus grande avidité les excellens mémoires dont vous avés enrichi les derniers volumes de Berlin et de Turin . . ." ["It is most flattering to me to have as my successor in Berlin the most outstanding mathematician of this century . . . I have avidly gone through your excellent contributions to the recent volumes...": *Eu.*IV A-5.504 = *Lag.*XIV.241|23 March 1775]. In that letter he makes no specific mention of the *Recherches*; but within a few months he had composed, under the title "*De insigni promotione scientiae numerorum*" ("On some famous progress in the science of numbers": *Eu.*I-4.163–196 = E 598|1775) a kind of commentary on that memoir, which he presented to the Academy on 26 October 1775; it appeared posthumously in 1785.

Not unlike many great discoveries, Lagrange's basic idea was simplicity itself; he had already described it briefly in his *Additions* to Euler's *Algebra* (*Lag.*VII.125–127 = *Eu.*I-1.603–605, no.70); but at that time Euler, while acknowledging (rather perfunctorily: *Eu.*IV A-5.496 = *Lag.*XIV.235|1773) receipt of that volume, had failed to take notice of this brilliant idea. In the *Recherches* of 1775 it is carried out in 4 pages (*loc.cit.*pp.697–700); for the present the following will suffice. If $-N$ is a quadratic residue modulo an integer e (not necessarily a prime), then, writing $-N = f^2 - eg$, we see that $-4N$ is the so-called "discriminant" of the quadratic form $F(X, Y) = eX^2 + 2fXY + gY^2$, while $F(1, 0) = e$ (in general the discriminant of a form $aX^2 + bXY + cY^2$ is defined as $\delta = b^2 - 4ac$). "It may seem that nothing much is gained by this", writes Euler ("*parum hinc lucri ad nostrum institutum afferri videtur*": *loc.cit.*p.165); "and so it would be, were it not for the illustrious Lagrange's discovery that the infinitely many forms" ("*infinitam . . . formularum multitudinem*") with the discriminant δ can, for each value of δ, be reduced to "a small number" ("*ad exiguum numerum*"). In fact, consider a form

$$F(X, Y) = aX^2 + bXY + cY^2$$

with the discriminant $\delta = b^2 - 4ac$; here we merely assume that δ is neither 0 nor a square; in particular, we assume $ac \neq 0$. Suppose that either $|a|$ or $|c|$ is $< |b|$; then, by sub-

stituting $X - rY$ for X if $|b| > |a|$ (resp. $Y - rX$ for Y if
$|b| > |c|$) and choosing r suitably, one can obtain a form

$$F_1(X \ Y) = a_1 X^2 + b_1 XY + c_1 Y^2$$

with the same discriminant δ, but where $|b_1| \leqslant |a| < |b|$ (resp.
$|b_1| \leqslant |c| < |b|$), so that in any case $|b_1| < |b|$. Obviously F and
F_1 are "equivalent" in the sense that they "represent properly"
the same integers; this means that the set of values taken
by $F(\alpha, \beta)$ for α prime to β is the same as the similar set for
F_1 (for more details, cf. Chap.IV, §IV). Clearly the process
of "Lagrangian reduction" by which F_1 has been derived
from F can be repeated, but no more than a finite number
of times since $|b|$ decreases at every step, until one reaches
a form

$$\phi(X, Y) = pX^2 + qXY + rY^2,$$

still equivalent to F and still with the same discriminant
$\delta = q^2 - 4pr$, for which $|p|$ and $|r|$ are both $\geqslant |q|$. An easy
calculation shows that $q^2 \leqslant \frac{1}{3}|\delta|$ if $\delta < 0$, $q^2 \leqslant \frac{1}{5}\delta$ if $\delta > 0$; as
$4pr = q^2 - \delta$, this implies that for a given δ there are only
finitely many such forms ϕ. Moreover, as $a = F(1, 0)$, we
must have $a = \phi(\alpha, \beta)$ with α prime to β, and therefore

$$4pa = 4p(p\alpha^2 + q\alpha\beta + r\beta^2) = (2p\alpha + q\beta)^2 - \delta\beta^2.$$

If, for a given δ, one lists all the forms

$$\phi_i(X, Y) = p_i X^2 + q_i XY + r_i Y^2 \quad (i = 1, 2, ..., m)$$

satisfying the above conditions for ϕ, this shows that one
of the integers $4p_i a$ can be represented by the form $X^2 -
\delta Y^2$. The question still arises whether all the forms ϕ_i are
needed for this purpose; this point had been examined in
full detail by Lagrange (*loc.cit.*pp.723–740; cf. Chap.IV,
§IV and App.II, III); Euler fails to mention it; either he
found it less interesting, or else Lagrange's rather laborious
calculations for solving that question were more than the
blind Euler could follow.

  In particular, take N squarefree as before, and let again
e be a divisor of $X^2 + NY^2$, i.e. a positive integer such that

$-N$ is a quadratic residue modulo e. Put $-N = f^2 - eg$ and apply the foregoing process of "Lagrangian reduction" to the form

$$F(X, Y) = eX^2 + 2fXY + gY^2$$

with the discriminant $\delta = -4N$; ϕ being as above, we have $q^2 - 4pr = -4N$, so that q is even. Writing $e = \phi(\alpha, \beta)$, we get now

$$pe = \left(p\alpha + \frac{q}{2}\beta\right)^2 + N\beta^2.$$

Thus, if we call again $\phi_1, ..., \phi_m$ the forms such as ϕ for the discriminant $\delta = -4N$, one of the integers $p_i e$ will be representable by the form $X^2 + NY^2$. Take for instance $N = 1$; then there is no other form ϕ than $\phi_1 = X^2 + Y^2$ (disregarding $-\phi_1$ which takes no positive values); therefore every divisor of that form is representable by it, and in particular every odd divisor of $X^2 + Y^2$ is $\equiv 1 \pmod 4$. In substance, this is nothing else than the proof that Goldbach had so laboriously concocted in 1743.

As this example suggests, many of Euler's old conjectures could now be deduced from the above results; some of them are treated in Lagrange's *Recherches;* more are added in a continuation of that memoir published in 1777 (*Lag.*III.759–795|1775). As a further illustration we may quote the case $N = 5$, $\delta = -20$. Here we have the forms $\phi_1 = X^2 + 5Y^2$, $\phi_2 = 2X^2 \pm 2XY + 3Y^2$; therefore, if p is an odd prime divisor of $X^2 + 5Y^2$, either it can be written as $p = a^2 + 5b^2$, in which case $p \equiv 1$ or $9 \pmod{20}$, or $2p$ can be so written, in which case $p \equiv 3$ or $7 \pmod{20}$. This argument fails to show the converse, which had been the other part of Euler's findings about that form, viz., that every prime $p \equiv 1, 3, 7, 9 \pmod{20}$ is a divisor of $X^2 + 5Y^2$. Nevertheless Euler was delighted and took great pleasure in adding several examples to Lagrange's, or rather, perhaps, in having them worked out by his assistants.

What was still beyond the scope of Lagrange's new methods was essentially the law of quadratic reciprocity. In his old age Euler succeeded at least in disengaging it from his early conjectures; it makes up the conclusion of a paper of 1772 on quadratic residues (*Eu*.I-3.510–512 in *E* 552|1772). His formulation for it, in terms of two odd primes p and s, amounts to writing the relation

$$\left(\frac{s}{p}\right) = \left(\frac{\pm p}{s}\right)$$

where the sign is taken so that $\pm p \equiv 1 \pmod 4$ (cf. Appendix I). In1773 he again attracted Lagrange's attention on such matters: "*Je suis fort assuré que la considération de ces circonstances pourra conduire à des découvertes très importantes*", he wrote or rather dictated ("I am sure that this will lead to very important discoveries": *Eu*.IV A-5.498 = *Lag*.XIV.238|1773). In his memoir of 1775 described above, after dealing with Lagrange's reduction theory, he pointed out how this confirmed in part his former guesses and concluded: "Thus all the "theorems" which I formulated long ago in vol.XIV of the old [Petersburg] *Commentarii* have acquired a much higher degree of certainty . . . and there seems to be no doubt that whatever in them is still to be desired will soon receive a perfect proof" ["*Nunc igitur omnia theoremata, quae . . . olim in Comment.veter. Tomo* XIV *dederam, multo maiorem gradum certitudinis sunt adepta . . . atque nullum dubium esse videtur, quin mox, quod in hoc genere adhuc desideratur, perfecta demonstratione muniatur*": *Eu*.I-4.195 in E 598|1775].

He was not far wrong. But mathematics in those days was moving ahead at a more leisurely pace than it does now. Proving the law of quadratic reciprocity still took more than twenty years; and it took Gauss to do it.

§X.

Fermat had stated, not only that every prime $p = 4m + 1$ is a sum of two squares, but also that it can be written thus in only one way; and a converse to this can easily be extracted from his further statements on this subject (cf. Chap.II, §§VIII and IX). Thus those results contain implicitly a criterion for an integer $4m + 1$ to be a prime.

Euler became aware of this long before he could prove it. In 1742 he wrote to Goldbach that $4m + 1$ is prime if and only if it can be written as $a^2 + b^2$ in one and only one way ("*unico modo*": *Corr*.I.134–135), illustrating this by the example of the "Fermat number" $2^{2^5} + 1$ (not a prime, as he knew since 1732; cf. §IV) which can indeed be written as

$$(2^{16})^2 + 1 = 62264^2 + 20449^2.$$

In his first memoir on sums of two squares, Euler qualified his earlier assertion by saying that $4m + 1$ is prime "if and only if it can be written in one and only one way as $a^2 + b^2$ with a prime to b " (*Eu*.I-2.314 in E 228|1749). Thus, if we again call a representation $n = a^2 + b^2$ of an integer n by a sum of two squares a *proper* one if a is prime to b (cf. above, Chap.II, §IX), Euler appears to say that $4m + 1$ is a prime if it has one and only one proper representation as such a sum. Neither one of his formulations is exact; as he must surely have known, $45 = 36 + 9$, $25 = 16 + 9$, $125 = 121 + 4$ offer counterexamples. In the fashion of his time he must have regarded such exceptions as "proving the rule", or rather as occurring only in cases too obvious to deserve mention. To be quite precise, he should have said that an integer $4m + 1$, with $m > 0$, is prime if and only if it has one and only one representation as a sum of two squares, and that representation is proper; in this formulation one must regard $a^2 + 0^2$ as a representation of a^2, but not a proper one.

In that memoir, Euler applied his criterion to 82421, 100981, 262657 which are prime, and 1000009, 233033, 32129 which are not (*loc.cit.*pp.321–327); here he notes that 32129 has the unique representation $95^2 + 152^2$ (but this is not proper), while in the case of 233033, which is not a sum of two squares, he was using a result for which he had no proof as yet, viz., that every prime $p = 4m + 1$ is a sum of two squares; as we have seen, this was proved shortly after that paper was written (cf. above, §§Vb and IX).

Of course, given a large number $n = 4m + 1$, it is impractical to search for its representations $n = a^2 + b^2$ by mere trial and error; shortcuts have to be devised. This is done in two papers ($Eu.$I-3.1–45 = E 283|1760, $Eu.$I-3.112–130 = E 369|1765) devoted chiefly to the search for large primes by means of the above criterion; Euler's idea is to use congruences. For instance, if $n \equiv 2 \pmod 3$, a and b must be prime to 3; if $n \equiv 2 \pmod 5$, a and b must be $\equiv \pm 1$ (mod 5); if $n \equiv 5 \pmod 8$, a or b must be $\equiv 2 \pmod 4$; etc. Using such remarks, Euler is able, in many cases, to reduce the amount of trials to manageable size.

In particular, if $n = a^2 + b^2$ is odd, one of the numbers a, b must be odd and the other even, so that one can write $n = a^2 + 4c^2$. Consequently the primality criterion applies equally well to the form $X^2 + 4Y^2$, provided we agree to call *proper* a representation $n = \mu a^2 + \nu b^2$ of an integer n by the form $\mu X^2 + \nu Y^2$ if and only if a is prime to νb and b is prime to μa. Similarly, assume that $n = a^2 + (3b)^2$ with a prime to $3b$; then n is $\equiv 1 \pmod 3$; if at the same time we have $n = c^2 + d^2$, either c or d must be $\equiv 0 \pmod 3$; from this it follows that the primality criterion is also valid for the form $X^2 + 9Y^2$. Using congruences modulo 8 and modulo 5, one finds in the same way that it applies also to $X^2 + 16Y^2$ and to $X^2 + 25Y^2$. In part, these observations are implicit in the paper of 1765 quoted above.

Eventually Euler found that the reach of his criterion extends well beyond the form $X^2 + Y^2$. One main point in that criterion was that no prime can be a sum of two squares in two different ways; this fact depends upon the result which appears as Theorem 2 in Euler's letter of 1747 to Goldbach ($Corr.$I.416), as Proposition 1 in his memoir of 1749 ($Eu.$I-2.302), and which has been reproduced above as lemma 2 of Chap.II, §VIII. As Euler discovered when he directed his attention to the forms $X^2 + 2Y^2, X^2 + 3Y^2$, most of the theory of the form $X^2 + Y^2$, including that theorem and its proof, can be rather easily extended to those two forms ($Eu.$I-2.459–492 = E 256|1753, $Eu.$I-2.556–575 = E 272|1759; cf. above, §IX, and Chap.II, §XII). In particular, they both give rise to a primality criterion similar to the one discussed above; Euler used this, for instance, for showing that 67579 is prime and that 40081 is not (*Eu*.I-2.490–491).

At some point Euler must have become aware that the proof for the crucial result quoted above (lemma 2 of Chap.II, §VIII) retains its validity, not only for all forms $X^2 + NY^2$, but even for all forms $\mu X^2 + \nu Y^2$, so that in particular, if μ and ν are positive, no integer can be a prime if it can be written as $\mu a^2 + \nu b^2$ in two different ways. In 1778 Euler's assistants wrote up several proofs or sketches of proofs for this, along with prescriptions for factoring an integer written as $\mu a^2 + \nu b^2$ in two different ways (*Eu*.I-3.422–423, *Eu*.I-4.271, *Eu*.I-4.305–310, etc.); they are clumsily written and hardly convincing. Actually Euler's proof of 1747, reproduced in Chap.II, §VIII, can easily be adapted to this general case. Take first $n = \mu a^2 + \nu b^2$; assume that $q = \mu x^2 + \nu y^2$ is a prime divisor of n, not dividing $\mu\nu$. In view of an identity well known to Euler (cf. e.g. Art.179 of his *Algebra*: *Eu*.I-1.424), we have

$$nq = (\mu ax \pm \nu by)^2 + \mu\nu(ay \mp bx)^2.$$

At the same time, q divides the integer

$$ny^2 - b^2q = \mu(ay - bx)(ay + bx);$$

as it is prime to μ, it must divide one of the integers $ay \mp bx$; as q^2 divides nq, this implies that q divides $\mu ax \pm \nu by$. Writing

$$\mu ax \pm \nu by = qu, \quad ay \mp bx = qv,$$

we get

$$\frac{n}{q} = u^2 + \mu\nu v^2.$$

In particular, for $\mu > 0$, $\nu > 0$, $\mu\nu > 1$, this shows that $n = q$ implies $u = \pm 1$, $v = 0$; therefore no prime can be written in two different ways as $\mu a^2 + \nu b^2$, as was to be proved. Of course the same conclusion would be reached more easily nowadays by using some ideal-theory in the field $\mathbf{Q}(\sqrt{-\mu\nu})$, or rather in the ring $\mathbf{Z}[\sqrt{-\mu\nu}]$.

Undoubtedly the observations referred to above, together
with Euler's experience about large primes, must have
prompted him to look for values of μ and ν for which the
last-mentioned result has a converse; such values would
then give rise to primality criteria, generalizing those he
knew already. This amounts to asking for the pairs (μ, ν)
such that every odd integer, prime to $\mu\nu$, is prime whenever
it has one and only one representation as $\mu a^2 + \nu b^2$ and
that representation is proper. This problem must have kept
Euler and his assistants busy for some time, until an answer
was found in 1778, and a list could be drawn up of 65
numbers from 1 to 1848 such that the pair (μ, ν) has the
property in question if the product $\mu\nu$ is one of the integers
in that list; he called such numbers "*numeri idonei*" ("suitable
numbers", i.e. suitable for testing the primality of large
integers). In May 1778 Euler sent a brief account of those
discoveries (*Eu.*I-3.418–420 = E 498|1778) to his friend
and former Berlin colleague, the Swiss mathematician Bé-
guelin, who communicated it to Lagrange; both asked for
more details. These were sent on 30 June 1778 by Euler's
young friend from Basel, Nicolas Fuss (*Eu.*I-3.420–428
= E 708a). Fuss had already sent a similar report to Daniel
Bernoulli in Basel. Bernoulli's less than enthusiastic answer,
of 18 March 1778, was as follows:

"... *Ce que vous me dites tant de votre part que de celle de M.
Euler est sans doute infiniment plus sublime; je veux parler du
beau théorème de M. Euler sur les nombres premiers et de sa nouvelle
méthode pour examiner tel nombre qu'on propose, quelque grand
qu'il puisse être, s'il est premier, ou non. Ce que vous vous êtes
donné la peine de me dire sur cette matière m'a paru fort subtil et
digne de notre grand maître. Mais ne trouvez-vous pas que c'est
presque faire trop d'honneur aux nombres premiers que d'y répandre
tant de richesses, et ne doit-on aucun égard au goût raffiné de
notre siècle? Je ne laisse pas de rendre justice à tout ce qui sort de
votre plume et d'admirer vos grandes ressources pour surmonter
les difficultés les plus épineuses; mais cette admiration se redouble
quand le sujet peut mener à des connaissances utiles. Je range dans
cette classe les profondes recherches dont vous me parlez, sur la*

force des poutres . . . " ["what you tell me, on your behalf and
that of Mr. Euler, is infinitely more exalted, no doubt; I
mean Mr. Euler's beautiful theorem on prime numbers and
his new method for deciding whether any given number,
however large, is a prime or not. What you have taken the
trouble to tell me about this question seems to me very
ingenious and worthy of our great master. But, pray, is it
not doing almost too much honor to prime numbers to
spread such riches over them, and does one owe no deference
at all to the refined taste of our time? I hold in due esteem
whatever comes from your pen and admire your high ability
to overcome the thorniest difficulties; but my admiration
is doubled when the topic leads to some useful piece of
knowledge. That includes, in my view, your deep researches
into the strength of beams . . .": *Corr*.II.676–677].

   This was nothing new to Euler; well he knew how little
interest his contemporaries, with the sole exception of Lagrange, had taken in his arithmetical work (cf. Chap.II,
§XVII). Anyway, on 16 March 1778 (almost the day of Bernoulli's letter to Fuss), he had three papers on his "*numeri
idonei*" and their use presented to the Petersburg Academy
(*Eu*.I-4.269–289 = E 708; *Eu*.I-4.303–328 = E 715; *Eu*.I-
4.360–394 = E 719); one more was added on 20 April
(*Eu*.I-4.395–398 = E 725). They are so ill coordinated with
one another, and some of the formulations and proofs in
them are so confused and defective, that one is tempted to
attribute them to a variety of assistants over whom Euler
was not keeping close control. To try to reconstruct from
these papers a full theory of the "*numeri idonei*" would be
a hopeless task[6]; for our purposes the following brief account
will suffice.

   Euler's list consists of sixty-five integers, out of which

---

   [6] Such a theory can be found in the conscientiously written paper of
F. Grube, *Zeitschr. für Math. u.Phys.*XIX (1874), pp.492–519, along with
an exposition and a detailed criticism of Euler's papers; it is based on
Gauss's theory of quadratic forms. Cf. also J. Steinig, *Elem.d.Math.*XXI
(1966), pp.73–88.

thirty-seven are squarefree, the rest consisting of products of some of those with 4, 9, 16, 25 or 36; as to the latter, specific rules are given by Euler (*Eu.*I-4.280–288, Theor.1 to 9; cf. F. Grube[6], §10). As Gauss was to observe incidentally some twenty years later in the *Disquisitiones* (Art.303), Euler's numbers are precisely those integers N for which every "properly primitive" positive form $aX^2 + 2bXY + cY^2$ with the "determinant" $b^2 - ac = -N$ (i.e. every such form where a, $2b$, c have no common divisor) is "*anceps*", this being tantamount to saying that it is equivalent either to a form $\mu X^2 + \nu Y^2$ with $\mu\nu = N$, or to a form

$$\tfrac{1}{2}\mu(X+Y)^2 + \tfrac{1}{2}\nu(X-Y)^2$$

with $\mu\nu = N$ and $\mu \equiv \nu \pmod 2$. For N squarefree, an equivalent definition would be to say that N, if squarefree, is "*idoneus*" if and only if every prime divisor p of $X^2 + NY^2$, prime to $2N$, is such that either p or $2p$ (or both) can be expressed as $\mu a^2 + \nu b^2$ with $\mu\nu = N$. One may well wonder whether Euler became aware of this last property, which would surely have appealed to him in view of his wide experience with forms such as $\mu X^2 + \nu Y^2$. In any case there can be no doubt that he must have accumulated a good deal of numerical evidence before singling out the "*numeri idonei*" for his purposes.

We shall not attempt to examine the definitions proposed for the "*numeri idonei*" in Euler's papers, nor the criteria described there for deciding which numbers are such; the latter, involving the examination of the integers $N + a^2$ for $a^2 \leqslant 3N$, must have been based on some process of descent, akin perhaps to Lagrange's method of reduction. The fact that the numbers on Euler's list are precisely those which answer Gauss's simple definition shows that, on this occasion also, his instinct was sound.

What particularly pleased Euler about his numbers is that they allowed him to discover quite large new primes; for instance, using the largest one 1848, he could easily establish the primality of $18518809 = 197^2 + 18480000$ (*Eu.*I-3.420, *Eu.*I-4.385–389). He also tried to extend his list beyond 1848, testing, as he tells us, all integers up to 10000 and beyond. To his utter astonishment, he drew a blank; that such a naturally defined set of numbers should apparently be finite, as he now came to conjecture, was a novel experience to him ("*Insigne istud paradoxon . . . circa nullam adhuc aliam seriem observatum esse memini*": *Eu.*I-4.396 in E 725; cf. *Eu.*I-3.420 in E 498). To us it comes as no surprise, since we know that the number of "classes", for a "determinant" $-N$, increases with N faster than the number of decompositions $N = \mu\nu$ of N into two factors. For a formal proof that the number of "*numeri idonei*" is finite, cf. W. E. Briggs and S. Chowla, *Can. J. of Math.* 6 (1954), pp. 463–470. It does not seem to be known whether there is any such number beyond Euler's list.

§XI.

Euler's interest in number theory had first been awakened by Fermat's statement about sums of four squares (cf. above, §IV); but he had sensibly postponed taking this matter up until he had made sufficient progress with sums of two squares. From 1747 onwards, we find more and more references to sums of three and of four squares in his correspondence (cf. §V*b*); but no real progress could be made until he discovered (perhaps by sheer guesswork: cf. *Eu.*I-6.312 in E 407|1770) the famous identity for products of sums of four squares: if

$$m = a^2 + b^2 + c^2 + d^2, \quad n = p^2 + q^2 + r^2 + s^2,$$

then

$$mn = A^2 + B^2 + C^2 + D^2$$

where one has put

$$A = ap + bq + cr + ds, \quad B = aq - bp - cs + dr,$$
$$C = ar + bs - cp - dq, \quad D = as - br + cq - dp;$$

obviously, then, one has also:

$$\frac{m}{n} = \left(\frac{A}{n}\right)^2 + \left(\frac{B}{n}\right)^2 + \left(\frac{C}{n}\right)^2 + \left(\frac{D}{n}\right)^2$$

(*Corr.*I.452|1748; *Eu.*I-2.368–369 in E 242|1751).

In the following year, after reaching complete success with sums of two squares (cf. above, §§Vb and IX), Euler tried to apply the same method, combined with the above identity, to sums of four squares (*Corr.*I.495–497|1749, 521|1750, 527|1750, and *Eu.*I-2.338–372 = E 242|1751); clearly this required the following two steps.

Firstly he had to show that every prime divides a sum of four squares. As to this he proved more, showing that, if p is any prime, the congruence $x^2 + y^2 + z^2 \equiv 0 \pmod{p}$ has a non-trivial solution, i.e. one other than $(0, 0, 0)$; later on, in his definitive paper on the same subject (*Eu.*I-3.218–239 = E 445|1772), he even extended this to an arbitrary congruence $\lambda x^2 + \mu y^2 + \nu z^2 \equiv 0 \pmod{p}$, where λ, μ, ν are prime to p; in fact, as he observed in substance, the function $x \to x^2 \pmod{p}$ takes $(p+1)/2$ distinct values, viz., 0 and the quadratic residues modulo p; the same must therefore be true of the functions $x \to \lambda x^2$ and $x \to -\mu x^2 - \nu$, so that those two sets must have at least one element in common.

Secondly it had to be shown that any prime which divides a sum $a^2 + b^2 + c^2 + d^2$ (where a, b, c, d have no common divisor, while some of them may be 0) is itself such a sum. Here Euler was unable to adapt the proof he had used for sums of two squares; all he could do was to show that every prime, and consequently every integer, is at any rate a sum of four rational squares. In fact, assume that this is not so, and let p be the smallest prime which cannot be expressed as such a sum. Let (a, b, c) be a non-trivial solution of $x^2 + y^2 + z^2 \equiv 0 \pmod{p}$; here one may assume that a, b, c are ≥ 0 and $< p/2$. Put

$$N = a^2 + b^2 + c^2 = pN',$$

so that $N < \frac{3}{4}p^2$, $N' < p$. As all prime factors of N' are $<p$, they are sums of four rational squares, and so is N' in view of Euler's identity; for the same reason, $p = N/N'$ is then a sum of four rational squares. Between that result and Fermat's theorem, there was still, as Euler realized, a wide gap which he found himself unable to bridge (cf. Appendix II).

There the matter rested for more than twenty years, until, in 1772, he received from Lagrange the memoir where Fermat's assertion was finally proved in full (*Lag*.III.189–201; cf. *Eu*.IV A-5.492); that proof was based on Euler's work on sums of two squares. Euler must soon have taken up the matter again; in a brilliant paper, presented to the Petersburg Academy on 21 September 1772 and first published in the *Nova Acta Eruditorum* of 1773 (*Eu*.I-3.218–239 = E 445), he begins by congratulating Lagrange on his achievement, then, rightly describing Lagrange's proof as "far-fetched and laborious" ("*nimis longe repetita et vehementer operosa*": *loc.cit.*p.218), he proceeds to give a new and elegant variant of the proof for sums of two squares (*ibid.*pp.222–224), showing finally that it applies equally well, not only to the forms $X^2 \pm 2Y^2$ (*ibid.*pp.225–226 resp. *Eu*.I-3.275 in E 449|1772) and $X^2 + 3Y^2$ (*ibid.*pp.226–227), but also to $X^2 + Y^2 + Z^2 + T^2$ (*ibid.*,pp.230–231). In the latter case, the proof is as follows.

Let p be an odd prime divisor of Σa_i^2, where the a_i, for $i = 1,2,3,4$, are integers without a common divisor which may be assumed to be ≥ 0 and $<p/2$. Putting $\Sigma a_i^2 = pm$, we have $m < p$. If $m = 2$, two of the a_i, say a_1 and a_2, must be odd, the others being even; then we have

$$p = \left(\frac{a_1+a_2}{2}\right)^2 + \left(\frac{a_1-a_2}{2}\right)^2 + \left(\frac{a_3+a_4}{2}\right)^2 + \left(\frac{a_3-a_4}{2}\right)^2,$$

and p is a sum of four squares; thus we may assume $m > 2$. For each i, put $a_i = b_i + mc_i$ with $|b_i| \leq m/2$; then $\Sigma b_i^2 \equiv 0$ (mod m), and we can write $\Sigma b_i^2 = mn$. As the a_i have no common divisor, not all the b_i can be 0, and not all can be $\pm m/2$, so that $0 < \Sigma b_i^2 < m^2$ and $0 < n < m$. Now apply Euler's identity to Σa_i^2 and Σb_i^2; this gives $m^2 pn = \Sigma A_i^2$, with $A_1 = \Sigma a_i b_i$, and A_2, A_3, A_4 given by

$$A_2 = a_1 b_2 - a_2 b_1 - a_3 b_4 + a_4 b_3 = m(c_1 b_2 - c_2 b_1 - c_3 b_4 + c_4 b_3)$$

and two similar formulas for A_3 and A_4. As $m^2 pn$ and A_2^2, A_3^2, A_4^2 are multiples of m^2, so is A_1^2; therefore we can put $A_i = mB_i$ for all i, and we get $pn = \Sigma B_i^2$. Call d the g.c.d. of the B_i, and put $B_i = da_i'$ for each i; then d^2 divides pn. As $0 < n < m < p$, d must be prime to p, and d^2 must divide n; put $n = d^2m'$. Thus we have $pm' = \Sigma a_i'^2$; this is a sum of four squares with no common divisor which is a multiple of p smaller than pm. Proceeding in this manner, one gets p itself expressed as a sum of four squares, as was to be proved.

More than a century later, Hurwitz was to write out a proof for the same theorem, based on the non-commutative ring of integral quaternions and on the fact that this ring has a euclidean algorithm (cf. his *Math. Werke*, II.303–330); whether he was aware of it or not, this was in substance nothing else than a transcription of Euler's proof in the language of quaternions.

§XII.

Euler's interest in "Pell's equation" $x^2 - Ny^2 = 1$ had been aroused in 1730, apparently by his correspondence with Goldbach (cf. above, §IV). Already at that time he had observed that the successive solutions make up "recurrent" sequences "conflated out of two geometric progressions" (*Corr.*I.37|1730; cf. *infra*, §XIII); in the case $N = 8$, for instance, this means that the successive integers $y_0 = 0, y_1 = 1, y_2 = 6$, etc., whose squares are "triangular numbers" (i.e. of the form $\frac{1}{2}n(n+1)$, or $\frac{1}{8}(x^2-1)$ if one puts $x = 2n+1$) are given by

$$y_n = \frac{(3+2\sqrt{2})^n - (3-2\sqrt{2})^n}{4\sqrt{2}}$$

(*Corr.*I.36) and satisfy the recurrence relation

$$y_{n+1} = 6y_n - y_{n-1}$$

(*Corr.*I.30). Soon he was publishing a short paper on the subject (*Eu.*I-2.6–17 = E 29|1733) devoted chiefly to the more general equation $y^2 = ax^2 + bx + c$ (cf. *infra*, §XIII), but including a description of Brouncker's method of so-

lution for "Pell's equation" (cf. Chap.II, §XIII), which he
attributes there, as he always did, to Pell and also, for no
apparent reason, to Fermat; he illustrates it by treating the
example $x^2 - 31y^2 = 1$ (*loc.cit.*p.13) and adds the important
observation that the solutions of $p^2 - Nq^2 = 1$ provide good
rational approximations p/q for $\sqrt{N}$ (*ibid.*p.15); this had already been known to Archimedes (cf. Chap.I, §VIII).

As to continued fractions, they first occur in Euler's correspondence in an altogether different context (*Corr.*I.58–
59|1731), in connection with Riccati's differential equation;
soon, however, he becomes interested in them for their own
sake (cf. *Eu.*I-14.187–215 = E 71|1737), observing inci-
dentally that rational numbers have finite continued fractions
(obtained by a process identical with the euclidean algorithm),
that periodic continued fractions represent quadratic irrationalities, and noting also that the expansion of any real
number into a continued fraction supplies the best rational
approximations for that number. Some of this had been
known to Huygens (cf. above, Chap.II, §XVII), but Euler
was clearly unaware of Huygens's work on that subject.

Since Brouncker's solution of Pell's equation $x^2 - Ny^2 = 1$,
and the continued fraction for $\sqrt{N}$, both supply good
rational approximations for $\sqrt{N}$, it had to occur to Euler to
compare them with one another. When he did so (*Eu.*I-
3.73–111 = E 323|1759), he discovered that the two al-
gorithms are in fact identical.

To verify that fact, consider Brouncker's method (the
"*methodus Pelliana*", as Euler always called it) as it has been
described in Chap.II, §XIII. It consisted in constructing a
sequence F_0, F_1, ... of quadratic forms

$$F_i(X_i, X_{i+1}) = A_i X_i^2 - 2B_i X_i X_{i+1} - C_i X_{i+1}^2,$$

beginning with

$$F_0(X_0, X_1) = NX_0^2 - (nX_0 + X_1)^2$$

and defined as follows. Denoting by $[\xi]$, for every real number
ξ, the integer m such that $m \leq \xi < m + 1$, one takes $n = [\sqrt{N}]$; then, calling ξ_i, for each $i \geq 0$, the positive root of

$F_i(\xi, 1)$, and putting $m_i = [\xi_i]$, one takes for F_{i+1} the transform of $-F_i$ by the substitution $X_i = m_i X_{i+1} + X_{i+2}$. This gives, for all $i \geqslant 0$:

$$B_i^2 + A_i C_i = N, \quad \xi_i = \frac{B_i + \sqrt{N}}{A_i}, \quad m_i = \left[\frac{B_i + n}{A_i}\right], \quad C_{i+1} = A_i.$$

At the same time, in view of the definition of the F_i, we have

$$\sqrt{N} = n + \frac{1}{\xi_0}; \qquad \xi_i = m_i + \frac{1}{\xi_{i+1}} \qquad (i = 0, 1, 2, ...)$$

so that $\sqrt{N}$ is given by the "continued fraction"

$$\sqrt{N} = n + 1/(m_0 + 1/(m_1 + 1/(m_2 + \cdots))).$$

Thus Brouncker's method includes the construction of the continued fraction for $\sqrt{N}$.

In Euler's presentation, on the other hand, one has recurrence formulas (*loc.cit.*p.82) which, in our notation, can be written thus:

$$m_i = \left[\frac{B_i + n}{A_i}\right], \qquad B_{i+1} = A_i m_i - B_i,$$

$$A_{i+1} = \frac{N - B_{i+1}^2}{A_i} = A_{i-1} + m_i(B_i - B_{i+1}),$$

with the initial values $B_0 = n, A_0 = N - n^2$; they are at once seen to be equivalent to those given above.

Once Brouncker's method has been re-interpreted in this way, the facts proved in Chap.II, §XIII, have some further consequences. Firstly they imply that the continued fraction for $\sqrt{N}$ is periodic; more precisely, there is $p > 0$ such that $m_{p+i} = m_i$ for all $i \geqslant 0$; this occurs when p is taken so that $F_p = F_0$; then $A_{p-1} = C_p = 1$. Write now $F_i = (A_i, B_i, C_i)$, as in Chap.II, §XIII, and put again $G_i = (C_i, B_i, A_i)$; it has been shown there that G_i can be derived from G_{i+1} just as F_{i+1} was derived from F_i, with the same value for m_i; this amounts to saying that if m_i is as before, and if η_{i+1} is the positive root of G_{i+1}, we have

$$m_i = [\eta_{i+1}] = \left[\frac{B_{i+1} + n}{C_{i+1}} \right].$$

As we have $F_p = F_0 = (N - n^2, n, 1)$, we have

$$G_p = (1, n, N - n^2), \qquad \eta_p = n + \sqrt{N}, \qquad m_{p-1} = 2n,$$

so that $F_{p-1}(X_{p-1}, X_p)$ is the transform of $-F_p(X_p, X_{p+1})$ by the substitution $X_{p+1} = X_{p-1} - 2nX_p$; this gives $F_{p-1} = G_p$. This shows that G_{p-1} is the same as F_0, therefore G_{p-2} the same as F_1, and, by induction on i, G_{p-i} the same as F_{i-1}; from this we get $\eta_{p-i} = \xi_{i-1}$, $m_{p-i-1} = m_{i-1}$, for $1 < i < p$. This is the so-called "palindromic" property of the sequence (m_i), by which is meant that we have:

$$(m_0, m_1, m_2, ..., m_{p-2}) = (m_{p-2}, m_{p-3}, m_{p-4}, ..., m_0).$$

While Euler drew attention to these properties of the continued fractions for square roots $\sqrt{N}$, as well as to their use in solving Pell's equation, there is no sign that he ever sought to back up his findings by anything more than experimental evidence. He did mention that the values obtained by his process for the integers B_i, A_i, m_i are necessarily bounded (with $B_i \le n$, $A_i \le B_i + n$, $m_i = 2n$ or $m_i \le n$ for all i: *loc.cit.* p.83); from this he could at least have derived the conclusion that the sequence (m_i) is periodic from a certain point onwards, but he failed to mention this, or did not bother to do so.

Much later he came back to this topic (*Eu.*I-3.310– 334 = E 454|1772), extending his observations to the continued fractions for other quadratic irrationalities and in particular for square roots $\sqrt{r}$ where r is any rational number >1; as to the latter, one can in fact prove the periodicity and the palindromic character of the continued fraction just as above, by taking as starting point for the sequence (F_i) the quadratic form

$$F_0(X_0, X_1) = bX_0^2 - a(nX_0 + X_1)^2$$

where one has put $r = b/a$ and $n = [\sqrt{r}]$. In that same paper Euler also applied continued fractions to the determination of the minimum value of $|F(a, b)|$ for integral values of a and b, not both 0, when F is any indefinite binary quadratic form.

Just for once he was coming too late. In 1768, taking up number theory for the first time, Lagrange had proved, still somewhat clumsily, all the basic theorems on Pell's equation (*Lag*.I.671–731); not much later he gave what may be regarded as a definitive treatment of the subject, based on the continued fraction algorithm (*Lag*.II.494–496); it is essentially the one we have described in Chap.II, §XIII; it is contained in a long memoir (*Lag*.II.377–535|1768) published in Berlin in 1769, which reached Euler in 1769 or early in 1770 (cf. *Eu*.IV A-5.466), along with a number of other contributions to which Euler must have paid more attention. Lagrange's solution, along with some further results on the same subject, was inserted by him in his *Additions* (composed in 1771, but published only in 1773; cf. *infra*, Chap.IV, §II) to Euler's *Algebra*; there he also gave a detailed treatment of the same minimum problem for quadratic forms which Euler was to discuss in his paper E 454. This, too, received scant attention from Euler (cf. Chap.IV, §III, and above, §IX). When in his later years Euler came back to the topic of Pell's equation (*Eu*.I-4.76–90 = E 559|1773), he added nothing of substance to what by that time was already public knowledge on that subject.

§XIII.

Already in 1730, and in all his subsequent work on "Pell's equation" (cf. above, §XII), Euler used to discuss that topic within the broader framework of the diophantine equations of degree 2 in two unknowns, to be solved either in integers or, as Diophantus had done, in rational numbers. For many years he confined his attention to the type $y^2 = \alpha x^2 + \beta x + \gamma$ which had been studied by Diophantus (cf. Chap.I,

§X). In 1759 (or perhaps in1763, since it seems that the manuscript was revised once) he indicates briefly that he has also extended his investigations to the more general type

$$F(x, y) = Ax^2 + 2Bxy + Cy^2 + 2Dx + 2Ey + F = 0$$

(*Eu.*I-3.76 in E 323|1759); but then the topic is dropped until 1772, when that equation is taken up for its own sake in a paper (*Eu.*I-3.297–309 = E 452|1772) perhaps intended as a supplement to Lagrange's work on the same subject (cf. Chap.IV, §II).

Except perhaps on one occasion, also in 1772 (cf. above, §IX), Euler always made clear that he could do nothing with a problem of that type unless at least one solution is somehow known beforehand (mostly "by guesswork", "*quasi divinando*": *Eu.*I-2.610). In the geometric interpretation which had been known to Newton (cf. above, Chap.II, §XV), the method for solving rationally an equation $F(x, y) = 0$ of degree 2, once a solution (a, b) is given, is to intersect the conic $F = 0$ with an arbitrary straight line going through (a, b); as we have seen (Chap.I, §X), this goes back to Diophantus. Euler does not mention that interpretation, but it may well have been at the back of his mind when he described that method in algebraic terms, at first in 1758 for the case $y^2 = \alpha x^2 + \beta x + \gamma$, then in 1772 in the general case. Taking (a, b) to be a solution, and intersecting the curve $F = 0$ with the line

$$\frac{x - a}{p} = \frac{y - b}{q}$$

(*Eu.*I-2.578 in E 279|1758; *Eu.*I-3.299 in E 452|1772), he obtains a point (a', b') given by formulas

$$a' = \lambda a + \mu b + \sigma,$$
$$b' = \nu a + \rho b + \tau,$$

where the coefficients are rational functions of p and q. In modern language, it is easily verified that the linear substitution

$$S = \begin{pmatrix} \lambda & \mu & \sigma \\ \nu & \rho & \tau \end{pmatrix}$$

defined by

$$(x, y) \rightarrow (\lambda x + \mu y + \sigma, \nu x + \rho y + \tau)$$

transforms F into itself; consequently its "homogeneous part"

$$S_0 = \begin{pmatrix} \lambda & \mu \\ \nu & \rho \end{pmatrix},$$

i.e. the substitution

$$(X, Y) \rightarrow (\lambda X + \mu Y, \nu X + \rho Y)$$

transforms into itself the homogeneous part

$$F_0(X, Y) = Y^2 - \alpha X^2 \text{ resp. } F_0(X, Y) = AX^2 + 2BXY + CY^2$$

of F. In other words, Euler's formulas determine automorphisms of the form F_0.

As Euler knew, nothing need be added to the above formulas if the question is to obtain all rational solutions of $F = 0$; but finding integral solutions is another matter. As to this, his method (already in 1733: Eu.I-2.7 in E 29|1733; cf. $Corr$.I.36|1730) can be described as trying to construct a linear substitution Σ with integral coefficients, transforming F into itself; if this can be done, then Σ, applied repeatedly to one solution in integers, may be expected to produce, if not all, at least infinitely many such solutions, provided Σ is not of finite order. Incidentally, the substitution S defined above cannot be used for that purpose, since it is of order 2; the failure to perceive this circumstance (a failure one may ascribe to Euler's assistants rather than to himself, since he had not committed it in 1758; cf. Eu.I-2.582, Coroll.2, in E 279|1758) vitiates the first part of his paper of 1772 (Eu.I-3.301–303).

In modern language, if Σ is to be such as Euler requires, its homogeneous part Σ_0 must belong to the group Γ_0 of the automorphisms of F_0, i.e. of the substitutions with integral coefficients and determinant $\lambda\rho - \mu\nu = 1$ which transform F_0 into itself; given such a Σ_0, there is a unique Σ (not necessarily with integral coefficients) which transforms F into itself. Following Euler, we exclude the essentially trivial case where the "determinant" $\Delta = B^2 - AC$ of F_0 is 0, i.e. where the conic $F = 0$ is a parabola. Then Σ is deduced from Σ_0 by adjusting σ, τ so that Σ leaves invariant the center of the conic $F = 0$.

Take first the case of the equation $y^2 = \alpha x^2 + \beta x + \gamma$; with Euler we assume, not only that α is not 0, but also that it is >0 and not a square, since otherwise the problem has only finitely many solutions which can be found by trial and error. Then Euler's calculations (*Eu*.I-2.8–9 in E 29|1733) show that the group Γ_0 of the automorphisms of $F_0 = Y^2 - \alpha X^2$ consists of the substitutions

$$\Sigma_0 = \begin{pmatrix} p & q \\ \alpha q & p \end{pmatrix}$$

where (p, q) is any solution of "Pell's equation" $p^2 - \alpha q^2 = 1$; as he was to observe (*Eu*.I-2.600–601 in E 279|1758), this fact is closely related to "Brahmagupta's identity", which he may well have first noticed precisely in that context (cf. above, §VIII).

With Σ_0 as shown above, the corresponding substitution Σ is given by

$$\Sigma = \begin{pmatrix} p & q & (p-1)\beta/2\alpha \\ \alpha q & p & \beta q/2 \end{pmatrix};$$

iterated, this gives:

$$\Sigma^2 = \begin{pmatrix} p^2+\alpha q^2 & 2pq & \beta q^2 \\ 2\alpha pq & p^2+\alpha q^2 & \beta pq \end{pmatrix}$$

which has integral coefficients. Thus, using either the group generated by Σ, or at any rate the group generated by Σ^2, one can always, from one solution in integers, derive infinitely many others. Lagrange was later to prove, in substance, that all the solutions can be derived thus from only finitely many (*Lag*.II.512–522, 704–705); Euler never raised this question. What aroused his interest was a different matter: what is the nature of the sequence (a_n, b_n) generated by applying a substitution such as Σ, and its iterates Σ^n, to a given solution (a_0, b_0)? The answer, as he had discovered quite early (*Eu*.I-2.10 in E 29|1733; cf. *Corr*.I.30,36|1730) was that (a_n), (b_n) are "recurrent sequences".

This concept had been studied in some detail by Daniel Bernoulli in a paper written in 1728 or 1729 (and published in 1732: *D.Bern*.II.49–64), following work on the same subject by A. de MOIVRE and others including Daniel's cousin Nicolas Bernoulli and Goldbach (cf. *ibid*.p.49). Euler must of course have been familiar with Daniel's paper from the time of its composition; he was to give a masterly exposition of the whole topic in Chapters IV, XIII and XVII of his *Introductio* of 1748 (*Eu*.I-8). An infinite sequence $(A_0, A_1, A_2, ...)$ is called "recurrent" if it satisfies a linear relation

$$A_{n+m} = L(A_{n+m-1}, A_{n+m-2}, ..., A_n)$$

for all $n \geqslant 0$; this is equivalent to saying that its "generating series" $\Sigma_0^\infty A_n x^n$ is the expansion of a rational function $P(x)/Q(x)$ at $x = 0$, with $Q(x) = 1 - L(x, x^2, ..., x^m)$, $P(x)$ being a polynomial of degree $< m$; that statement has to be slightly modified if L, instead of being homogeneous, has a constant term. In particular, if Q has no multiple root and has the simple roots $\xi_1, \xi_2, ..., \xi_m$, one can write

$$\frac{P(x)}{Q(x)} = \sum_1^m \frac{\lambda_\mu}{1 - \xi_\mu x}$$

and consequently

$$A_n = \sum_1^m \lambda_\mu \xi_\mu^n$$

so that, in Euler's words, the sequence (A_n) can be said to be "conflated out of m geometric progressions" (cf. above, §XII).

Now, coming back to the equation $y^2 = \alpha x^2 + \beta x + \gamma$, take first, as Euler does (*Eu*.I-2.589–590 in E 279|1758) the case $\beta = 0$. Then the solutions (a_n, b_n) derived from a given one (a_0, b_0) are obtained by applying to it the iterates Σ_0^n of the substitution Σ_0 defined above. This gives, for $n \geqslant 0$:

$$a_{n+1} = pa_n + qb_n, \qquad b_{n+1} = \alpha qa_n + pb_n,$$
$$a_{n+2} = pa_{n+1} + qb_{n+1}, \qquad b_{n+2} = \alpha qa_{n+1} + pb_{n+1},$$

and therefore, since $p^2 - \alpha q^2 = 1$:

$$a_{n+2} = 2pa_{n+1} - a_n, \qquad b_{n+2} = 2pb_{n+1} - b_n.$$

From the theory of recurrent sequences ("*ex doctrina serierum recurrentium*": *ibid.*p.584) it follows now that they can be written in the form

$$a_n = (r+s\sqrt{\alpha})(p+q\sqrt{\alpha})^n + (r-s\sqrt{\alpha})(p-q\sqrt{\alpha})^n$$

with suitable rational r and s; there is a similar formula for b_n. For $\beta \neq 0$ one gets a corresponding result (*ibid.* p.584–585).

As to the general case of an equation $F(x, y) = 0$ of degree 2, Euler assumes that the "determinant" $\Delta = B^2 - AC$ of F_0 is >0 and not a square; otherwise, as he probably knew, the problem has only finitely many solutions. Then, with the same notations as above, it turns out that the form F_0 admits the automorphism

$$\Sigma_0 = \begin{pmatrix} p + Bq & Cq \\ -Aq & p - Bq \end{pmatrix}$$

where (p, q) is a solution of $p^2 - \Delta q^2 = 1$; here again this can be extended to an automorphism Σ of F itself (*Eu.*I-3.76 in E 323|1759; *Eu.*I-3.306 in E 452|1772) where the coefficients σ, τ of Σ appear as fractions with the denominator Δ, and need not be integers. Did Euler know that in this case too the coefficients of Σ^2 are integers? This is not mentioned. On the other hand, he proves at length that the successive transforms (a_n, b_n) of a given solution (a_0, b_0) by the iterates of Σ are expressed by formulas of the type

$$a_n = (r+s\sqrt{\Delta})(p+q\sqrt{\Delta})^n + (r-s\sqrt{\Delta})(p-q\sqrt{\Delta})^n - 2r + a_0.$$

This time, thanks to a newly acquired technique (cf. *infra*, §XIV), the proof makes no appeal to "the theory of recurrent series" (*Eu*.I-3.306–309 in E 452|1772).

§XIV.

Already Fermat had applied the theory of quadratic forms to diophantine problems (cf. Chap.II, §XII). This same idea was bound to occur to Euler when he took up the study of Fermat's observations on Diophantus, around 1748 (cf. above, §IV). One first instance of this is given by a paper (*Eu*.I-2.223–240 = E 167|1748; cf. above, §IV) where he makes use of results on the form $X^2 - 2Y^2$ (which he was far from being able to prove at the time) in order to deal with a diophantine problem proposed by Fermat. At that same time his attention was attracted to Fermat's statement on $x^n + y^n = z^n$ (*Corr*.I.445|1748); in 1753 he could announce to Goldbach that he had solved the case $n = 3$ (*Corr*.I.618). As he indicated some years later (*Eu*.I-2.557 in E 272|1759), this was based on the theory of the form $X^2 + 3Y^2$, which he had not yet completed but with which he had already made decisive progress (cf. *Eu*.I-2.481–482, Scholion, in E 256|1753, and above, §IX). Undoubtedly his proof for the case $n = 3$ of Fermat's equation must have been in substance the one which has been described in Chap.II, §XVI. Again in 1754 he applied the theory of the same form to the equation $x^3 + y^3 + z^3 = t^3$ (*Eu*.I-2.428–458 = E 255|1754; cf. also *Eu*.I-2.557); his main result there, translated into modern language, is that the surface defined by that equation in projective space is rational; from our point of view it belongs to algebraic geometry.

His proof of 1753 for $x^3 + y^3 = z^3$ was never published; but, by the time he decided to insert it into the last section of his *Algebra* (cf. above, §Vc), he had worked out a new technique for dealing with quadratic irrationalities and using them in the theory of binary quadratic forms. Similar ideas, and even more far-reaching ones, had just occurred to Lagrange in connection with his algebraic work (cf. *Lag*.II.522–

537). This is how Euler wrote to Lagrange on discovering this coincidence:

"J'ai aussi fort admiré votre méthode d'employer les nombres irrationels et même les imaginaires dans cette espèce d'analyse, qui est uniquement attachée aux nombres rationels. Il y a dejà quelques années que j'ai eu de semblables idées; . . . ayant publié ici une Algebre complette en langue russe, j'y ai developpé cette matière fort au long, où j'ai fait voir que, pour résoudre l'équation

$$xx + nyy = (pp + nqq)^\lambda,$$

on n'a qu'à résoudre celle-ci:

$$x + y\sqrt{-n} = (p + q\sqrt{-n})^\lambda,$$

Cet ouvrage s'imprime actuellement aussi en allemand . . . Mais je n'y ai pas poussé mes recherches au delà des racines quarrées; et l'application aux racines cubiques et ultérieures vous a été réservée uniquement. C'est de là que j'ai tiré cette formule très remarquable

$$x^3 + ny^3 + nnz^3 - 3nxyz$$

dont les trois facteurs sont $x + y\sqrt[3]{n} + z\sqrt[3]{n^2}$; d'où l'on voit qu'on peut toujours aisément déterminer les lettres x, y et z pour que cette formule devienne un quarré, ou un cube . . . ou quelque plus haute puissance" ["I have greatly admired your method of using irrational and even imaginary numbers in this kind of analysis which deals with nothing else than rational numbers. Already for several years I have had similar ideas; . . . having published here a complete Algebra in Russian, I have developed this matter quite at length, showing that, in order to solve

$$x^2 + ny^2 = (p^2 + nq^2)^\lambda,$$

it is enough to solve

$$x + y\sqrt{-n} = (p + q\sqrt{-n})^\lambda.$$

A German translation is being printed . . . But I did not go beyond square roots; and the application to cubic and higher roots is yours only. That is where I found the remarkable formula

$$x^3 + ny^3 + n^2z^3 - 3nxyz$$

whose factors are $x + y\sqrt[3]{n} + z\sqrt[3]{n^2}$; from which one sees that x, y, z can always easily be determined so as to make that formula equal to a square, a cube . . . or any higher power": *Eu.*IV A-5.467|1770; cf. Art.191 of his *Algebra*, *Eu.*I-1.431]. Similarly, commenting upon the same method in 1772, he wrote: "*Haec methodus . . . eo magis est notatu digna, quod ex doctrina irrationalium est petita, cuius alioquin nullus videtur esse usus in Analysi Diophantea. Eximium autem huius doctrinae usum iam pridem in Algebra mea Ruthenice et Germanice edita fusius ostendi*" ["This method is all the more remarkable since it comes out of the theory of irrationals which otherwise seems to be of no use in diophantine analysis. It has such use, however, and an eminent one, as I have shown more at length some time ago in the Algebra I have published in Russian and in German": *Eu.*I-3.309 in E 452|1772].

So far as Euler was concerned, the new method consisted in splitting a quadratic form $x^2 + ny^2$ into its factors $x \pm y\sqrt{-n}$, or a form $ax^2 + cy^2$ into the factors $x\sqrt{a} \pm y\sqrt{-c}$. As long as this is done only in order to derive algebraic identities, the use of this method is of course unimpeachable. Such is the case when it is applied to the proof of "Brahmagupta's identity" (Art.176 of the *Algebra*, *Eu.*I-1.422; cf. Chap.I, §VIII) or the more general identity

$$(ap^2 + cq^2)(r^2 + acs^2) = a(pr \pm cqs)^2 + c(qr \mp aps)^2$$

(*ibid.*, Art.178, p.424). In the same way, by putting

$$x\sqrt{a} \pm y\sqrt{-c} = (p\sqrt{a} \pm q\sqrt{-c})^3$$

(*ibid.*, Art.188, p.429), Euler concludes rightly that, for

$$x = ap^3 - 3cpq^2, \qquad y = 3ap^2q - cq^3,$$

the number $ax^2 + cq^2$ becomes a cube, viz., $(ap^2 + cq^2)^3$.

What he is after with such identities, however, and what he needs for his treatment of the equation $x^3 + y^3 = z^3$ (cf. *ibid.*, Art.247, pp.492, 493), is the converse, a converse which he has no hesitation in spelling out. According to him, if the form $aX^2 + cY^2$ does not split into rational linear factors (i.e. if $-ac$ is not a square), and if $ax^2 + cy^2$ is a cube while

x is prime to y, then there must be two integers p, q such that x and y are given by the above formulas "because then $x\sqrt{a} + y\sqrt{-c}$ and $x\sqrt{a} - y\sqrt{-c}$ have no common factor and so must be cubes" (*ibid.*, Art.191, p.431). For instance, if $x^2 + 3y^2$ is a cube while x is prime to y, then there are integers p, q such that

$$x = p^3 - 9pq^2, \qquad y = 3p^2q - 3q^3$$

(*ibid.*, Art.189, p.430).

This last assertion is true; it is no other than the content of our lemma 5 in Chap.II, §XII; it must have been known to Fermat (cf. Chap.II, §§XII and XVI) and is at any rate implicit in Euler's theory of the form $X^2 + 3Y^2$ as contained in his memoir E 272|1759 (*Eu*.I-2.556–575) which is expressly designed to supply the foundation for Euler's earlier treatment of the equation $x^3 + y^3 = z^3$; in fact, the result in question is all that is needed for that purpose. In his seeking to justify it by his new "method", his instinct had not wholly misled him, but this was due to the fact that the field $\mathbf{Q}(\sqrt{-3})$ admits a euclidean algorithm (cf. Chap.II, Appendix I), so that the laws of ordinary arithmetic retain their validity for the integers in that field. In this matter, too, it took Gauss to set things right.

Nevertheless a momentous event had taken place. Algebraic numbers had entered number theory — through the back-door.

§XV.

What we now call "elliptic curves" (i.e. algebraic curves of genus 1) were considered by Euler under two quite different aspects, without his ever showing an awareness of the connection between them, or rather of their substantial identity.

On the one hand, he must surely have been familiar, from the very beginning of his career, with the traditional methods for handling diophantine equations of genus 1 (cf. Chap.I, §X, Chap.II, §XV, and above, §V*d*). Also in connection with such equations he eventually came to reconstruct some of Fermat's proofs by descent (cf. Chap.II, §XVI, and above, §§V*d* and XIV). Finally, late in life, he took up such problems more systematically, including the "double equations" which had been a favorite both for Diophantus and for Fermat, and the equations $\Phi(x,y) = 0$ of degree 2 in x and 2 in y which were perhaps his most original contribution to the whole subject.

On the other hand he had inherited from his predecessors, and notably from Johann Bernoulli, a keen interest in what we know as "elliptic integrals" because the rectification of the ellipse depends upon integrals of that type; they were perceived to come next to the integrals of rational functions in order of difficulty. That the latter can always be integrated by means of logarithmic and inverse trigonometric functions, and of course of rational functions, came gradually to be recognized in the seventeenth and eighteenth centuries until it was accepted as an established fact (cf. *Corr.*I.50–57 and 170–171); eventually Euler derived it from the so-called "fundamental theorem of algebra" (*Eu.*I-17.70–194 = E 162,163|1748; cf. above, §VII). Functions depending rationally upon x and a square root $\sqrt{ax^2 + bx + c}$ could then be integrated by the "diophantine method", i.e. by the same change of variable

$$t = \sqrt{ax^2 + bx + c} - x\sqrt{a}, \qquad x = \frac{t^2 - c}{b - 2t\sqrt{a}}$$

which is used for the solution of the diophantine equation $y^2 = ax^2 + bx + c$ in rational numbers when a is a square (cf. *Corr.*II.190 and *J.Bern.*III.393). Consequently, when Euler, in his younger days, tried (vainly, of course) to "integrate" the differential $dx/\sqrt{1 - x^4}$ (*Corr.*I.47|1730), he must have been aware that this was almost the same problem as finding rational solutions for $y^2 = 1 - x^4$, or (what amounts to the same) integral solutions for $Z^2 = X^4 - Y^4$; thus, when he reconstructed Fermat's proof for the

impossibility of the latter equation (*Eu.*I-2.38–58 = E
98|1738), this must have helped to persuade him that the
differential in question could not be integrated in terms of
known functions. In the same year, in fact, while sending
to his old teacher Johann Bernoulli his latest discovery about
integrals of that type, he expressed himself as follows:

"*Observavi nuper insignem elasticae rectangulae proprietatem,*

$$in\ qua\ si\ abscissa\ ponatur\ x,\ est\ applicata = \int \frac{xx dx}{\sqrt{(a^4 - x^4)}}\ et\ longitudo$$

$$curvae = \int \frac{aa dx}{\sqrt{(a^4 - x^4)}}\ quae\ expressiones\ ita\ sunt\ comparatae,\ ut$$

inter se comparari nequeant. At si abscissa sumatur = a, *inveni
rectangulum sub applicata et arcu comprehensum aequale esse areae
circuli cuius diameter sit abscissa* = a; *quae observatio mihi quidem
notatu maxime digna videtur*" ["I have newly discovered a remarkable property of the elastic curve given by the equation
$y = f(x)$, with

$$f(x) = \int_0^x \frac{x^2 dx}{\sqrt{a^4 - x^4}};$$

its arc-length from 0 to (x,y) is

$$s(x) = \int_0^x \frac{a^2 dx}{\sqrt{a^4 - x^4}};$$

these two indefinite integrals have no relation to one another,
but I have found that $f(a)s(a) = \frac{1}{4}\pi a^2$, which seems to me
most noteworthy": *Bibl.Math.*(III)5 (1904), p.291|1738; cf.
*Eu.*I-14.268 in E 122|1739, and Euler's letter to Clairaut,
*Eu.*IV A-5.114|1742]. That discovery was nothing else than
Legendre's relation between the periods of elliptic integrals
of the first and of the second kind, specialized to the lemniscatic case.

As to Euler's later discoveries about elliptic integrals, that
story is well known (cf. e.g. A. Enneper, *Elliptische Functionen,*

Theorie und Geschichte, Halle 1876, and *Eu.*I-20, pp.VII–VIII). Here we shall deal with the aspects most relevant to our main theme. For proper background, cf. *infra,* Appendix III.

Euler had closely followed contemporary work on those integrals, including more particularly that of MACLAURIN and of d'ALEMBERT (cf. e.g. his letter to d'Alembert, *Eu.*IV A-5.252|1746). On 23 December 1751 the two volumes of Fagnano's *Produzioni Matematiche,* just published, reached the Berlin Academy and were handed over to Euler; the second volume contained reprints of pieces on elliptic integrals (*Fag.*II.287–318) which had appeared between 1714 and 1720 in an obscure Italian journal and had remained totally unknown. On reading those few pages Euler caught fire instantly; on 27 January 1752 he was presenting to the Academy a memoir (*Eu.*I-20.80–107 = E 252) with an exposition of Fagnano's main results, to which he was already adding some of his own.

The most striking of Fagnano's results concerned transformations of the "lemniscatic differential"

$$\omega(z) = \frac{dz}{\sqrt{1 - z^4}};$$

how he had reached them was more than even Euler could guess. "Surely his discoveries would shed much light on the theory of transcendental functions", Euler wrote in 1753, "if only his procedure supplied a sure method for pursuing these investigations further; but it rests upon substitutions of a tentative character, almost haphazardly applied . . ." ["*Hinc certe theoriae quantitatum transcendentium insigne lumen accenderetur, si modo via, qua FAGNANUS est usus, certam methodum suppeditaret in huiusmodi investigationibus ulterius progrediendi; sed . . . tota substitutionibus precario factis et quasi casu fortuito adhibitis nititur . . .*": *Eu.*I-20.110 in E 263|1753).

But, as he had written the year before: "To this end [the progress of Analysis] such observations are particularly valuable as have been made almost by chance and discovered *a posteriori,* while no *a priori* explanation, no direct way of reaching them, has been perceived . . . Of that kind are a few observations . . . which I have found in Count Fagnano's newly published volume . . . " ["*Ad hunc scopum imprimis accomodatae videntur eiusmodi observationes, quae cum quasi casu sint factae et a posteriori detectae, ratio ad easdem a priori ac per viam directam perveniendi . . . neutiquam est perspecta . . . Huiusmodi autem observationes . . . nonnullas deprehendi in opere Ill. Comitis FAGNANI nuper in luce edito*": *Eu.*I-20.81 in E 252|1752].

With characteristic generosity Euler never ceased to acknowledge his indebtedness to Fagnano (cf. e.g. his letter to Heinsius, *PkU.*109|1764); but surely none but Euler could have seen in Fagnano's isolated results the germ of a new branch of analysis. His first contribution was to extend Fagnano's duplication formula for the lemniscate to a general multiplication formula (*Eu.*I-20.100 in 252|1752). If, as he says in substance, $z \to u = f(z)$ transforms the differential $\omega(z)$ into $\omega(u) = n\omega(z)$, then the substitution

$$
\text{(A)} \qquad z \to v = \frac{z\sqrt{\dfrac{1-uu}{1+uu}} + u\sqrt{\dfrac{1-zz}{1+zz}}}{1 - uz\sqrt{\dfrac{(1-uu)(1-zz)}{(1+uu)(1+zz)}}}
$$

transforms $\omega(z)$ into $\omega(v) = (n+1)\omega(z)$. Soon he was to rewrite this in the equivalent form

$$
\text{(A')} \qquad v = \frac{z\sqrt{1-u^4} + u\sqrt{1-z^4}}{1 + uuzz}
$$

(*Eu.*I-20.65–66 in E 251|1753) and to recognize that this implies

$$
\omega(v) = \omega(z) + \omega(u)
$$

when z, u are treated as independent variables, so that it is nothing less than the addition formula for the lemniscatic integral $\int \omega(z)$.

In this, his second paper on that subject, we already find a clear indication of the main ideas which he was to develop later about elliptic integrals. Of course he also explored some blind alleys; as such we may describe his persistent efforts to give geometric interpretations of his results in terms of the arc-lengths of conic sections and other curves (cf. e.g. *Corr.*I.568–569|1752, E 264|1755, E 639|1776); such are also the elaborate calculations in which he sought "a shorter and more natural way" ("*ein kürzerer und natürlicherer Weg*": *PkU.*109|1764; cf. *Eu.*I-20.305–311 in E 345|1765) for proving what in substance was the addition theorem. On the latter attempt he has this comment, however: "I cannot deny that this is done so circuitously that one could hardly expect those operations to come spontaneously to anyone's mind" ["*Diffiteri non possum, hoc per multas ambages esse praestitum, ita ut vix sit expectandum cuiquam has operationes in mentem venire potuisse*": *Eu.*I-20.311].

What seems to have struck Euler in the first place is that Fagnano's work supplied him with one algebraic integral of the differential equation $\omega(x) = \omega(y)$ (besides the trivial one $x = y$), this being given by $x^2y^2 + x^2 + y^2 - 1 = 0$. Soon he must have noticed that his formula (A), or equivalently (A'), contained implicitly the general integral of the same equation; after replacing v, z and u, in (A'), by y, x and a constant c, the relation between x and y can be written as

(B) $$c^2x^2y^2 + x^2 + y^2 = c^2 + 2xy\sqrt{1 - c^4}$$

(*Corr.*I.567|1752; *Eu.*I-20.63 in E 251|1753), which reduces to Fagnano's for $c = 1$.

Perhaps Fagnano had somehow been guided by the analogy with the equation

$$\frac{dx}{\sqrt{1 - x^2}} = \frac{dy}{\sqrt{1 - y^2}}.$$

At any rate Euler used it quite consciously and systematically,

observing that this last equation had the general integral
(the "complete" integral, as he preferred to call it):

$$x^2 + y^2 = c^2 + 2xy\sqrt{1 - c^2}$$

(*Corr.*I.567|1752); this is obtained of course by putting both
sides of the equation equal to dt, then taking $x = \sin t, y = \sin(t+t_0), c = \sin t_0$. As he observes at once, a similar treatment for $\omega(x)$, $\omega(y)$ "is not possible or at least not avail-
able" ["*huiusmodi comparatio in formulis transcendentibus
$\int dx/\sqrt{1 - x^4}$ et $\int dy/\sqrt{1 - y^4}$ locum non habet seu saltem non
constat*": *Eu.*I-20.61]. Had he pursued this idea further, he
would, as we know now, have discovered Jacobi's elliptic
functions, or at least, as Gauss did, the "lemniscatic" sine
and cosine. Instead of this, after obtaining the desired integral "not by any sure method but by guess-work" ["*nulla
certa methodo . . . sed . . . tentando vel divinando*": *ibid.* p.61],
he merely verified (B) a posteriori by differentiating it and
at the same time using it to express x in terms of y and y in
terms of x.

From then on, Euler's handling of elliptic integrals always
rested upon the consideration of relations $\Phi(x,y) = 0$ of
degree ≤ 2 in x and ≤ 2 in y. Such a relation can be written

$$\Phi(x,y) = P_0(x)y^2 + 2P_1(x)y + P_2(x) = 0$$

or also as

$$\Phi(x,y) = Q_0(y)x^2 + 2Q_1(y)x + Q_2(x) = 0$$

where the P_i and Q_i are polynomials of degree ≤ 2. By solving
this either for y or for x, one sees that the curve defined by
$\Phi = 0$ is birationally equivalent to either one of the curves
$z^2 = F(x)$ and $w^2 = G(y)$, with

$$F = P_1^2 - P_0P_2, \qquad G = Q_1^2 - Q_0Q_2.$$

Thus a relation $\Phi = 0$ can serve to establish an isomorphism
between two curves $z^2 = F(x)$, $w^2 = G(y)$; this idea has been
occasionally used by Euler (cf. *Eu.*I-20.69–71,77, 248–249).

Moreover, we have here:

$$\tfrac{1}{2}\frac{\partial \Phi}{\partial x} = Q_0(y)x + Q_1(y) = \pm \sqrt{G(y)},$$

$$\tfrac{1}{2}\frac{\partial \Phi}{\partial y} = P_0(x)y + P_1(x) = \pm \sqrt{F(x)},$$

so that, differentiating the relation $\Phi = 0$, one gets

$$\pm \frac{dx}{\sqrt{F(x)}} = \pm \frac{dy}{\sqrt{G(y)}}$$

(this being the differential of the first kind on the curve $\Phi = 0$ in the "general case" when F and G are of degree 3 or 4). We may also say that the isomorphism defined as above between the curves $z^2 = F(x)$ and $w^2 = G(y)$ transforms the differentials dx/z and dy/w into each other if signs have been chosen properly.

Almost invariably Euler confines himself to the case when Φ is symmetric in x and y; then he writes it as

$$\alpha + 2\beta(x+y) + \gamma(x^2+y^2) + 2\delta xy + 2\varepsilon xy(x+y) + \zeta x^2 y^2 = 0$$

and calls it "the canonical equation" (*Eu.*I-20.321 in E 347|1765; cf. *Eu.*I-20.71 in E 251|1753 and *Eu.*I-20.155 in E 261|1755). Here F and G are the same polynomial, and, in Euler's language, $\Phi = 0$ is an integral of the differential equation

$$\pm \frac{dx}{\sqrt{F(x)}} = \pm \frac{dy}{\sqrt{F(y)}}$$

His main concern is then to determine Φ when F is given; he does this by the method of indeterminate coefficients, which gives him for Φ a polynomial containing an arbitrary constant, i.e., in his language, a "complete" (or general) integral for the differential equation determined by F. In this he includes the case when F is of degree 2 or 1, even devoting a whole memoir to that subject (an elementary one, since $dx/\sqrt{F(x)}$ can then be integrated by logarithmic or inverse trigonometric functions; cf. *Eu.*I-20.110–152 =

E 263|1753. In that case, as he shows, it is enough to consider relations $\Phi = 0$ of degree 2 in x *and* y, putting $\varepsilon = \zeta = 0$ in the above formula. In doing this he is merely pushing further the analogy he had noticed between $dx/\sqrt{1 - x^4}$ and $dx/\sqrt{1 - x^2}$, an analogy which to us is best explained by saying that he is dealing with the various kinds of algebraic groups of dimension 1, defined over the field of real numbers (the circle being of course one of them). His real concern, however, is with the case when F is a polynomial of degree 3 or 4, and more specifically with the case when F is an even polynomial of degree 4:

$$F(x) = 1 + mx^2 + nx^4.$$

As he perceived, the general case can always be reduced to that one; in fact, this can always be done by a suitable substitution

$$x \rightarrow \frac{\lambda x + \mu}{\nu x + \rho},$$

as he once sought to prove (*Eu.*I-20.303–304 in E 345|1765); his proof remained incomplete, since it only established the validity of this result over complex numbers, while real numbers are clearly what he had in mind. The matter was later cleared up by Legendre (cf. his *Mémoire sur les transcendantes elliptiques*, Paris, An II [= 1793], pp.9–10).

For $F(x) = 1 + mx^2 + nx^4$ (or, what amounts to the same except in number-theoretical contexts, $F(x) = A + Cx^2 + Ex^4$), Euler found that one may take for Φ an even polynomial in x, y (for an *a priori* justification of this fact, cf. Appendix III). Then the method of indeterminate coefficients leads directly to the "complete integral"

$$0 = c^2 - x^2 - y^2 + nc^2x^2y^2 + 2xy\sqrt{1 + mc^2 + nc^4}$$

where c is the arbitrary constant (*Eu.*I-20.67 in E 251|1753; *Eu.*I-20.155–159 in E 261|1755; *Eu.*I-20.310 in E 345|1765), and then to the addition formulas (formulas (3) and (4) of Appendix III; *Eu.*I-20.69 in E 251; *Eu.*I-20.159 in E 261; *Eu.*I-20.311 in E 345).

In 1765 Euler went one step further, applying the method of indeterminate coefficients to the calculation of the "canonical equation" $\Phi = 0$ in the general case

$$F(x) = A + 2Bx + Cx^2 + 2Dx^3 + Ex^4.$$

From our modern point of view, this amounts to defining, in terms of a suitable parameter, the group of automorphisms of the curve Γ given by $y^2 = F(x)$. It is now well-known that this group can be identified (over the groundfield over which Γ is defined) with the so-called "Jacobian" of Γ; this is an elliptic curve

$$Y^2 = 4X^3 - iX - j$$

birationally equivalent to Γ over some algebraic extension of the groundfield but not necessarily over the groundfield itself; i, j are no other than the invariants of the quartic polynomial F (cf. e.g. A.Weil, *Coll.Papers* II.111–116). Of course such concepts lay far beyond Euler's horizon. It is nonetheless remarkable that his calculations (*Eu.*I-20.321–326 in E 347|1765; cf. *Eu.*I-20.71–76 in E 347|1753) contain all the facts which lie behind the above statements, including the formulas for the invariants i, j in terms of A, B, C, D, E.

Finally it must be pointed out that in this § we discussed only the algebraic aspects of Euler's work on elliptic integrals. Euler himself, however, just as Fagnano before him, always put much stress on the relations between integrals which accompanied his algebraic results. The arc-lengths of the lemniscate and of the ellipse and hyperbola had provided him with typical examples of so-called integrals "of the first kind" and "of the second kind", respectively. Eventually he extended his investigations even to the most general case of an integral of the third kind, i.e. one of the form $\int (Z\,dz)/\sqrt{F(z)}$, where F is as before and Z is an arbitrary rational function of z (*Eu.*I-21.39–56 = E 581|1775). Actually all such results are nothing more than special cases of Abel's theorem combined with the algebraic addition formulas (cf.

Appendix III), so that we need not enter into that topic here.

§XVI.

During the years when Euler was pursuing most actively the subject of elliptic functions he seems to have left in abeyance the topic of diophantine equations of genus 1 (a closely allied one in our eyes). He only came back to it when composing the last section of his *Algebra* of 1770. From that time on, diophantine problems, those of genus 1 and a variety of others, were to become one of his favorite pastimes.

Nine papers on such problems, composed from 1770 onwards, appeared during his lifetime; 24 others, all written from 1777 to 1783, were printed posthumously. "I must confess", he wrote in 1772, "that I derive nearly as much pleasure from investigations of this kind as from the deepest speculations of higher mathematics. I have indeed devoted most of my efforts to questions of greater moment, but such a change of subject always brings me welcome relaxation. Moreover, higher Analysis owes so much to the Diophantine method, that one has no right to dismiss it altogether" ["... *fateri cogor me ex huiusmodi investigationibus tantundem fere voluptatis capere quam ex profundissimis Geometriae sublimioris speculationibus. Ac si plurimum studii et laboris impendi in quaestionibus gravioribus evolvendis, huiusmodi variatio argumenti quandam mihi haud ingratam recreationem affere solet. Ceterum Analysis sublimior tantum debet Methodo Diophanteae, ut nefas videatur eam penitus repudiare*": *Eu.*I-3.174 in E 427|1772].

At first sight these papers, which take up so much space in volumes III, IV and V of Euler's *Opera Omnia*, have the appearance of a collection of difficult exercises, more or less haphazardly assembled. One can admire Euler's virtuosity in performing the most abstruse calculations, or rather (since he was practically blind at the time) in directing his assistants' handling of them, without feeling tempted to scrutinize them in detail. On closer inspection, one notices

Euler's continuing preoccupation with equations of genus 1, to which he had already reserved a place of honor in his *Algebra* (cf. §V*d*). In E 405|1770 and E 451|1772, for instance (*Eu*.I-3.148–171 and 282–296), some problems of little intrinsic interest are solved by means of equations of the form

$$y^2 = A + Bx + Cx^2 + Dx^3 + Ex^4$$

where A and E are squares; on each occasion he describes anew Fermat's methods for finding infinitely many solutions of such equations (cf. Chap.II, §XV, and Euler's *Algebra*, *Eu*.I-1.396), calling them "the usual methods" ("*methodos consuetas*", *Eu*.I-3.158); perhaps his purpose there was merely to teach those methods to his assistants. He takes up repeatedly two of Fermat's most notable problems (problems (A) and (B) of Chap.II, §XV; cf. ibid., §XVI and Appendix V), amounting, as we have seen, to the equations $2X^4 - Y^4 = \pm Z^2$, or also (as Euler, following a hint from Leibniz, prefers to formulate it) to the pair of equations $X + Y = U^2$, $X^2 + Y^2 = V^4$; in 1773 he outlines a solution by descent (*Eu*.I-4.96–98 = E 560^{IV}, published in 1783), thus anticipating to some extent Lagrange's definitive treatment of the subject in a paper of 1777 (*Lag*.IV.377–398); in 1780, quoting Lagrange, he seeks to improve still upon his own and Lagrange's treatment, but, failing to use descent, his new proof remains incomplete (*Eu*.I-5.77–81 = E 769).

In 1777, and again in 1782, he seeks criteria for an equation $X^4 + mX^2Y^2 + Y^4 = Z^2$, and for a "double equation"

$$X^2 + Y^2 = Z^2, \qquad X^2 + NY^2 = T^2$$

to have infinitely many solutions (*Eu*.I-4.235–244 = E 696|1777; *Eu*.I-5.35–47 = E 755|1782; *Eu*.I-4.255–268 = E 702|1777); not surprisingly, he fails (it is still an open problem) but does obtain non-trivial sufficient conditions. In 1780 he extends the latter question to double equations

$$X^2 + mY^2 = Z^2, \qquad X^2 + nY^2 = T^2$$

(*Eu*.I-5.48–60 = E 758|1780); here his interest and our own turns to proofs of impossibility for the cases $m = 1$,

$n = 3$ or 4 and others equivalent to these two; one may well admire his beautiful technique (rather different from Fermat's) for applying the infinite descent to such problems.

More important to us are some papers, written in 1780 (*Eu*.I-5.82–115, 146–181 = E 772, 773, 777, 778), the first one of which bears the characteristically enthusiastic title *"De insigni promotione Analysis Diophanteae"* ("On some famous progress in Diophantine Analysis": E 772). Along with a number of other papers by Euler on various subjects, they were all published in 1830 in volume XI of the *Mémoires de l'Académie des Sciences de Saint Pétersbourg* (which had changed their title from Latin to French in 1809). It is this publication which gave Jacobi the occasion for writing his note of 1835 with the no less characteristic title *"De usu theoriae integralium ellipticorum et integralium abelianorum in analysi diophantea"* ["On the use of elliptic and abelian integrals in diophantine analysis": *Jac*.II.53–55; cf. Chap.II, §XVII], and for expressing the view that Euler must have been aware of the connection between diophantine equations of the form $y^2 = F(x)$, with F of degree 3 or 4, and the elliptic integral $\int dx/\sqrt{F(x)}$.

To us it is indeed obvious that the addition formulas (formulas (3) and (4) of Appendix III) determine a rational solution (x', y') of the equation $y^2 = 1 + mx^2 + nx^4$ when two such solutions (x, y), (a, b) are known; it would indeed be surprising if this had never occurred to Euler as he repeatedly wrote those same formulas and the multiplication formulas derived from them (cf. *Eu*.I-20.67,159,311, and above, §XV). Actually he gives no hint that he perceived such facts; luckily his papers of 1780 give some clue to his way of thinking.

As we have seen in §XV, his way of dealing with elliptic integrals had been entirely based on his "canonical equations" of the form $\Phi(x, y) = 0$, with Φ of degree 2 in x and 2 in y. Now his "famous progress" in the paper E 772 consisted precisely in the introduction of such equations as models for elliptic curves, to replace the equations and "double equations" previously in use. That he did so in full awareness of the connection between the two topics seems proved by his use of the same name (the "canonical equation") for this equation $\Phi = 0$ in the new context (*Eu.*I-5.158 in E 778) and by his using for it precisely the same notations (*Eu.*I-5.153 in E 777, to be compared with *Eu.*I-20.321).

To our eyes it is the perception of this connection (later made explicit by Jacobi) which was Euler's "famous progress", and it is something of an anticlimax to consider his further steps. Of course the question arises how to obtain a "canonical" model (if one exists) when the problem is given otherwise. Algebraic geometry supplies a criterium for this; for a given curve Γ of genus 1, there is a model $\Phi(x, y) = 0$, with Φ of degree 2 in x and 2 in y, if and only if there are on Γ two non-equivalent rational divisors of degree 2 (cf. Chap.II, Appendix II). For Euler, the curve is given by an equation $y^2 = F(x)$, and the question, which he can only solve by trial and error, is to find polynomials P, Q, R of degree 2 such that $F = P^2 + QR$, after which he can take

$$\Phi(x, y) = Q(x)y^2 + 2P(x)y - R(x)$$

(*Eu.*I-5.158). Once this is done, he uses the equation $\Phi = 0$ to perform what we called "the ascent" (cf. Chap.II, §XV), i.e. to construct new solutions out of any that may already be known, rightly claiming the new method to be far more efficient than the traditional one, which, as he says, not only "requires most tedious calculations" but "soon produces such enormous numbers that hardly anyone would undertake so much labor" ["*nimis taediosas ambages requirit ... mox ad tam enormes numeros pervenitur, ut vix quisquam tantum laborem suscipere voluerit*": *Eu.*I-5.82 in E 772]. His procedure (an example of which has been given above in Chap.II, Appendix V) is as follows.

As before, we can write:

$$\Phi(x, y) = P_0(x)y^2 + 2P_1(x)y + P_2(x)$$
$$= Q_0(y)x^2 + 2Q_1(y)x + Q_2(y)$$

where the P_i, Q_i are polynomials of degree 2. If $M = (x, y)$ is a solution of $\Phi = 0$, there is another one $N = (x, y')$, where y' is given by $y + y' = -2P_1(x)/P_0(x)$, or alternatively by $yy' = P_2(x)/P_0(x)$; then there is a solution $M' = (x', y')$ given by $x + x' = -2Q_1(y')/Q_0(y')$ or $xx' = Q_2(y')/Q_0(y')$. Repeating this process, one gets thus an infinite sequence of solutions M, M', M'', M''', etc., which in general are all different. Another similar sequence can of course be obtained by exchanging the roles of x and y in this procedure.

To understand what goes on here, it is best to use the concepts explained in Chap.II, Appendix II. On the curve of genus 1 defined by $\Phi = 0$, let A, A' be the points where $x = \infty$ and B, B' those where $y = \infty$; put $\mathfrak{a} = A + A'$, $\mathfrak{b} = B + B'$; these are the divisors of poles of x and of y, respectively; therefore they are rational. With the above notations, x takes the same value at M and at N, so that we have $M + N \sim \mathfrak{a}$; similarly we have $N + M' \sim \mathfrak{b}$. Therefore, putting $\mathfrak{m} = \mathfrak{b} - \mathfrak{a}$, we have $M' \sim M + \mathfrak{m}$, and similarly $M'' \sim M + 2\mathfrak{m}$, $M''' \sim M + 3\mathfrak{m}$, etc. The modern theory of heights shows that the height of the n-th point of this sequence, i.e. the size of the integers needed to express its coordinates (cf. Chap.II, Appendix IV), is of the order of magnitude of n^2, in contrast with the points produced by the "traditional method", whose heights increase exponentially since those methods rest upon the duplication formulas. This accounts for Euler's findings.

§XVII.

[For brevity we will write $\zeta(s)$ for the series $\Sigma_1^\infty n^{-s}$ or rather, as Euler would put it, for

$$1 + \frac{1}{2^s} + \frac{1}{3^s} + \frac{1}{4^s} + \text{etc.},$$

$L(s)$ for $\Sigma_0^\infty (-1)^n(2n+1)^{-s}$ or

$$1 - \frac{1}{3^s} + \frac{1}{5^s} - \text{etc.},$$

and $n!$ for $1.2.3...n$].

At the time of Euler's apprenticeship under Johann Bernoulli, the summation of the series $\zeta(n)$, when n is an integer $\geqslant 2$, was a classical problem which had exercised Leibniz and the Bernoulli brothers (cf. *Corr.*II.15|1737). The summation of $L(1)$ in 1674 had been perhaps the most striking early discovery of the young Leibniz (one which, as Huygens predicted, would be "for ever famous among geometers": *Huy.* VII.394|1674), but it depended upon the power-series expansion

$$\text{arc tan } x = \frac{x}{1} - \frac{x^3}{3} + \frac{x^5}{5} - \cdots,$$

which gives $L(1) = \pi/4$; this could not be extended to $\zeta(n)$ nor to $L(n)$ for any $n > 1$. Even the numerical evaluation of these series was no mean problem, in view of their slow convergence.

In 1728 we find Daniel Bernoulli writing to Goldbach that the sum of the series $\zeta(2)$ is "very nearly $\frac{8}{5}$" (*Corr.*II.263|1728); Goldbach, using an elementary method, answers that $\zeta(2) - 1$ lies between 16223/25200 and 30197/46800 (and thus between 0,6437 and 0,6453: *Corr.*II.282|1729). At that time Euler was already in Petersburg, and in daily contact with Daniel; he must have been aware of those letters. Soon he was presenting to the Academy a paper ending up with a much better evaluation for $\zeta(2)$, viz., 1,644934, obtained by an ingenious application of integral calculus (*Eu.*I-14.41 in E 20|1731).

There can be no doubt that these and similar problems supplied Euler with a powerful motivation in his discovery of the so-called "Euler-MacLaurin summation formula" (it was published independently by MacLaurin in his *Treatise of Fluxions*, Edinburgh 1742). This is already stated in a paper presented to the Academy in 1732 (*Eu.*I-14.42–72 = E 25); three years later he gave a full exposition of the same topic (*Eu.*I-14.108–123 = E 47|1735; cf. *Eu.*I-14.435–439 in E 130|1739, and Chap.V of his *Institutiones Calculi Differentialis* of 1755, Part II, *Eu.*I-10.309–336). His procedure may be summarized as follows.

The problem, as he sees it, is to find a formula for a sum

$\Sigma f(i)$ taken over the range $1 \leqslant i \leqslant n$, $f(x)$ being what we should call an analytic function of x for all $x > 0$; from this is to be derived the value of $\Sigma_1^x f(i)$. The answer is sought in the form $S(n)$, S being also an analytic function for $x > 0$. Clearly S will have the required property if it satisfies the difference equation

$$S(x) - S(x-1) = f(x)$$

and if $S(0) = 0$, or at any rate if it satisfies $S(v) = \Sigma_1^v f(i)$ for some integer v. As to the difference equation, Euler, using Taylor's theorem (for which he refers to Taylor's publication of 1715, although he must have known that it had been obtained previously by Leibniz and by Johann Bernoulli; cf. *Eu.*I-14.109), rewrites it in the form

$$f(x) = \frac{dS}{dx} - \frac{1}{2!}\frac{d^2S}{dx^2} + \frac{1}{3!}\frac{d^3S}{dx^3} - \text{etc.}$$

and seeks to solve it by putting

$$S(x) = \alpha \int f dx + \beta f + \gamma \frac{df}{dx} + \delta \frac{d^2f}{dx^2} + \text{etc.}$$

with indeterminate coefficients α, β, γ, etc.; these are to be determined by substituting back this tentative formula into the preceding equation and identifying both sides. The additive constant implicit in the indefinite integral $\int f dx$ can then be determined by taking for x any integral value, e.g. $x = 0$. This gives him at once the values of α, β, γ, etc. by means of a recurrence formula; he gets $\alpha = 1$, $\beta = 1/2$, then for instance

$$\delta = \frac{\gamma}{2!} - \frac{\beta}{3!} + \frac{\alpha}{4!},$$

etc., and hence (but "with a great deal of work", "*multo labore*": *Eu.*I-14.436) the numerical values for the first fourteen of those coefficients; those of even order, other than β, all turn out to be 0.

In this Euler says that he had proceeded according to "a known method" ("*methodo cognita*"; *Eu*.I-14.112), by which he probably meant no more than the "method" of indeterminate coefficients. Actually he had been bold to the point of rashness; but, as the Latin proverb says, "*audaces fortuna iuvat*" (fortune favors the bold). Formally speaking, his procedure can be clarified by using the operator $D = d/dx$; then one may write:

$$S(x-1) = e^{-D}S(x), \qquad f(x) = S(x) - S(x-1) = (1-e^{-D})S(x)$$

and, formally of course,

$$S(x) = (1-e^{-D})^{-1}f(x) = (\alpha D^{-1}+\beta+\gamma D+\delta D^2 + \cdots)f(x)$$

where α, β, γ, etc., appear now as the coefficients of the Taylor expansion of $z/(1-e^{-z})$ at $z = 0$ (*Eu*.I-14.436 in E 130|1739). From this fact, which had escaped Euler's notice at first, one deduces easily that the coefficients of the even powers of D in the above formula are all 0, as his previous numerical computations had indicated (cf. *ibid*.). If now one puts

$$\frac{z}{1 - e^{-z}} = \sum_{m=0}^{\infty} \frac{1}{m!} b_m z^m,$$

one gets $b_0 = 1$, $b_1 = \frac{1}{2}$, $b_{2m-1} = 0$ for $m > 0$; as to the b_{2m} for $m > 0$, they have the successive values

$$\frac{1}{6}, \; -\frac{1}{30}, \; \frac{1}{42}, \; -\frac{1}{30}, \; \frac{5}{66}, \; -\frac{691}{2730}, \text{ etc. };$$

and Euler's result, formally speaking, amounts to the formula

$$(EM) \quad S(x) = \sum_{i=1}^{v} f(i) + \int_{v}^{x} f(t)dt$$

$$+ \sum_{m=1}^{\infty} \frac{1}{m!} b_m \left[f^{(m-1)}(x) - f^{(m-1)}(v) \right]$$

where the $f^{(i)}$, for $i \geq 0$, denote f and its successive derivatives.
As Euler observed at once (*Eu*.I-14.43–44 in E 25|1732;

Eu.I-14.115–117 in E 47|1735), this result is valid at any rate when $f(x)$ is a polynomial in x. Then the series given above for $S(x)$ reduces to a finite sum and defines a polynomial which is a solution of the difference equation $S(x) - S(x-1) = f(x)$. For instance, taking $f(x) = x^r, r \geqslant 1$, and $\nu = 0$, one gets:

$$1^r + 2^r + \cdots + n^r = S_r(n)$$

where S_r is the polynomial given by

$$S_r(x) = \frac{x^{r+1}}{r+1} + \frac{x^r}{2} + \sum_{1 \leqslant m < \frac{r+1}{2}} \binom{r}{2m-1}\frac{1}{2m}b_{2m}\,x^{r+1-2m}$$

(*Eu*.I-14.116–117), or more simply by:

$$\frac{dS_r(x)}{dx} = B_r(x) = \sum_{m=0}^{r}\binom{r}{m}b_m x^{r-m}, \qquad S_r(0) = 0.$$

This result had been anticipated, partly by Fermat (cf. above, Chap.II, §II), and more fully by Jacob Bernoulli (cf. pp.96–97 of his *Ars Conjectandi* published posthumously in Basel in 1713); the latter's priority was eventually perceived by Euler, who then adopted A. de Moivre's designation "Bernoulli numbers" for the numbers b_{2m} (cf. A. de Moivre, *Miscellanea Analytica*, Londini 1730, Complementum, pp.6 and 19–21): "these are the numbers", Euler wrote, "that are called Bernoullian from their inventor Jacob Bernoulli, since he used them in his *Ars Conjectandi* to sum the powers of natural integers" ["*numeri ab inventore Bernoulliani vocati, quippe quibus olim Jacobus Bernoulli in Arte Conjectandi est usus ad progressiones potestatum numerorum naturalium summandas*": *Eu*.I-15.92 in E 393|1768; cf. the similar comment in his *Institutiones Calculi Differentialis* of 1755, *Eu*.I-10.321].

What Euler had chiefly in mind at this juncture, however, was the numerical evaluation of such series as $\zeta(n)$ for $n = 2, 3, 4$, etc., and also, to begin with, of the partial sums of the series $\zeta(1)$. In such cases a modern analyst would say at once that the summation formula (*EM*) breaks down because

(as Euler himself did not fail to discover: *Eu.*I-14.357 in
E 125|1739; cf. *Eu.*I-14.119 in E 47|1735 and, in the *Insti-*
tutiones Calculi Differentialis, the passage *Eu.*I-10.327) the
numbers $|b_{2m}|$ increase so rapidly that the series involved
are always divergent. The correct way to use them for purposes of numerical computation is to cut them down to
finite sums and evaluate the remainder. Instead of this,
Euler's prescription is to sum the terms "until they begin
to diverge" ("*quoad termini divergere incipiant*": *Eu.*I-14.357);
perhaps one should rather say that he allows himself to be
guided by his instinct (not a bad guide, truly, in such matters).
It is thus that he is able to compute $\zeta(2)$ with 20 decimal
figures:

$$\zeta(2) = 1,64493406684822643647... ,$$

$\zeta(3)$ with 15 figures, $\zeta(4)$ with 16 figures, the so-called "Euler
constant" with 16 figures (*Eu.*I-14.119–122 in E 47|1735),
then π with 15 figures (*Eu.*I-14.359 in E 125|1739); eventually
he used the same method to compute $\zeta(n)$ with 18 figures
for $2 \leqslant n \leqslant 16$ (*Corr.*I.207|1743; cf. *Eu.*I-14.440 in
E 130|1739, and Chap.XI of the *Introductio* of 1748, *Eu.*I-
8.201–205). There can be no doubt that he took much
pleasure in such calculations, even apart from the theoretical
conclusions which he was sometimes able to derive from
them.

§XVIII.

"So much work has been done on the series $\zeta(n)$", Euler
writes in 1735, "that it seems hardly likely that anything
new about them may still turn up . . . I, too, in spite of
repeated efforts, could achieve nothing more than approximate values for their sums . . . Now, however, quite
unexpectedly, I have found an elegant formula for $\zeta(2)$,
depending upon the quadrature of the circle [i.e., upon π]"
["*Tantopere iam pertractatae et investigatae sunt series reciprocae*
potestatum numerorum naturalium, ut vix probabile videatur de
iis novi quicquam inveniri posse . . . Ego etiam iam saepius . . .

has series diligenter sum persecutus neque tamen quicquam aliud
sum assecutus, nisi ut earum summam . . . proxime veram definiverim
. . . Deductus sum autem nuper omnino inopinato ad elegantem
summae huius seriei $1 + \frac{1}{4} + \frac{1}{9} + \frac{1}{16} +$ etc. *expressionem, quae*
a circuli quadratura pendet": *Eu.*I-14.73–74 in E 41|1735].

When he wrote these lines, he had obviously just discovered
his famous result $\zeta(2) = \pi^2/6$. His paper on that subject,
and on various closely related summations, was presented
to the Academy on 5 December 1735; it was not printed
until 1740 (*Eu.*I-14.73–86 = E 41), but by that time everyone
who counted in the international mathematical community
had been apprised of it: first of all, perhaps, Euler's friend
Daniel Bernoulli in Basel (cf. *Corr.*II.435|12 Sept. 1736 and
*Corr.*II.15–16|1737); also STIRLING in Edinburgh (*Eu.*IV
A-1, no.2621|8 June 1736), EHLER and KÜHN in Dantzig
(*PkU.*345|3 April 1736), POLENI in Padova (*PkU.*216|13
March 1736), NAUDÉ in Berlin (*PkU.*193–199|1740), and,
directly or through these, many others including of course
Johann Bernoulli (cf. *Corr.*II.15–16|1737). As to Goldbach,
he was in Petersburg at the time of the discovery and surely
must have heard of it at once. In 1742, noticing (not without
a shade of disappointment) that CLAIRAUT was insuffi-
ciently informed of the matter, he promptly sends him full
details, prefacing them with the words: *"M' Jacq. Bernoulli*
. . . parle de ces suites, mais . . . il avouë, que malgré toutes les
peines, qu'il s'etoit données, il n'avoit pû venir à bout, de sorte que
M^{rs} [Jean] Bernoulli, de Moivre et Stirling, grands Maitres dans
cette matiere, ont été extrêmement surpris, quand je leur annonçois
que j'avois trouvé la somme de cette serie

$$1 + \frac{1}{4} + \frac{1}{9} + \frac{1}{16} + \frac{1}{25} + \text{etc.}$$

et même de celle-ci

$$1 + \frac{1}{2^n} + \frac{1}{3^n} + \frac{1}{4^n} + \frac{1}{5^n} + \text{etc.}$$

quand n est un nombre pair" ["Jacob Bernoulli does mention those series, but confesses that, in spite of all his efforts, he could not get through, so that Joh. Bernoulli, de Moivre and Stirling, great authorities in such matters, were highly surprised when I told them that I had found the sum of $\zeta(2)$, and even of $\zeta(n)$ for n even": *Eu*.IV A-5.120]. These are almost the same words which Fermat had once used in describing one of his own discoveries (*"c'est une de mes inventions qui a quelquesfois estonné les plus grands maistres"*: cf. Chap.II, §XV). From 1736 onwards, and for the next eight or ten years, Euler's result, and the degree of validity of his proof for it, become a recurrent topic in his correspondence with his friends and colleagues all over Europe.

Here again Euler had been rash, applying Newton's theorem on the sums of the n-th powers of the roots of algebraic equations to transcendental equations of the form $y = \sin s$, where y is given and s is the unknown. Using the power-series for the sine, and calling A, B, C, etc., the roots of the equation, he writes it as

$$0 = 1 - \frac{s}{y} + \frac{s^3}{1.2.3.y} - \frac{s^5}{1.2.3.4.5.y} + \text{etc.}$$

$$= \left(1 - \frac{s}{A}\right)\left(1 - \frac{s}{B}\right)\left(1 - \frac{s}{C}\right) \cdot \text{etc} .$$

and concludes from this, firstly, that

$$\frac{1}{y} = \frac{1}{A} + \frac{1}{B} + \frac{1}{C} + \text{etc.,}$$

then, comparing terms in s^2, that

$$0 = \frac{1}{AB} + \cdots = \frac{1}{2}\left(\frac{1}{A} + \frac{1}{B} + \cdots\right)^2 - \frac{1}{2}\left(\frac{1}{A^2} + \frac{1}{B^2} + \cdots\right),$$

which gives $\Sigma 1/A^2 = 1/y^2$, and similarly for higher powers. Taking in particular $y = 1$, he states that the roots A, B, C, etc., are $q, q, -3q, -3q, 5q, 5q$, etc., where q is one fourth of the circumference of a circle of radius 1 (i.e. $\pi/2$ in the notation he was soon to adopt). This gives:

$$1 = 2 \times \frac{2}{\pi}\left(1 - \frac{1}{3} + \frac{1}{5} - \cdots\right),$$

which was Leibniz's result, and further:

$$1 = 2 \times \left(\frac{2}{\pi}\right)^2 \left(1 + \frac{1}{3^2} + \frac{1}{5^2} + \cdots\right).$$

It was of course obvious to Euler that one has

$$1 + \frac{1}{3^2} + \frac{1}{5^2} + \cdots = \left(1 + \frac{1}{2^2} + \frac{1}{3^2} + \cdots\right) - \left(\frac{1}{2^2} + \frac{1}{4^2} + \frac{1}{6^2} + \cdots\right)$$

$$= \left(1 + \frac{1}{2^2} + \frac{1}{3^2} + \cdots\right)\left(1 - \frac{1}{2^2}\right) = \frac{3}{4}\zeta(2),$$

so that the above conclusions give $\zeta(2) = \pi^2/6$. The same procedure, pushed further, gives to Euler (somewhat laboriously, "*multo labore*": *Eu*.I-14.85) the values of $\zeta(n)$ for $n = 4, 6, 8, 10, 12$, and those of $L(n)$ for $n = 3, 5, 7$. Applied to other values of y than $y = 1$, it gives the value of still other series (Dirichlet series in modern terminology). All this is rapidly sketched in Euler's article of 1735; one can almost feel in it the feverish excitement of the writer.

Euler's method had been open to grave objections, as he perceived himself (cf. *Eu*.I-14.79 in E 41|1735, and *Eu*.I-14.139–140 in E 61|1743) even before they were pointed out by his correspondents: Johann Bernoulli in 1737 (*Corr*.II.16–17), as well as Daniel Bernoulli (cf. *Corr*.II.477|1741), Daniel's cousin Nicolas Bernoulli (cf. *Corr*.II.683|1742), Cramer in Geneva (cf. *Corr*.II.477,683) and perhaps others. Apart from the question of the convergence, or possible divergence, of the series and products involved (a question raised, somewhat clumsily, by Nicolas Bernoulli: *Corr*.II.683–684 and 691|1742), one could well ask whether the transcendental equation

$$1 = s - \frac{1}{1.2.3}s^3 + \cdots$$

had not, besides its "visible" real roots, other "imaginary" roots which would vitiate the whole calculation (*Corr.*II.16; cf. *Eu.*IV A-5.121); nor was this objection limited to the possible existence of roots of the form $a + b\sqrt{-1}$, even if this could have been meaningfully asked before Euler defined the sine for such values of the argument (cf. *infra*, §XIX); for the so-called fundamental theorem of algebra, which at least had the status of a plausible conjecture for algebraic equations (cf. above, §VII) could not even be reasonably formulated for equations "of infinite degree" such as the one Euler was trying to handle. More specifically, it was objected to him that if an ellipse was substituted for the circle in the geometric definition of the sine, his conclusions would become obviously false, while the visible real roots of his equation would remain the same (*Corr.*II.477).

At first Euler held his doubts in abeyance by observing that his method agreed with Leibniz's result in the case of $L(1)$ (so that, as he says, if the equation $1 = \sin x$ had "imaginary roots" besides the visible real ones, at least the sum of their reciprocals would be 0), and also by comparing his new values for $\zeta(2)$, $\zeta(4)$, etc., with the numerical values he had computed previously (cf. above, §XVII). Thus, he says, "I did not hesitate to publish these sums as being perfectly true"("*non dubitavi istas summas tanquam verissimas producere*": *Eu.*I-14.140 in E 61|1743).

Nevertheless, for the next ten years, Euler never relaxed his efforts to put his conclusions on a sound basis. At first this led him only to an analytical proof for $\zeta(2) = \pi^2/6$ (or rather for the equivalent result $1 + 1/3^2 + 1/5^2 + \cdots = \pi^2/8$) which did not seem capable of being extended even to $\zeta(4)$ (cf. his letter of 1737 to Johann Bernoulli, *Bibl.Math.*(III) 5 (1904), pp.258–259, and P. Stäckel's article, *Eu.*I-14.156–176). But now his attention had been drawn by his heuristic proof of 1735, as well as by his earlier work on what is now called the gamma function (cf. *Corr.*I.3–6|1729, etc.) to infinite products in their relation to infinite series, and this, too, led him to a momentous discovery. In a paper presented to the Academy in 1737 (*Eu.*I-14.216–

244 = E 72), after discussing some rather queer series suggested to him by Goldbach, he goes on to consider some
infinite products "no less admirable", he says, "since in them
the factors proceed according to the prime numbers, whose
progression is no less irregular [than the terms in Goldbach's
series]" (". . . *neque minus erunt admirabilia . . . cum . . . in his
. . . termini progrediantur secundum numeros primos, quorum
progressio non minus est abstrusa*": *Eu.*I-14.227). What he means
by this is the famous "eulerian" product for $\zeta(s)$:

$$\zeta(s) = \prod_p (1-p^{-s})^{-1}$$

extended to all primes p, or in his notation

$$1 + \frac{1}{2^n} + \frac{1}{3^n} + \frac{1}{4^n} + \text{etc.} = \frac{2^n \cdot 3^n \cdot 5^n \cdot 7^n \cdot \text{etc.}}{(2^n-1)(3^n-1)(5^n-1)(7^n-1) \cdot \text{etc.}}$$

(*Eu.*I-14.230), and the similar product

$$\frac{\pi}{4} = 1 - \frac{1}{3} + \frac{1}{5} - \frac{1}{7} + \frac{1}{9} - \text{etc.} = \frac{3 \cdot 5 \cdot 7 \cdot 11 \cdot 13 \cdot \text{etc.}}{4 \cdot 4 \cdot 8 \cdot 12 \cdot 12 \cdot \text{etc.}}$$

or in modern notation

$$L(1) = \prod_{p \neq 2} \frac{p}{p \pm 1}$$

where the sign is determined by $p \pm 1 \equiv 0 \pmod 4$ (*Eu.*I-
14.233; cf. *Corr.*I.577–578|1752). From the former result
he deduces that the sum $\Sigma \, 1/p$ is infinite "as the logarithm
of the sum of the harmonic series $\Sigma \, 1/n$"; as he knew that
the sum $\Sigma_1^N \, 1/n$ is of the order of magnitude of $\log N$ (cf.
e.g. *Eu.*I-14.87–100 = E 43|1734), this meant that the sum
$\Sigma_{p<N} \, 1/p$ has the order of magnitude of $\log \log N$. From
the result about $L(1)$ he was later to conclude that the sum
$\Sigma_{p \neq 2} \pm 1/p$ (where the sign is determined as before, and
the odd primes are arranged in increasing order) has a
finite value, approximately 0,334980 (*Corr.*I.587|1752; cf.
*Eu.*I-4.146–153 in E 596|1775). From these observations it
follows obviously that the sum $\Sigma 1/p$ extended to the primes
of the form $4n + 1$, and the similar sum extended to the

primes of the form $4n - 1$, are both infinite and of the same order of magnitude.

Some years after thus introducing "eulerian products", Euler devoted to the same topic a whole chapter of his *Introductio in Analysin Infinitorum* (Chap.XV = *Eu*.I-8.284–312), where he not only writes the products for $\zeta(s)$ and $L(s)$ in almost modern form, writing for instance for $\zeta(n)$ the following expression

$$\frac{1}{\left(1-\frac{1}{2^n}\right)\left(1-\frac{1}{3^n}\right)\left(1-\frac{1}{5^n}\right)\left(1-\frac{1}{7^n}\right) \text{etc.}},$$

but extends this to what we would now call L-series with Dirichlet characters modulo 3 (*Eu*.I-8.308–309) and modulo 8 (*Eu*.I-8.310–311). He even shows how such series (and "innumerable others": *Eu*.I-8.306) can be summed for $n = 1$, using the results of the preceding chapter X (cf. *infra*, §XIX). One may well regard these investigations as marking the birth of analytic number-theory.

§XIX.

In 1735 Euler had been unable to prove his results on $\zeta(2n)$ to his own and his colleagues' full satisfaction. In 1745 he had already completed in manuscript his *Introductio in Analysin Infinitorum,* presenting in its Chapters VIII to XI a masterly exposition of a theory of trigonometric functions and their series and product expansions (*Eu*.I-8.133–212), with the results in question as one of its main applications. Some of it had been published in Berlin some time before (*Eu*.I-14.138–155 = E 61|1742); most of it had apparently been discovered in the first year or two of his stay in Berlin (cf. *Eu*.IV A-5.115,120–124|1742 and *Corr*.I.131–132|1742).

Since his younger days he had known that e^x is the limit of $(1+x/n)^n$ for $n \to \infty$ (a fact which he chose to express by writing

$$e^x = \left(1+\frac{x}{n}\right)^n$$

"for n infinite" ("*existente n numero infinito*": *Eu*.I-14.143 in

INTRODUCTIO

IN ANALYSIN

INFINITORUM.

AUCTORE

LEONHARDO EULERO,

Profeſſore Regio BEROLINENSI, *& Academiæ Im-*
perialis Scientiarum PETROPOLITANÆ
Socio.

TOMUS PRIMUS.

LAUSANNÆ,

Apud MARCUM-MICHAELEM BOUSQUET & Socios.

MDCCXLVIII.

E 61|1742; cf. his *Introductio*, *Eu*.I-8.132, 147, etc.). He was also familiar with the power-series for e^x, $\sin x$, $\cos x$; in particular, he knew that the power-series for e^x can be obtained by expanding $(1+x/n)^n$ by the binomial formula when n is taken to be a (finite) integer, and then taking the limit, term by term, for $n \to \infty$ (or, as he preferred to put it, substituting for n an infinite number: *Eu*.I-8.123–124).

In 1740, in an exchange of letters with Joh. Bernoulli on the solutions of linear differential equations with constant coefficients (among other topics), he stated for the first time the celebrated formula

$$2 \cos x = e^{+x\sqrt{-1}} + e^{-x\sqrt{-1}},$$

deriving it from the identity between the power-series expansions of both sides (*Bibl.Math*.(III) 6 (1905),pp.76–77); implicitly he was using the power-series for e^x in order to extend its definition to complex values of the argument. The similar formula for $\sin x$ must of course have been discovered at the same time (cf. *Eu*.I-14.142,144 in E 61|1742). Combining this with the above formula for e^x, Euler could then conclude that $\sin x$ is the limit for $n \to \infty$ (or, as he said, "the value for n infinitely large") of the polynomial

$$P_n(x) = \frac{1}{2\sqrt{-1}} \left[\left(1 + \frac{x\sqrt{-1}}{n}\right)^n - \left(1 - \frac{x\sqrt{-1}}{n}\right)^n \right].$$

Now P_n (for finite n, as he would say) can easily be split into its real factors of degree 1 and 2 (cf. above, §VII). For an odd $n = 2p + 1$, this is effected by using the identity

$$X^n - Y^n = (X-Y) \prod_{\nu=1}^{p} (X - e^{+2\pi i\nu/n}Y)(X - e^{-2\pi i\nu/n}Y)$$

$$= (X-Y) \prod_{\nu=1}^{p} \left(X^2 - 2XY \cos \frac{2\pi\nu}{n} + Y^2\right).$$

This gives, again for $n = 2p + 1$:

$$P_n(x) = x \prod_{\nu=1}^{p} Q_{n,\nu}(x), \qquad Q_{n,\nu}(x) = 1 - \frac{1 + \cos \dfrac{2\pi\nu}{n}}{1 - \cos \dfrac{2\pi\nu}{n}} \frac{x^2}{n^2}.$$

Clearly, for any given ν, and "for n infinite", $Q_{n,\nu}(x)$ becomes $1 - x^2/\nu^2\pi^2$; from this Euler draws the conclusion

$$\sin x = x \prod_{\nu=1}^{\infty} \left(1 - \frac{x^2}{\nu^2\pi^2}\right),$$

and a similar calculation gives him the corresponding infinite product for the cosine:

$$\cos x = \prod_{\nu=0}^{\infty} \left(1 - \frac{4x^2}{(2\nu+1)^2\pi^2}\right)$$

(Eu.I-14.141–146 in E 61|1742; cf. the $Introductio$, Chap. IX = Eu.I-8,153–176), showing, in his own words (Eu.I-14.146), that the equations of infinite degree which he had previously applied to the calculation of the series $\zeta(2)$, $\zeta(4)$, etc., "have indeed no other roots" than the visible real ones. As noted above, all this is expressed by Euler in the language of "infinitely large" and "infinitely small" numbers (a language that has been revived in our days by the adepts of so-called "non-standard analysis"). The evaluations which, from our present point of view, would legitimize his reasonings are of course lacking; but in the present case any tiro could easily fill them in.

From these infinite products, as Nicolas Bernoulli rightly observed ($Corr$.II.689|1742; cf. ibid.,p.694), one deduces at once, by logarithmic differentiation, further formulas, such as

$$\frac{\pi \cos \pi x}{\sin \pi x} = \frac{1}{x} + \sum_{\nu=1}^{\infty} \left(\frac{1}{\nu+x} - \frac{1}{\nu-x}\right),$$

which Euler writes as follows:

$$\frac{\pi \cos \pi x}{\sin \pi x} = \frac{1}{x} - \frac{1}{1-x} + \frac{1}{1+x} - \frac{1}{2-x} + \frac{1}{2+x} - \frac{1}{3-x} + \text{etc.}$$

(without the parentheses which in our eyes are to guarantee the absolute convergence of the series). Combining this with the formula which can be similarly derived from the product for cos x, one obtains, in Euler's notation

$$\frac{\pi}{\sin \pi x} = \frac{1}{x} + \frac{1}{1-x} - \frac{1}{1+x} - \frac{1}{2-x} + \frac{1}{2+x} + \frac{1}{3-x} - \frac{1}{3+x} - \text{etc.}$$

(cf. e.g. *Eu.*I-14.148). For this Euler has two other proofs, one based on integral calculus (*Eu.*I-14.146–148 in E 61|1742; cf. his letter to Clairaut, *Eu.*IV A-5.115|1742), the other based directly on the extension of the above infinite products to functions of the form cos x + cos a, sin x + sin a (*Eu.*I-8.170–176, 184–187 in *Introductio*, Chap.IX–X). From this it is easy to derive by repeated differentiation not only the values of $\zeta(n)$ for even $n > 0$ and of $L(n)$ for odd $n > 0$, but also that of infinitely many other "Dirichlet series" (as they would now be called) (cf. e.g. *Corr.*I.132|1742 or *Eu.*I-14.151–152). In particular one finds that $\zeta(n)$ is of the form $r(2\pi)^n$ for every even $n > 0$, and $L(n)$ of the form $s(2\pi)^n$ for every odd $n > 0$, the numbers r, s being rational.

At first sight there is no connection between those results and the Euler-MacLaurin formula. Euler, however, could not but notice fairly soon the relation between the values of $\zeta(2n)$, which he had calculated already in 1735 up to

$$\zeta(12) = \frac{691\pi^{12}}{6825 \times 93555}$$

(*Eu.*I-14.85 in E 41|1735), and the numbers b_{2n} (not yet called by him Bernoullian) which had turned up in his work on the summation formula (cf. *Eu.*I-14.114 in E 47|1735); in particular the occurrence of the unusual prime 691 (a kind of "tracer" for Bernoulli numbers) both in $\zeta(12)$ and in b_{12} could not have failed to strike him. Once this had been observed, Euler had no difficulty in establishing the general relation

$$\zeta(2n) = \frac{1}{2}(-1)^{n-1}\frac{b_{2n}}{(2n)!}(2\pi)^{2n}$$

(*Eu.*I-14.434–439 in E 130|1739).

§XX.

The mystery of the values of $\zeta(n)$ for odd $n > 1$ still remained intact; indeed it has remained so down to the present day. Knowing of course that $\zeta(s)$ is related to the alternating series

$$\varphi(s) = 1 - \frac{1}{2^s} + \frac{1}{3^s} - \frac{1}{4^s} + \text{etc.}$$

by the formula

$$\varphi(s) = \left(1 - \frac{2}{2^s}\right)\zeta(s),$$

Euler apparently thought of clearing up the mystery by first investigating the values to be attributed (according to his views on divergent series: cf. $Eu.$I-14.585–617 = E 247|1746, and the passage $Eu.$I-10.81–82 in his *Institutiones Calculi Differentialis* of 1755) to the divergent alternating series

$$\varphi(-m) = 1 - 2^m + 3^m - 4^m + \text{etc.}$$

for odd values of $m > 0$ (cf. $Eu.$I-14.443 in E 130|1739). To him this could be done by taking $\varphi(-m) = R_m(1)$, where R_m is the rational function defined (for $|x| < 1$, as one would insist on saying now) by the series

$$R_m(x) = 1 - 2^m x + 3^m x^2 - 4^m x^3 + \text{etc.}$$

"Differential calculus", as he says, "gives us an easy way of summing such series" ["*Le calcul différentiel nous fournit un moyen fort aisé de trouver la somme de ces séries*": $Eu.$I-15.71 in E 352|1749]. In fact, one has $R_0(x) = 1/(1+x)$, and, for every integer $m \geqslant 0$:

$$R_{m+1}(x) = \frac{d}{dx}[xR_m(x)],$$

so that, for $m > 0$, R_m is of the form

$$R_m(x) = \frac{P_m(x)}{(1+x)^{m+1}}$$

where P_m is a polynomial of degree $m - 1$ with integral coefficients. Moreover, by writing down P_m for a few values of m, as Euler must have done, one notices at once that R_m

satisfies the condition

$$R_m\left(\frac{1}{x}\right) = (-1)^{m+1}x^2 R_m(x),$$

a fact which can then be verified easily by induction on m. Thus or otherwise Euler found that $\varphi(-m) = R_m(1) = 0$ for every even $m > 0$ (*Eu.*I-14.442 in E 130|1739).

A far more surprising result comes out when one computes even the first few values of $\varphi(-m)$ for odd $m \geqslant 1$; when Euler did this, it must have become immediately clear to him that those values are related to those of $\zeta(m+1)$, which in turn he had found to be related to the Bernoulli numbers b_{m+1}. Actually the relation between them can be written as

$$\varphi(-m) = R_m(1) = (2^{m+1}-1)\frac{b_{m+1}}{m+1},$$

this being valid for all odd $m > 0$, and also for all even $m > 0$ since then both sides are 0. In 1739, Euler, having made the computation for $m = 1, 3, 5, 7$, chooses to write the result as

$$1 - 2^m + 3^m - 4^m + \text{etc.}$$
$$= \frac{\pm\, 2\cdot1\cdot2\cdot3\cdots m}{\pi^{m+1}}\left(1+\frac{1}{3^{m+1}}+\frac{1}{5^{m+1}}+\text{etc.}\right)$$

with alternating signs $+, -, +, -$ for $m = 1, 3, 5, 7$; here the series in the right-hand side has clearly the value $(1-2^{-m-1})\zeta(m+1)$.

There Euler let the matter rest for the next ten years, until at last he looked for a general proof and of course found it, using, as he said, "a most peculiar method" ["*Il faut employer une méthode toute particulière pour démontrer cette harmonie*": *Eu.*I-15.75 in E 352|1749]. The "peculiar" method consisted in a wholly illegitimate application of the Euler-MacLaurin formula to the series $\varphi(-m)$, so hazardous that there would be little point in describing it here; even Euler might have hesitated to put it forward, had he not had two other methods at his disposal for checking the results in each specific case, one of these being the one indicated above;

for the other, cf. *Eu.*I-14.594–595 in E 247|1746, and *Eu.*I-10.222–224 in Part II, Chap.I, of the *Institutiones Calculi Differentialis.* Anyway he did obtain the correct value as shown above, thus verifying the validity of the above formulas for all odd $m > 0$.

This time, however, Euler did not stop there. Substituting $n - 1$ for m, and rewriting the result in terms of the function φ, one gets

$$\frac{\varphi(1 - n)}{\varphi(n)} = -C_n \frac{1 \cdot 2 \cdot 3 \ldots (n - 1)(2^n - 1)}{(2^{n-1} - 1)\pi^n}$$

with $C_n = (-1)^{n/2}$, this being valid for all even $n > 0$. As Euler observes, the left-hand side has the value 0 for all odd $n > 1$, so that the formula is also valid for those values of n provided one takes then $C_n = 0$.

Ever since his early days in Petersburg he had been interested in the interpolation of functions and formulas given at first only for integral values of the argument; that is how he had created the theory of the gamma function (cf. *Corr.*I.3–4|1729). Not surprisingly, he sought to do the same for the above formula by substituting $\cos \pi n/2$ for C_n and the gamma function, i.e. $\Gamma(n)$ in the notation later introduced by Legendre, for $(n-1)!$. The result is equivalent to what we know as the functional equation of the zeta function. In view of what Euler knew about the gamma function, it holds for all integral values of n, positive or negative (including $n = 0$ and $n = 1$), also for $n = \frac{1}{2}$, and it could be checked numerically, within the limits of accuracy of his calculations, for some other values of n. Well could he entitle his paper *"Remarques sur un beau rapport entre les séries de puissances tant directes que réciproques"* (*Eu.*I-15.70–90 = E 352|1749). For good measure he threw in the functional equation for the function

$$L(s) = 1 - \frac{1}{3^s} + \frac{1}{5^s} - \frac{1}{7^s} + \text{etc.},$$

arrived at in the same way. "This last conjecture", he wrote, "is expressed by a simpler formula than the preceding one;

as it is equally certain, one may hope that efforts to find a
perfect proof for it will be more successful; this will not fail
to throw much light on a number of similar investigations"
["*Cette dernière conjecture renferme une expression plus simple
que la précédente; donc, puisqu'elle est également certaine, il y a
à espérer qu'on travaillera avec plus de succès à en chercher une
démonstration parfaite, qui ne manquera pas de répandre beaucoup
de lumière sur quantité d'autres recherches de cette nature*": Eu.I-
15.90].

Euler's paper, written in 1749, appeared in 1768 in the
Memoirs of the Berlin Academy; it was never reprinted.
Some of its conclusions, in the far less suggestive form in
which they had been stated in Euler's memoir of 1739
(E 130, published only in 1750 in the Petersburg *Commentarii*)
were repeated in 1773 under a less than promising title
("*Exercitationes Analyticae*": Eu.I-15.131–167 = E 432|1772).
For a hundred years after their discovery, Euler's functional
equations were utterly forgotten. Even in 1849, when both
Schlömilch and Malmstén did mention the equation for $L(s)$
(cf. Eu.I-16^2, pp.LXXXIV–LXXXV, note 2, and A.
Weil,*Coll.Papers* III.293), they clearly regarded it as a mere
curiosity. In 1859 Riemann resurrected the subject; its further history does not concern us here.

§XXI.

In September 1740, Euler, who was already contemplating
his move to Berlin (cf. Corr.II.461|1740), received from his
future colleague Philippe Naudé a letter which raised a
number of mathematical questions (Eu.IV A-1, no.1903|29
Aug.1740).

Naudé, a mathematician of protestant French origin, was
at that time active in Berlin, where he died in 1745, aged
only 61 years. As natural at the time, his letter to Euler
asked first of all for details about Euler's summation of the
series $\zeta(2n)$. Another question concerned an equation of
Pell's type. Finally Naudé proposed a combinatorial problem:
in how many ways can a given number m be written as a

sum of μ (positive) integers, all distinct from one another? in how many ways can it be written as a sum of μ (positive) integers, distinct or not? what is for instance the answer for $m = 50$, $\mu = 7$?

Euler answered Naudé on 23 September "new style" (*PkU*.193–205, dated 12 Sept.1740 "old style"). Letters between Berlin and Petersburg were then normally no less than 10 to 12 days in transit; thus Euler's answer was prompt enough, even though he apologizes for its tardiness ("due to bad eyesight", he writes, "from which I have been suffering for some weeks": *PkU*.193; cf. *Corr*.I.102|1740, and above, §III). Nevertheless he is not only able to give full details on $\zeta(2n)$, but he can already outline the solution to Naudé's last problems, thereby creating in effect a new branch of analysis to which he gave its name, "*Partitio Numerorum*". Soon he was giving a full exposition of his results in the last memoir which he presented personally to the Petersburg Academy before leaving for Berlin (*Eu*.I-2.163–193 = E 158|6 April 1741). He was to come back upon the same topic in greater detail in Chap.XVI of the *Introductio in Analysin Infinitorum* (*Eu*.I-8.313–338) and in a memoir also composed in Berlin (*Eu*.I-2.254–294 = E 191|1750), later adding a few more results in a paper written in Petersburg (*Eu*.I-3.131–147 = E 394|1768).

As he had perceived at once (possibly even before hearing from Naudé: cf. *PkU*.203), the key to such questions lies in the consideration of suitable formal power-series in one or more indeterminates and in the relation between products (finite or infinite) of binomials and their expansions into sums or into series; as to the latter aspect, he was also encountering it in his investigations on trigonometric functions on the one hand, and on "eulerian products" on the other (cf. above, §§XVIII–XIX). Now, in order to solve Naudé's problems, he introduces first of all the infinite product $\Pi_i (1 + x^i z)$; in modernized notation, his calculation is as follows. Put:

$$P(x, z) = \prod_{i=1}^{\infty} (1 + x^i z) = \sum_{\mu=0}^{\infty} A_\mu(x) z^\mu = \sum_{m,\mu=0}^{\infty} N_{m,\mu} x^m z^\mu;$$

as he observes, $N_{m,\mu}$ is the number of ways in which m can be expressed as a sum of μ distinct integers, i.e. the number occurring in Naudé's first question. One has now

$$P(x, z) = (1 + xz)P(x, xz),$$

hence, identifying both sides:

$$A_\mu(x) = A_\mu(x)x^\mu + A_{\mu-1}(x)x^\mu.$$

By induction on μ, this gives:

$$A_\mu(x) = \frac{x^{\mu(\mu+1)/2}}{(1-x)(1-x^2) \cdots (1-x^\mu)}.$$

Now put:

$$\frac{1}{(1-x)(1-x^2) \cdots (1-x^\mu)} = \prod_{i=1}^{\mu} (1 + x^i + x^{2i} + \cdots) = \sum_{m=0}^{\infty} M_{m,\mu}x^m;$$

$M_{m,\mu}$ is the number of ways in which m can be expressed as a sum of integers (in any number) belonging to the set $\{1, 2, ..., \mu\}$. This gives:

$$A_\mu(x) = x^{\mu(\mu+1)/2} \sum_{m=0}^{\infty} M_{m,\mu}x^m, \qquad N_{m,\mu} = M_{m-\mu(\mu+1)/2,\mu}.$$

On the other hand, Euler also introduces the product:

$$Q(x, z) = \prod_{i=1}^{\infty} (1 - x^i z)^{-1} = \prod_{i=1}^{\infty} (1 + x^i z + x^{2i}z^2 + \cdots)$$

$$= \prod_{\mu=0}^{\infty} B_\mu(x)z^\mu = \prod_{m,\mu=0}^{\infty} N'_{m,\mu}x^m z^\mu;$$

here $N'_{m,\mu}$ is the number in Naudé's second question, i.e. the number of ways in which m can be written as a sum of μ integers, distinct or not. Here one has

$$Q(x, xz) = (1 - xz)Q(x, z)$$

and then, proceeding just as before:

$$B_\mu(x) = \frac{x^\mu}{(1-x)(1-x^2) \cdots (1-x^\mu)}, \qquad N'_{m,\mu} = M_{m-\mu,\mu}.$$

From these facts Euler is able to derive a number of recurrence formulas between the coefficients $M_{m,\mu}$ and to construct extensive tables for these numbers. Another noteworthy identity concerning partitions is the following:

$$\prod_{i=1}^{\infty} (1+x^i) = \frac{\prod_{i=1}^{\infty} (1-x^{2i})}{\prod_{i=1}^{\infty} (1-x^i)} = \frac{1}{\prod_{i=1}^{\infty} (1-x^{2i-1})}$$

which, by the same argument as before, connects the number of ways an integer can be written as a sum of distinct integers with the number of ways of expressing it as a sum of odd integers, distinct or not.

It was of course at once obvious to Euler that his numbers $M_{m,\mu}$ have a property of "stability" when μ increases; more precisely, they are independent of μ for $\mu \geqslant m$, and one has

$$\prod_{i=1}^{\infty} (1-x^i)^{-1} = \sum_{m=1}^{\infty} M_{m,m} x^m.$$

Already in 1741 this suggested to him to calculate the product $\prod_{i=1}^{\infty} (1-x^i)$. "Let me end this paper", he writes, "with a noteworthy observation, which however I have not yet been able to prove with geometric rigor" ["*Finem huic dissertationi faciat observatio notatu digna, quam quidem rigore geometrico demonstrare mihi nondum licuit*": *Eu.*I-2.191 in E 158|1741]. The result, conjectured after Euler had carried out the product in question for no less than 51 factors:

$$(1-x)(1-x^2) \cdots (1-x^{51})$$
$$= 1 - x - x^2 + x^5 + x^7$$
$$- x^{12} - x^{15} + x^{22} + x^{26} - x^{35} - x^{40} + x^{51} + \text{etc.}$$

was indeed astonishing; all coefficients in the series are 0 or ± 1, and in the exponents Euler recognized at once the "pentagonal numbers" $\frac{1}{2}n(3n+1)$ for $n = 0, -1, +1, \ldots,$ $-5, +5, -6$, suggesting the celebrated formula

$$\prod_{i=1}^{\infty} (1-x^i) = \sum_{n=-\infty}^{+\infty} (-1)^n x^{n(3n+1)/2}$$

(*Eu.*I-2.191–192 in E 158|1741; cf. *Corr.*II.467–468|28

Jan.1741, *Eu*.I-2.280–281 in E 191|1750 and *Eu*.I-8.332 in
Chap.XVI of the *Introductio*). With its generalizations it was
to play a central role in the next century in Jacobi's *Fundamenta Nova* of 1829 and in all his later work on theta
functions (cf. *Jac*.I.234–236, and *Eu*.I-2.191–192, note 1).
Of course Euler's gifts of divination did not extend as far
as that; to him the formula was no more than a formal
identity amounting to an arithmetical result.

Some years later he transformed that formula into a
statement about the arithmetical function that he denoted
by $\int n$ (cf. above, §VI); it is the sum of the divisors of n,
including 1 and n, and is now commonly denoted $\sigma_1(n)$; his
interest in it had lately been aroused by an attempt to deal
with "amicable numbers" (*Eu*.I-2.59–61 = E 100|1747, *Eu*.I-
2.86–162 = E 152|1747). Taking the logarithmic derivative
of the product $\Pi\,(1-x^i)$, one gets formally

$$\frac{d}{dx}\log\prod_{i=1}^{\infty}(1-x^i)=-\sum_{i=1}^{\infty}\frac{ix^{i-1}}{1-x^i}$$

$$=-\sum_{i=1}^{\infty}\left(i\sum_{\nu=1}^{\infty}x^{\nu i-1}\right)=-\sum_{N=1}^{\infty}(\int N)x^{N-1};$$

identifying this with the logarithmic derivative of the series
in Euler's identity, one obtains

$$\sum_{n=-\infty}^{+\infty}(-1)^n\int(N-\tfrac{1}{2}n(3n+1))=0$$

provided one puts $\int m=0$ for all $m<0$, and provided one
replaces $\int 0$ by N in that relation if such a term occurs in it,
i.e. if N is one of the "pentagonal numbers" $\tfrac{1}{2}n(3n+1)$. Euler
lost no time in sending this to Goldbach, calling it *"eine sehr
wunderbare Ordnung"* (*Corr*.I.407–408|1 April 1747); very
soon he was writing it up for publication under the title
*"Découverte d'une loi toute extraordinaire des nombres par rapport
à la somme de leurs diviseurs"* (*Eu*.I-2.241–253 = E 175|22
June 1747). Of its validity he had no doubt: *"elle appartient
à ce genre dont nous pouvons nous assurer de la vérité, sans en
donner une démonstration parfaite"* ("it belongs to the kind of

results whose truth is certain even though we cannot give a perfect proof for it": *Eu*.I-2.242).

Of course this did not stop him from looking for a "perfect proof", and (of course, one is tempted to say) he found one (*Corr*.I.522–524|1750; *Eu*.I-2.390–398 = E 244|1751). This proof, another dazzling display of algebraic virtuosity but quite elementary, rests upon the construction of a sequence of formal power series $P_0 = \Pi(1 - x^n)$, P_1, P_2, etc., such that the following relations hold for all $n > 0$:

$$P_{n-1} = 1 - x^{2n-1} - x^{3n-1} P_n,$$

$$P_n = \sum_{\nu=0}^{\infty} x^{\nu n}(1-x^n)(1-x^{n+1}) \cdots (1-x^{n+\nu}).$$

The case $n = 1$ is obtained by a straightforward application, to the product $\Pi(1 - x^n)$, of the obvious identity

$$\prod_{n=1}^{\infty}(1-\alpha_n) = 1 - \sum_{n=1}^{\infty} \alpha_n(1-\alpha_1)(1-\alpha_2) \cdots (1-\alpha_{n-1}).$$

Then, assuming the P_i to have been constructed up to P_n, we have:

$$P_n = 1 + x^n(1-x^{n+1}) + \sum_{\nu=2}^{\infty} x^{\nu n}(1-x^{n+1}) \cdots (1 - x^{n+\nu})$$

$$-x^n - \sum_{\nu=1}^{\infty} x^{\nu n+n}(1 - x^{n+1}) \cdots (1 - x^{n+\nu}).$$

Replacing ν by $\nu + 2$ in the former sum, by $\nu + 1$ in the latter, and then combining the two, one finds indeed that this is nothing else than $1 - x^{2n+1} - x^{3n+2}P_{n+1}$, so that one can proceed by induction on n. This gives:

$$P_0 = 1 - x - x^2(1-x^3) + x^{2+5}(1-x^5) - \cdots$$

$$+ (-1)^n x^{2+5+\cdots+(3n-1)} P_n$$

for all $n > 0$, which completes the proof.

With another brilliant idea, arisen in the same context,

Euler was less successful. Having come to regard Fermat's famous assertion about "three triangular numbers, four squares, five pentagonal numbers, etc." (cf. above, §Vb) as belonging by rights to the subject of *"partitio numerorum"*, he naturally thought of applying to it the method he had introduced for such purposes, viz., the method of formal power-series (*Corr.*I.531–532|1750; cf. *Eu.*I-3.132, 144–145 in E 394|1768). For instance, as he observes, the statement about four squares amounts to saying that if one writes s for the formal power-series

$$s = 1 + x^1 + x^4 + x^9 + \text{etc.}$$

all powers of x occur in the series s^4 with non-zero coefficients; well he knew, of course, that the coefficient of x^n in s^4 would indicate in how many ways n can be written as a sum of four squares. "This", he writes to Goldbach, "seems to me the most natural way of attaining to the proof of Fermat's theorems" ["*Dieser Weg däucht mir noch der natürlichste zu seyn, um zum Beweis der theorematum Fermatianorum zu gelangen*": *Corr.*I.532|1750]. Except for the substitution of the "theta series"

$$2s - 1 = \sum_{n=-\infty}^{+\infty} x^{n^2}$$

for s, this is precisely how Jacobi proved the theorem in question as a worthy conclusion to his *Fundamenta*, thus fulfilling Euler's prediction. Nor was Euler content with the mere consideration of s as a formal series; he would not have been himself if he had not made it an occasion for numerical computations; what attracted him was its behavior as x approaches -1, or, what amounts to the same, the behavior of the series

$$t = 1 - x^4 + x^9 - x^{16} + \text{etc.}$$

near $x = 1$ (*Corr.*I.529–531|1750). Here, substituting at first $1 - y$ for x, he obtains formally

$$\sum_{n=0}^{\infty} (-1)^n (1-y)^{n^2} = \tfrac{1}{2} + 0 \cdot y + 0 \cdot y^2 + 0 \cdot y^3 + \text{etc.},$$

this being an easy consequence of his earlier results according to which $1^m - 2^m + 3^m -$ etc. is 0 for even $m > 0$ and has the value $\frac{1}{2}$ for $m = 0$ (cf. above, §XX). It suggests that, for x approaching 0, $2t - 1$ tends to 0 faster than any power of $1 - x$, an expectation which Euler seeks to confirm by computing t with 10 decimal figures, successively for $x = 0, \frac{1}{2}, \frac{2}{3}, \frac{7}{10}, \frac{9}{10}$ (no mean feat, in view of the slow convergence of the series t for x close to 1). "But I have vainly looked for a sure way of summing this series numerically for $x = 1 - \omega$ when ω is very small", he writes to Goldbach; "a method for doing that would be of high value" ["*es wäre eine Methode hoch zu schätzen, vermittelst welcher man im Stande wäre den Werth von t proxime zu bestimmen, wenn ω ein sehr kleiner Bruch ist*": *Corr.*I.531]. Here again one must admire Euler's sureness of judgment; the answer to his question is contained in the functional equation of the theta series; this, too, was given to the mathematical world by Jacobi in his *Fundamenta*.

§XXII.

Euler's arithmetical work, although it extends over some fifty years, makes up only a small part of his immense production. Nevertheless some of the main characteristics of his scientific temperament emerge clearly from the above survey.

Perhaps its most salient feature is the extraordinary promptness with which he always reacted even to casual suggestions or stimuli: a question from Goldbach about the primality of Fermat's numbers $2^{2^n} + 1$ (cf. above, §§IV, Va and VI); Fermat's statement on sums of four squares, which at first he did not even read through (§§IV, Vb, XI); an elementary combinatorial problem proposed by Naudé (§§Vh, XXI); a few isolated theorems in a book by a little-known Italian nobleman (§§Ve, XV); a theorem by Lagrange on congruences modulo a prime (§VI); Lagrange's somewhat less than felicitous proof for Fermat's theorem on sums of four squares (§XI); etc. Nor was he neglecting chance ob-

servations such as his calculation of $\Pi(1-x^n)$ (cf. §XXI);
"such observations are all the more valuable", he had written
after receiving Fagnano's volume (*loc.cit.* §XV). Every occasion was promptly grasped; each one supplied grist to
his mill, often giving rise to a long series of impressive
investigations.

Hardly less striking is the fact that Euler never abandoned
a problem after it had once aroused his insatiable curiosity.
Other mathematicians, Hilbert for instance, have had their
lives neatly divided into periods, each one devoted to a
separate topic. Not so Euler. All his life, even after the loss
of his eyesight, he seems to have carried in his head the
whole of the mathematics of his day, both pure and applied.
Once he had taken up a question, not only did he come
back to it again and again, little caring if at times he was
merely repeating himself, but also he loved to cast his net
wider and wider with never failing enthusiasm, always ex-
pecting to uncover more and more mysteries, more and
more "*herrliche proprietates*" lurching just around the next
corner. Nor did it greatly matter to him whether he or
another made the discovery. "*Penitus obstupui*", he writes ("I
was quite flabbergasted": *Eu.*I-21.1 in E 506|1777; cf. his
last letter to Lagrange, *Eu.*IV A-5.505|1775) on learning of
Lagrange's additions to his own work on elliptic integrals;
after which he proceeds to improve upon Lagrange's
achievement. Even when a problem seemed to have been
solved to his own satisfaction (as happened with his first
proof of Fermat's theorem $a^p \equiv a \bmod p$, or in 1749 with
sums of two squares) he never rested in his search for better
proofs, "more natural" (*Eu.*I-2.510 in E 262|1755; cf. §VI),
"easy" (*Eu.*I-3.504 in E 552|1772; cf. §VI), "direct" (*Eu.*I-
2.365 in E 242|1751; cf. §VI); and repeatedly he found
them.

Some problems eluded him all his life. Most conspicuous
among these perhaps, so far as number theory is concerned,
are his conjectures related to the law of quadratic reciprocity,
conjectures first formulated in 1742 and soon thereafter,
and whose importance he was still emphasizing at the end

of his life (cf. §§V*i*, VIII, IX); by that time he could scarcely hope to obtain a proof himself but was expressing his confident expectation that their proof would be "forthcoming soon" ("*quam . . . mox expectare licebit*": *Eu.*I-4.191 in E 598|1775). How pleased he would have been to see even the partial proof given by Legendre in 1785, not to mention the full proof found by Gauss in 1796 and published in 1801! It was not given to him to reach the promised land; but his eyes, now turned inward, had been allowed a glimpse into its splendor, or, to use his own word, its *Herrlichkeit*.

Appendix I

The Quadratic Reciprocity Law

As explained above in §VIII (cf. Chap.II, §XII), a prime p, assumed to be odd and not a divisor of N, is in Euler's sense a "prime divisor of the form $X^2 + NY^2$" if it divides some integer $a^2 + Nb^2$ with a prime to Nb; this is equivalent to saying, either that $-N$ is a quadratic residue modulo p, or alternatively that p splits into two prime ideal factors in the field $\mathbf{Q}(\sqrt{-N})$. Our purpose here is to describe the facts about those primes in the form best suited for comparison with Euler's empirical results.

Without restricting the generality we shall assume that N is squarefree and neither -1 nor 0; we shall put $d = -N$ or $d = -4N$ according as $N \equiv -1 \pmod 4$ or not, and $D = |d|$. Then there is a function ω of integers with the following properties:

(I) $\omega(n) = \pm 1$ *if n is prime to D; otherwise* $\omega(n) = 0$.

(II) ω *is periodic of period D*, i.e. it satisfies

$$\omega(n) = \omega(n+D)$$

for all n; this can also be expressed by saying that ω is a function of integers modulo D.

(III) $\omega(mn) = \omega(m)\omega(n)$ *for all m and n*. This is expressed by saying that ω is a "Dirichlet character" modulo D. In particular, $\omega(n^2) = 1$ for all n prime to D.

(IV) $\omega(-1)$ *is* $+1$ *or* -1 *according as N is* <0 *or* >0.

(V) *There is no divisor D' of D other than D such that* $\omega(m) = \omega(n)$ *whenever m, n are prime to D and* $m \equiv n \pmod{D'}$. This is expressed by saying that the character ω is *primitive* modulo D; in view of (I) and (III), it amounts to saying that there is no divisor D' of D, other than D, such that $\omega(n) = 1$ whenever n is prime to D and $\equiv 1 \pmod{D'}$.

Finally, the relevance of ω to the question of the prime divisors of $X^2 + NY^2$ is given by the following:

(VI) *An odd prime p is a prime divisor of $X^2 + NY^2$ if and only if $\omega(p) = +1$.*

In view of (I), (II) and (III), ω determines a character of order 2 (in the sense now usual in group-theory) on the multiplicative group $G_D = (\mathbf{Z}/D\mathbf{Z})^\times$ of integers prime to D modulo D; thus its kernel, i.e. the set of congruence-classes modulo D on which it takes the value 1 is a subgroup of G_D of index 2; according to (III) this kernel contains the group $(G_D)^2$ of quadratic residues modulo D.

It will now be shown that conditions (I) to (V) determine such a function ω and determine it uniquely. Take first the case when D is a prime; this will be so whenever $N = \pm q$, where q is an odd prime and the sign is determined by $N \equiv -1 \pmod 4$. In that case, as Euler had found out not later than 1751 (*Eu.*I-2.339–355 in E 242|1751; cf. nos. 284–307 of the *Tractatus, Eu.*I-5.227–230, and above, §VI), quadratic residues make up a subgroup of index 2 of $G_D = G_q$. Therefore $\omega(n)$ is $+1$ or -1 according as n is a quadratic residue or non-residue modulo q; in the notation in use since Legendre (cf. Chap.IV, §VI), this is written $\left(\dfrac{n}{q}\right)$, or more compactly (n/q). That this satisfies (IV) is nothing else than the determination obtained by Euler in 1749 of the quadratic residue character $(-1/q)$ of -1 modulo q (cf. above, §VI, and Chap.II, §VII).

In the general case, write $D = q_0 q_1 q_2 \ldots q_r$, where q_0 is 1, 4, or 8, and $q_1, q_2, \ldots, q_r$ are distinct odd primes. As Euler knew, at least from the time when he wrote the *Tractatus*, an integer is a quadratic residue modulo D if and only if it is such modulo each one of the q_i. In modern terms, $G_D/(G_D)^2$ is a group of type $(2, 2, \ldots, 2)$ with 2^r, 2^{r+1} or 2^{r+2} elements according as q_0 is 1, 4 or 8; each one of its characters can be written as the product of some or all of the characters

(n/q_i) for $i = 1, 2, ..., r$, and, if $q_0 = 4$ or 8, of a character modulo q_0; a primitive one such as ω is then necessarily given, for n prime to D, by a product

$$\omega(n) = \psi(n) \prod_{i=1}^{r} \left(\frac{n}{q_i}\right)$$

where $\psi = 1$ if $q_0 = 1$, and otherwise ψ is a primitive character modulo q_0. If $q_0 = 4$ there is only one such character ψ_0, given by $\psi_0(1) = 1$, $\psi_0(-1) = -1$. If $q_0 = 8$ we have $G_8 = \{\pm 1, \pm 3 \bmod 8\}$; a primitive character of G_8 must satisfy $\psi(-3) = -1$ since otherwise $\psi(n)$ would depend only upon the value of n modulo 4; thus there are two such characters ψ_1, ψ_2, respectively determined by $\psi_1(-1) = 1$, $\psi_2(-1) = -1$. If $q_0 = 8$, the character ψ occurring in the above formula for ω is therefore the one determined by condition (IV), i.e. the one given by

$$\psi(-1) = \mathrm{sgn}(-N) \prod_{i=1}^{r} \left(\frac{-1}{q_i}\right)$$

where $\mathrm{sgn}(-N)$ is 1 or -1 according as $-N$ is >0 or <0. The fact that condition (IV) is satisfied also for $q_0 = 1$ or 4, when ω is determined as above, is again an easy consequence of the determination of $(-1/q)$ for any odd prime q.

As to condition (VI), which can now be written as

$$\left(\frac{-N}{p}\right) = \omega(p)$$

(this being valid for all odd primes p, not dividing N), take first the case $N = 1$. Then $D = 4$; ω is the character ψ_0 modulo 4; and (VI) says that $(-1/p) = \psi_0(p)$, which is indeed the case. Take $N = \pm 2$; then $D = 8$; ω is ψ_1 if $N = -2$, ψ_2 if $N = +2$; then (VI) says that $(2/p) = \psi_1(p)$, $(-2/p) = \psi_2(p)$, as was found to be the case in §IX.

Take now $N = \pm q$, the sign being determined by $N \equiv -1 \pmod 4$, so that $D = q$. Then (VI) can be written as

$$\left(\frac{\mp q}{p}\right) = \left(\frac{p}{q}\right)$$

where p, q are any two distinct odd primes, and the sign is such that $\mp q \equiv 1 \pmod 4$. In view of the previous determination of $(-1/p)$, this is tantamount to the law of quadratic reciprocity[1], as first formulated by Legendre in 1785 (cf. *infra*, Chap.IV, §VI). Thus the above properties (I) to (VI) of ω may be said to include that law. Now, assuming its validity, we will show that condition (VI) is satisfied for all values of N.

Put again $D = q_0 q_1 \dots q_r$ as before, and, for $i = 1, 2, \dots$, r, put $q'_i = \pm q_i$, the sign being determined by $q'_i \equiv 1 \pmod 4$. We can write $-N = q'_0 q'_1 \dots q'_r$ with $q'_0 = \pm 1$ or ± 2. Then we have:

$$\left(\frac{-N}{p}\right) = \left(\frac{q'_0}{p}\right) \prod_{i=1}^{r} \left(\frac{q'_i}{p}\right) = \left(\frac{q'_0}{p}\right) \prod_{i=1}^{r} \left(\frac{p}{q_i}\right).$$

As noted above, we have $(q'_0/p) = \psi'(p)$, where ψ' is 1, ψ_0, ψ_1 or ψ_2 according as q'_0 is $+1$, -1, $+2$ or -2; the proof will be complete if we show that ψ' is the same as the character ψ occurring in the definition of ω. This is so for $q'_0 = 1$, since in that case we have $-N \equiv 1 \pmod 4$, $D = |N|$, $q_0 = 1$, $\psi = 1$. Similarly, if $q'_0 = -1$, we have $N \equiv 1 \pmod 4$, $D = 4|N|$, $q_0 = 4$, $\psi = \psi_0$. Now take the case $q'_0 = \pm 2$; then $D = 4|N|$, $q_0 = 8$, $\psi = \psi_1$ or ψ_2, and we have to show that $\psi(-1) = \psi'(-1)$. But in that case we have

$$\text{sgn}(q'_i) = \left(\frac{-1}{q_i}\right), \ \text{sgn}(-N) = \text{sgn}(q'_0) \prod_{i=1}^{r} \left(\frac{-1}{q_i}\right)$$

and therefore

$$\psi(-1) = \text{sgn}(q'_0) = \psi'(-1),$$

which completes our proof.

---

[1] The most lucid exposition of this topic is the one contained in Chap.III of Dirichlet's *Vorlesungen über Zahlentheorie* (ed. R. Dedekind, Braunschweig 1894). For perhaps the simplest known proof of the "law" in question, which is Eisenstein's, cf. A. Weil, *Number Theory for Beginners*, Springer 1979, §XII. Cf. also J.-P. Serre, *Cours d'Arithmétique*, P.U.F., Paris 1970, Chap.I.

Finally, in order to facilitate the comparison of the above results with Euler's conjectures (cf. §VIII), it is convenient to introduce the kernel K_N of the character induced by ω on the multiplicative group modulo $4|N|$, i.e. the set of the congruence classes prime to $4|N|$ modulo $4|N|$ on which ω takes the value 1. Clearly this has the following properties:

(A) K_N is a subgroup of index 2 of the multiplicative group modulo $4|N|$.

(B) -1 belongs to K_N or not, according as $N < 0$ or $N > 0$.

(C) If r and s are prime to $4|N|$, and if $r \equiv s$ (mod D), then both belong to K_N whenever one of them does, and no divisor D' of D, other than D, has the same property.

Conversely, let K_N be a subgroup of index 2 of the multiplicative group modulo $4|N|$, satisfying conditions (B) and (C). Let ω' be the function of integers modulo $4|N|$ such that $\omega'(n) = 0$ if n is not prime to $4|N|$, $\omega'(n) = 1$ if n belongs to K_N, and $\omega'(n) = -1$ if n is prime to $4|N|$ and not in K_N. For $N \equiv 1$ or 2 (mod 4), ω' satisfies all the conditions (I) to (V) listed above, so that it is no other than ω. Now take $N \equiv -1$ (mod 4), so that $D = |N|$. Define a function ω'' of integers by putting $\omega''(n) = 0$ when n is not prime to N, and $\omega''(n) = \omega'(r)$ when n is prime to N, r is prime to $4|N|$, and $n \equiv r$ (mod $|N|$); in view of (C), this is independent of the choice of r, so that ω'' is well defined; it is easy to see that ω'' satisfies all the conditions (I) to (V), so that it is the same as ω. This shows that ω' in the former case, and ω'' in the latter case, satisfies condition (VI). Thus an odd prime, not a divisor of N, is a divisor of the form $X^2 + NY^2$ if and only if it belongs to K_N modulo $4|N|$, which is what Euler had conjectured in 1744.

Appendix II

An Elementary Proof For Sums of Squares

In 1751, Euler, having proved that every integer is a sum of (at most) four squares "*in fractis*", i.e. in rational numbers, wrote as follows:

"*In Analysi quidem Diophantea pro certo assumi solet nullum numerum integrum in quatuor quadrata fracta dispertiri posse, nisi eius resolutio in quatuor quadrata integra vel pauciora constet . . . Verum nusquam adhuc eiusmodi demonstrationem inveni . . .*" ["In diophantine analysis one usually takes for granted that no integer can be split into four rational squares unless it has an expression as a sum of four integral squares or less . . . but so far I have never found a proof of this . . .": *Eu*.I-2.372 in E 242|1751]. This is indeed a question which must occur naturally, concerning sums of 2, 3 or 4 squares, to any reader of Diophantus, and Fermat had already given it some thought, apparently without success, at the beginning of his career as a number-theorist (cf. Chap.II, §V). Euler himself had raised it, more particularly concerning sums of two squares, in 1745 (cf. *Corr*.I.312), and again repeatedly, concerning sums of four squares, in his correspondence with Goldbach (cf. *Corr*.I.521|1750, 527|1750, 559|1751).

Here we shall reproduce a proof due to L. Aubry (*Sphinx-Œdipe* 7 (1912), pp.81–84) which applies equally well to sums of 2, 3, or 4 squares and a few other quadratic forms. We first describe it in geometric terms for sums of 3 squares.

Points in $\mathbf{R}^3$ will be called *rational* (resp. *integral*) if they have rational (resp. integral) coordinates. To each point $a = (x, y, z)$ there is an integral point $a' = (x', y', z')$ at a euclidean distance less than 1 from a; for instance one can take for x', y', z' the integers respectively closest to x, to y and to z, in which case the distance from a to a' is $\leqslant \sqrt{3}/2$.

Now assume that an integer N is a sum of three rational squares; this is the same as to say that there is a rational point a_0 on the sphere S given by $N = x^2 + y^2 + z^2$. Let a_0' be the integral point (or one of the integral points, if there are two or more) closest to a_0; its distance from a_0 is <1. Call a_1 the second intersection of S with the straight line joining a_0 to a_0'; it is rational. Let a_1' be an integral point closest to a_1; let a_2 be the intersection of S with the line joining a_1 to a_1'; etc. It will now be shown that, for some n, a_n is an integral point.

Put $a_0 = (x, y, z)$; assume that it is not an integral point, and call m the lowest common denominator for x, y, z; write $x = n/m, y = p/m, z = q/m$; we have

$$Nm^2 = n^2 + p^2 + q^2.$$

Let $a_0' = (n', p', q')$ be an integral point closest to a_0, and put

$$r = n - mn', \qquad s = p - mp', \qquad t = q - mq',$$
$$N' = n'^2 + p'^2 + q'^2, \qquad M = 2(nn' + pp' + qq').$$

The squared distance between a_0 and a_0' is then

$$\frac{1}{m^2}(r^2 + s^2 + t^2) = N + N' - \frac{M}{m}$$

which can be written as m'/m, where m' is an integer; as it is <1, we have $0 < m' < m$, and at the same time:

$$r^2 + s^2 + t^2 = mm', \qquad M = m(N + N') - m'.$$

Now the line joining a_0 to a_0' consists of the points

$$(n' + \lambda r, p' + \lambda s, q' + \lambda t);$$

here the point a_1 is given by

$$
\begin{aligned}
0 &= (n' + \lambda r)^2 + (p' + \lambda s)^2 + (q' + \lambda t)^2 - N \\
&= mm'\lambda^2 + (M - 2mN')\lambda + N' - N \\
&= (m\lambda - 1)(m'\lambda + N - N').
\end{aligned}
$$

The root $\lambda = 1/m$ corresponds to the point a_0, so that the other root $\lambda = (N' - N)/m'$ corresponds to a_1. Thus m' is a common denominator for the coordinates of a_1; as it is $< m$, this shows that the lowest common denominators for the coordinates of a_0, a_1, a_2, etc. make up a decreasing sequence of positive integers, which proves our assertion.

The only property of the quadratic form

$$
F(X, Y, Z) = X^2 + Y^2 + Z^2
$$

which has been used in this proof is that, to every non-integral point (x, y, z), there is an integral point (x', y', z') such that

$$
0 < |F(x - x', y - y', z - z')| < 1.
$$

This applies equally well, for instance, to the forms $X^2 + Y^2$, $X^2 \pm 2Y^2$, $X^2 - 3Y^2$. Now, modifying the notation in an obvious manner, we will show how the same proof can be applied to some quadratic forms $F(x)$ in $\mathbf{R}^n$, taking integral values at all integral vectors x and such that, to every non-integral vector x in $\mathbf{R}^n$, there is an integral vector x' for which $0 < |F(x - x')| \leq 1$; this will be the case for instance for $X^2 + 3Y^2, X^2 + Y^2 + 2Z^2, X^2 + Y^2 + Z^2 + T^2$. Define the bilinear form $B(x, y)$ by

$$
F(\lambda x + \mu y) = \lambda^2 F(x) + \lambda \mu B(x, y) + \mu^2 F(y).
$$

As before, take a rational point a_0 such that $F(a_0) = N$, and an integral point a_0' such that

$$
0 < |F(a_0 - a_0')| \leq 1.
$$

Let m be the smallest integer such that $n = ma_0$ is an integral point; put $a_0' = n', r = n - mn', N' = F(n'), M = B(n, n')$.

We have

$$F(a_0 - a_0') = F(m^{-1}r) = F(m^{-1}n - n') = N + N' - \frac{M}{m};$$

writing this as m'/m, we have $0 < m' \leqslant m$, and m' is an integer.

The line joining a_0 to a_0' consists of the points $n' + \lambda r$, and its intersection a_1 with the hypersurface $F(x) = N$ is given by

$$0 = F(n' + \lambda r) - N = mm'\lambda^2 + (M - 2mN')\lambda + N' - N$$

$$= (m\lambda - 1)(m'\lambda + N - N').$$

As before, a_1 corresponds to the root $\lambda = (N' - N)/m'$, so that $m'a_1$ is an integral point, and m' is a common denominator for the coordinates of a_1; however, we have now to take into account the case $|m'| = m$, i.e. $|F(m^{-1}r)| = 1$. Taking for F one of the three forms listed above, we may assume that we have taken for a_0' the point, or one of the points, whose coordinates are the integers closest to those of a_0, so that the coordinates of $m^{-1}r$ are $\leqslant \frac{1}{2}$ in absolute value; then, for $F(m^{-1}r)$ to be 1, they must all be $\pm \frac{1}{2}$, and $2a_0$ must be integral, so that $m = 2$. In that case there are 2^n possible choices for a_0', and this choice can be made so that $N - N'$ is even; since at the same time m' must be 1 or 2, a_1 is then integral.

The above proof would have been easily understood by Euler; perhaps, with a little more effort, it would have been understood by Fermat, whose algebraic skills still fell somewhat short of the required level. That it was discovered so late may serve as an encouragement to those who seek elementary proofs for supposedly sophisticated results.

Appendix III

The Addition Theorem For Elliptic Curves

Our purpose here will be to provide an appropriate background for Euler's theory of elliptic functions. As in Chapter II, Appendix II, let Γ be an algebraic curve defined over a groundfield of characteristic 0. By a differential on Γ one understands any expression $\omega = f \cdot dg$, where f, g are rational functions on Γ; this may be identified with $f(dg/dx)dx$, where one can take for x any non-constant rational function on Γ.

To every point M on Γ one can attach a rational function x on Γ having a simple zero at M; one can then use it as a local coordinate at M, and every function f on Γ can be expressed in a neighborhood of M as a Laurent series $f = \Sigma_{i=n}^{\infty} a_i x^i$; if $a_n \neq 0$, n is the order of f at M; it is >0 if f has at M a zero of multiplicity n, and it is <0 if f has at M a pole of multiplicity $-n$. We recall that the number of zeros of a non-constant function f on Γ is equal to the number of its poles, both being counted with their multiplicities; if this number is N, f is said to be of order N, and the field of all rational functions on Γ is an algebraic extension of degree N of the field $K(f)$ of rational functions of f, K being the groundfield.

Let M and x be as above; then every differential ω on Γ can be expressed in the neighborhood of M as $\omega = \Sigma_{i=n}^{\infty} a_i x^i dx$; it is said to have a zero of multiplicity n at M

if $a_n \neq 0$, $n > 0$, and a pole of multiplicity $-n$ at M if $a_n \neq 0$, $n < 0$. It is said to be exact at M if it can be written there as $d(\Sigma_{i=m}^{\infty} b_j x^j)$; this is so if and only if $a_{-1} = 0$. A differential is said to be of the first kind if it has no pole, of the second kind if it is exact everywhere; otherwise it is said to be of the third kind.

From now on we shall take the groundfield to be the field $\mathbf{C}$ of complex numbers. Then differentials can be integrated along any (piecewise differentiable) path on Γ; such an integral, taken along a closed path, is called a period.

Let first Γ be of genus 0; then there is on Γ a rational function x of order 1; rational functions on Γ are those of the form $R(x) = P(x)/Q(x)$, where P, Q are polynomials; Γ can be identified with the "projective straight line" (over $\mathbf{C}$), also known as "the Riemann sphere", i.e. the plane of the complex variable x to which is added one point at infinity, written $x = \infty$. On Γ there is no differential of the first kind; if $\omega = R(x)dx$ is any differential on Γ, then (as was in substance proved by Euler; cf. above, §§VII and XV) its integral $\int\omega$, which is well-defined up to an additive constant, can always be expressed in terms of rational and logarithmic functions (rational, logarithmic and inverse trigonometric functions if the groundfield is the field of real numbers, as is always understood by Euler). This integral is itself a rational function, i.e. one can write $\omega = dS(x)$ with a rational function S, if and only if ω is of the second kind.

Next, let Γ be any algebraic curve; let $\Sigma_1^N A_i$, $\Sigma_1^N B_j$ be two equivalent divisors on Γ (cf. Chap.II, Appendix II); we assume that $A_i \neq B_j$ for all i, j. This means that there is on Γ a function x of order N, with the N zeros A_i and the N poles B_j (both being counted with their multiplicities). Then Γ can be identified with an N-sheeted Riemann surface spread over the "Riemann sphere" for x, and the field of rational functions on Γ can be written as $\mathbf{C}(x, y)$, where y is a function on Γ, algebraic of degree N over $\mathbf{C}(x)$. Take the irreducible equation for y over $\mathbf{C}(x)$:

$$P(x, Y) = \sum_0^N F_i(x) Y^{N-i} = F_0(x) \prod_1^N (Y - y_i) = 0,$$

where the F_i are polynomials, and $y_1 = y$. Let $\omega = R(x, y)dx$ be any differential on Γ; put

$$\eta = \sum_1^N R(x, y_i)dx.$$

This can be written as $S(x)dx$, where S is a rational function; it is a differential on the "projective straight line". It is of the first kind, and therefore 0, if ω is of the first kind; it is of the second kind if ω is of the second kind. Now take any path λ from $x = 0$ to $x = \infty$ in the x-plane; its inverse image on Γ consists of N paths λ_i going respectively from A_i to B_i on Γ (for $1 \leqslant i \leqslant N$) if the points A_i, B_i have been numbered suitably. We have in that case:

$$\int_\lambda \eta = \sum_1^N \int_{\lambda_i} \omega.$$

Here the left-hand side, and therefore also the right-hand side, must be 0 if ω is of the first kind; they can be expressed rationally in terms of the data if ω is of the second kind; in the general case the result is the same except that logarithmic terms may be needed. This, in substance, is Abel's theorem. If other paths than the λ_i are used in the right-hand side, its value is merely modified by some period of ω.

The relevant case for Euler's theory of elliptic integrals (and for Fagnano's earlier investigations on the same subject) is that of a curve Γ given by an equation $y^2 = F(x)$, where F is a polynomial of degree 3 or 4 without double roots (cf. Chap.II, Appendix II); then Γ is of genus 1, and x is of order 2 on Γ; up to a constant factor, there is on Γ one and only one differential of the first kind, given by $\omega = dx/y$; it has no zero. Let $A + N, B + M$ be equivalent divisors on Γ; then, by Abel's theorem, we have

$$(1) \qquad \int_A^M \omega = \int_B^N \omega,$$

or, what amounts to the same

$$(2) \qquad \int_A^N \omega = \int_A^M \omega + \int_A^B \omega$$

if the paths of integration are suitably chosen; otherwise both sides differ by a period of ω. If ω is replaced by a differential ω' of the second or third kind, then, for the same reason, the difference of both sides can be expressed, up to a period of ω', by rational and logarithmic functions in terms of the data; it can be so expressed as a rational function if ω' is of the second kind. As to the condition $A + N \sim B + M$, it can also be written as $N \sim M + B - A$, or, in the notation explained in Chap.II, Appendix II, as $N = M \dotplus B$ if A is taken as origin. As follows from the general theory which has been sketched there, N can be expressed rationally in terms of A, B, M; as will be seen presently, explicit formulas for this were obtained by Euler and were perhaps his main discovery on the subject of elliptic integrals.

In the classical theory of elliptic functions according to Jacobi, Abel and their successors, one takes as independent variable the value of the integral

$$u = \int_A^M \omega$$

along any path with the origin A and the end-point M on Γ; for a given M this is defined modulo a period of ω; one proves that the periods of ω make up a lattice L in the complex plane $\mathbf{C}$ and also that M is uniquely determined by the value of u modulo L; thus the mapping $u \rightarrow M$ determines an isomorphism of the group $\mathbf{C}/L$ onto the group of the points on Γ under the operation $\dotplus$. "Elliptic functions" are then the rational functions on Γ, viewed as functions (meromorphic functions, as their definition implies) of the independent variable u. The algebraic formulas which express $M \dotplus M'$ in terms of M and M' (resp. $nM = M \dotplus \cdots \dotplus M$ in terms of M) can then be viewed as addition (resp. multiplication) formulas for the elliptic functions belonging to Γ. This point of view, which has dominated the theory during the nineteenth and part of the twentieth century, is irrelevant to Euler's work on the subject.

Take again $N = M \dotplus B$; put $M = (x, y)$, $N = (x', y')$;

taking A, B to be constant, and differentiating (1), we get

$$\frac{dx}{y} = \frac{dx'}{y'};$$

this can be expressed by saying that, for given A and B, the mapping $M \to N = M \overset{.}{+} B$ is an automorphism of Γ transforming the differential ω into itself. Clearly such automorphisms make up a group $\mathcal{G}$, isomorphic to the group of the points of Γ under the operation $\overset{.}{+}$; this group is simply transitive, i.e. there is one and only one such automorphism mapping a given point P onto a given point Q, viz., the one given by $B \sim A + Q - P$.

More generally, let α be any analytic mapping of a neighborhood of a point P of Γ onto a neighborhood of a point Q; assume that it maps P onto Q and transforms ω into ω. Combining it with an automorphism $M \to M \overset{.}{+} B$ mapping Q onto P, we get a mapping α' of a neighborhood of P onto a neighborhood of P, leaving both P and ω invariant. Let t be a function on Γ with a simple zero at P; taking it as the local coordinate at P, we can write $\omega = \sum_{i=1}^{\infty} a_i t^{i-1} dt$, with $a_1 \neq 0$ since ω has no zero. This can also be written as $\omega = du$ with $u = \sum_{i=1}^{\infty} a_i t^i / i$; expressing the mapping α' in terms of the local coordinate u, we see that it must be the identity, since it leaves both du and the point $u = 0$ invariant. This shows that a mapping such as α must coincide everywhere locally with a mapping $M \to M \overset{.}{+} B$. Using either analytic continuation or the principle of conservation of algebraic identities, one concludes from this that every automorphism of Γ, transforming ω into itself, must belong to the group $\mathcal{G}$ defined above.

Define a function f on Γ by putting $f(M) = x$ whenever $M = (x, y)$; also, for $M = (x, y)$, write $M^- = (x, -y)$, so that $f(M) = f(M^-)$, and that $M \to M^-$ is an automorphism of Γ, changing ω into $-\omega$. If P is any point on Γ, other than A and A^-, the function

$$\varphi(M) = \frac{f(M) - f(P)}{f(M) - f(A)}$$

on Γ has the zeros P, P^- and the poles A, A^-, so that we have $P + P^- \sim A + A^-$ for all P. Now take again $N = M \dotplus B$, i.e. $N \sim M + B - A$, and put $M = (x, y)$, $N = (x', y')$ as before. As N depends rationally upon M, B and A, we can express $x' = f(N)$ as $R(x) + yS(x)$, where R, S are rational functions of x, with coefficients which depend rationally upon A and B. Call $\Phi(x, x') = 0$ the irreducible relation between $x = f(M)$ and $x' = f(N)$; it must be of degree 2 or 1 in x'. Moreover, we have

$$N^- \sim A + A^- - N \sim A + A^- - (M+B-A) \sim B^- + A - M;$$

therefore the mapping $M \to N^-$ is an automorphism of Γ, exchanging M and N^- and therefore exchanging $x = f(M)$ and $x' = f(N^-)$. Consequently the relation $\Phi(x, x') = 0$, for given A and B, must be symmetric in x and x', so that it can be written in Euler's "canonical form" ($Eu.$I-20.71 in E 251|1753; cf. above, §XV):

$$\Phi(x, x') = \alpha + 2\beta(x+x') + \gamma(x^2+x'^2) \\ + 2\delta xx' + 2\varepsilon xx'(x+x') + \zeta x^2 x'^2 = 0,$$

where the coefficients depend rationally upon A and B. It is easy to see that this cannot be of degree 1 in x', consequently in x, except for some special values of B (more precisely, for $2B \sim 2A$).

If this is written as

$$P_0(x)x'^2 + 2P_1(x)x' + P_2(x) = 0,$$

where P_0, P_1, P_2 are polynomials of degree 2, then, since it has a root x' of the form $R(x) + yS(x)$, the polynomial $P_1^2 - P_0 P_2$ must be of the form $\rho^2 F$, with ρ depending rationally upon A and B but not upon M. Conversely, assume that this is so. We have then:

$$\frac{1}{2}\frac{\partial\Phi}{\partial x'} = P_0(x)x' + P_1(x) = \pm\sqrt{P_1(x)^2 - P_0(x)P_2(x)} = \pm\rho\sqrt{F(x)}.$$

Since Φ is symmetric, we have also

$$\frac{1}{2}\frac{\partial\Phi}{\partial x} = \pm\rho\sqrt{F(x')}.$$

Therefore, differentiating the relation $\Phi(x, x') = 0$, we get

$$\pm \frac{dx}{\sqrt{F(x)}} = \pm \frac{dx'}{\sqrt{F(x')}},$$

which can also be written as

$$\frac{dx}{y} = \frac{dx'}{y'}$$

if the signs of y, y' are chosen properly. In Euler's language, this is expressed by saying that the relation $\Phi = 0$ is an integral of the above differential equation; it is a "complete" (i.e. general) integral, since it contains an arbitrary constant (viz., $f(B)$; A is taken as given once for all).

A noteworthy special case occurs when F is of the form

$$F(x) = 1 + mx^2 + nx^4;$$

as shown by Euler and Legendre, the general case can always be reduced to this one by a suitable substitution $x \rightarrow (\lambda x + \mu)/(\nu x + \rho)$ (cf. above, §XV). For every $M = (x, y)$, put $M^* = (-x, -y)$; as $M \rightarrow M^*$ is an automorphism of Γ, transforming ω into itself, it must be of the form $M \rightarrow M \dotplus B_0$, with some B_0 on Γ; as it is of order 2, we must have $B_0 \dotplus B_0 = 0$, i.e. $2B_0 \sim 2A$. As before, take $N = M \dotplus B$, $x = f(M), x' = f(N)$, so that we have $\Phi(x, x') = 0$. The automorphism $M \rightarrow M^*$ maps N onto $N^* = M^* \dotplus B$; as we have $f(M^*) = -x, f(N^*) = -x'$, we must have $\Phi(-x, -x') = 0$. Therefore Φ must be of the form

$$\Phi(x, x') = \alpha + \gamma(x^2 + x'^2) + 2\delta xx' + \zeta x^2 x'^2$$

as postulated by Euler (cf. above, §XV).

As the point A can be arbitrarily chosen, we may assume that we have taken $A = (0, 1)$. Put $B = (a, b)$; then the relation $\Phi = 0$ must be satisfied for $x = f(A) = 0, x' = f(B) = a$; this gives $\alpha = -\gamma a^2$. Also, if P_0, P_1, P_2 are again the coefficients of $x'^2, 2x', 1$ in Φ, we must have

$P_1^2 - P_0 P_2 = \rho^2 F$; this gives

$$\alpha\gamma = -\rho^2, \qquad \zeta\gamma = -n\rho^2, \qquad \delta^2 = \gamma^2 + \alpha\zeta + m\rho^2$$

(cf. $Eu.$I-20.158–161 in E 261|1755). Since $\alpha, \gamma, \delta, \zeta, \rho$ are defined only up to a common factor, we may assume that $\gamma = -1$. This gives:

$$\alpha = a^2 = \rho^2, \qquad \zeta = na^2, \qquad \delta^2 = 1 + ma^2 + na^4 = b^2,$$

$$\Phi(x, x') = a^2 - x^2 - x'^2 + 2\delta xx' + na^2 x^2 x'^2$$

as Euler had found (cf. above, §XV). Solving $\Phi(x, x') = 0$ for x', we get:

$$x' = \frac{\delta x \pm ay}{1 - na^2 x^2}, \qquad \delta = \pm b$$

and a similar formula for x in terms of x', y', since Φ is symmetric. The double signs can be determined by observing that N depends symmetrically upon M and B, and that it is B when one takes $M = A$. This gives $\delta = b$ and

$$(3) \qquad\qquad x' = \frac{bx + ay}{1 - na^2 x^2}.$$

Similarly, writing now the formula for x in terms of x', y', and observing again that this must hold for $M = A$, $N = B$, we get

$$(4) \qquad\qquad x = \frac{bx' - ay'}{1 - na^2 x'^2}$$

which of course can also be used for expressing y' in terms of x, y, a, b. These are the addition and subtraction formulas obtained by Euler ($Eu.$I-20.69 in E 251|1753, etc.; cf. above, §XV). From this, one can derive multiplication formulas in the usual manner. For instance, for $M_2 = M \dot{+} M$, i.e. $M_2 \sim 2M - A$, we have only to put $B = M$ in the addition formula, and we get

$$x_2 = f(M_2) = \frac{2xy}{1 - nx^4}.$$

Similarly, defining M_ν by $M_\nu = M \dotplus M_{\nu-1}$ for $\nu > 2$, or, what amounts to the same, by

$$M_\nu \sim \nu M - (\nu - 1)A$$

for all $\nu \geqslant 2$, and applying the addition formula repeatedly, one obtains the coordinates of M_ν in terms of x, y for all ν. One has then, by Abel's theorem

$$\int_A^{M_\nu} \omega = \nu \int_A^M \omega$$

if the paths of integration are taken suitably; otherwise both sides differ by a period of ω.

All this can be illustrated by considering the "lemniscatic case" $y^2 = 1 - x^4$, which played a decisive role in the work of Fagnano and in the early work of Euler on this subject (cf. above, §XV). Following these authors, we take now as groundfield the field of real numbers; all square roots are to be taken positively. The lemniscate is the curve

$$(x^2 + y^2)^2 = x^2 - y^2;$$

for convenience we consider only that part of the curve which lies in the first quadrant $x \geqslant 0$, $y \geqslant 0$; it is an arc of origin $0 = (0, 0)$ and end-point $P = (1, 0)$. This arc can be represented parametrically as

$$x = \sqrt{\tfrac{1}{2}(z^2 + z^4)}, \qquad y = \sqrt{\tfrac{1}{2}(z^2 - z^4)}, \qquad 0 \leqslant z \leqslant 1;$$

the arc-length on it is given by

$$ds = \sqrt{dx^2 + dy^2} = \frac{|dz|}{\sqrt{1 - z^4}}$$

(cf. *Eu.*I-20.91 in E 252|1752). Put $t = \sqrt{1 - z^4}$; $(x, y) \rightarrow (z, t)$ is a one-to-one correspondence between the arc in question on the lemniscate and the arc $0 \leqslant z \leqslant 1$, $t \geqslant 0$ on the curve Γ given by $t^2 = 1 - z^4$. All integrals on Γ will be taken along segments of that arc.

The addition and subtraction formulas on Γ are now given by

$$z' = \frac{\pm z\sqrt{1 - \zeta^4} + \zeta\sqrt{1 - z^4}}{1 + \zeta^2 z^2}.$$

(*Eu.*I-20.65 in E 251|1753). Thus, for any given ζ, the mapping $z \to z'$ transforms the arc-length ds into itself. For $\zeta = 1$, the subtraction formula gives the mapping

$$z \to z_1' = \sqrt{\frac{1 - z^2}{1 + z^2}}$$

which maps O and P onto one another, the arc OP onto itself, and preserves the arc-length. For $\zeta = z$, we get the duplication mapping

$$z \to z_2 = \frac{2z\sqrt{1 - z^4}}{1 + z^4}$$

which changes ds into $2ds$. These two mappings were Fagnano's discoveries (*Fag.*II.294,308; cf. *Eu.*I-20.92–95 in E 252|1752); the addition formula was first obtained by Euler (*Eu.*I-20.63–64 in E 251|1753; cf. *Eu.*I-20.100 in E 252|1752). Fagnano found that, if M is any point on the arc OP of the lemniscate, the mapping $z \to z_1'$ maps the arc OM onto an arc PM' of equal length; the point Q such that OQ and QP have equal length can be obtained by putting either $z = z_1'$ or $z_2 = 1$, and is given by $z^2 = \sqrt{2} - 1$; if M is on the arc OQ, the mapping $z \to z_2$ maps OM onto an arc of double length.

It is of some interest to note that Fagnano obtained the duplication formula $z \to z_2$, not of course by using the addition formula which he did not possess, but (*Fag.*II.305–307) by combining the two mappings

$$z \to u = \frac{z\sqrt{2}}{\sqrt{1 - z^4}}, \qquad u \to z_2 = \frac{u\sqrt{2}}{\sqrt{1 + u^4}}$$

for which we have

$$\frac{du}{\sqrt{1 + u^4}} = \frac{dz\sqrt{2}}{\sqrt{1 - z^4}}, \qquad \frac{dz_2}{\sqrt{1 - z_2^4}} = \frac{du\sqrt{2}}{\sqrt{1 + u^4}}.$$

These formulas define "Landen transformations", i.e. transformations (or, in modern language, isogenies) of degree 2 between the curves $t^2 = 1 - z^4$ and $w^2 = 1 + u^4$; replacing u by ηu, where η is the eighth root of unity $\eta = (1+i)/\sqrt{2}$, one obtains the formulas for the complex multiplication by $1 \pm i$ on the curve $t^2 = 1 - z^4$. This aspect of Fagnano's work, which was to acquire considerable importance at the hands of Abel, failed to attract Euler's attention.

Finally, consider a conic section C given by an equation $y^2 = ax^2 + b$ with $b \neq 0$, $a \neq 0$, -1; this is an ellipse for $a < 0$, a hyperbola for $a > 0$.

Its arc-length is given by

$$ds = \sqrt{\frac{1 + qx^2}{1 + px^2}} \, |dx|$$

with $p = a/b$, $q = (a+a^2)/b$. If we define a curve Γ by

$$z^2 = (1+px^2)(1+qx^2),$$

then ds is a differential of the second kind on Γ, and we have a one-to-one correspondence between the points (x, y) of C for which $x \geq 0$, $y \geq 0$ and the points of Γ for which $x \geq 0$, $ax^2 + b \geq 0$, $z \geq 0$. One can then proceed just as in the case of the lemniscate, except for the changes required, in view of Abel's theorem, by the fact that the arc-length is now of the second kind.

Take for instance the case $b > 0$, $0 > a > -1$, corresponding to an ellipse; one is dealing here with the arc $x \geq 0$, $y \geq 0$ of the ellipse, with origin $P = (0, \sqrt{b})$ and endpoint $Q = (\alpha, 0)$, where we have put $\alpha = \sqrt{-(b/a)}$; P and Q correspond respectively to the points $(0, 1)$ and $(\alpha, 0)$ on Γ. Apply the subtraction formula (4) to the points (x, z) and $(\alpha, 0)$ on Γ (writing x, x', α, 0, pq instead of x', x, a, b, n, respectively); we get

$$x' = \alpha \sqrt{\frac{1 + px^2}{1 + qx^2}},$$

which determines a mapping $x \to x'$ of the arc PQ of the ellipse onto itself, or rather onto QP since it maps P and Q onto one another; the relation between x and x' can also be written as

$$1 + p(x^2 + x'^2) + pqx^2x'^2 = 0,$$

this being of course a special case of the relation denoted above by $\Phi = 0$. Differentiating this, we get

$$(1 + qx'^2)xdx + (1 + qx^2)x'dx' = 0,$$

which implies that $x \to x'$ is, as expected, a one-to-one mapping of PQ onto QP. This can be written:

$$\frac{dx}{x'} + \frac{dx'}{x} + qd(xx') = 0.$$

At the same time the arc-length on PQ can be written as $ds = \alpha(|dx|/x')$ and therefore its transform under the mapping $x \to x'$ is $dx' = \alpha(|dx'|/x)$. Let $M = (x, y)$ be any point on PQ; let M' be its image under $x \to x'$; then the lengths λ, λ' of the arc PM and of its image QM' are respectively given by

$$\lambda = \alpha \int_0^x \frac{dx}{x'}, \qquad \lambda' = -\alpha \int_0^x \frac{dx'}{x};$$

in view of the relation obtained above, this gives

$$\lambda' - \lambda = q\alpha xx'.$$

This, too, was Fagnano's discovery (*Fag.*II.287–289; cf. *Eu.*I-20.82–90 in E 252|1752). Eventually it was extended by Euler to include the effect of the addition, subtraction and multiplication formulas on all elliptic integrals, not only those of the second kind (*Eu.*I-20.156–158 in E 261|1755, etc.), but even those of the third kind (*Eu.*I-21.39–56 = E 581|1775). All such results may of course be viewed as special cases of Abel's theorem.

Lagrange (by Delpech). *Courtesy, Archives de l'Académie des Sciences, Paris.*

Chapter Four

An Age of Transition: Lagrange and Legendre

§I.

In 1745 Euler had been hailed by his old teacher Johann Bernoulli as *"mathematicorum princeps"*, the first of mathematicians (cf. Chap.III, §III). By 1775 he clearly felt ready to pass the title on to Lagrange. "It is most flattering to me", he wrote to his younger colleague, "to have as my successor in Berlin the most outstanding geometer of this century" (*loc.cit.* Chap.III, §IX). Such was indeed by then the universal verdict of the scientific world. *"Le célèbre Lagrange, le premier des géomètres"* is how LAVOISIER referred to him in 1793 in an official request to the Convention in behalf of his friend (*Lag.*XIV.314–315) at the onset of the Terror which was soon to claim Lavoisier himself as a victim. In the next century the title of *"princeps mathematicorum"* was bestowed upon Gauss by the unanimous consent of his countrymen. It has not been in use since.

LAGRANGE was born in 1736 in Turin, where his birth was registered as that of Giuseppe Lodovico Lagrangia. On his father's side at least the family was of French origin, and Lagrange never used the Italianized form of his name; he always signed his letters as Louis (or Lodovico, or Luigi) de la Grange (or La Grange) and later as Joseph Louis Lagrange or simply Lagrange. Under the House of Savoy, Piedmont was largely bilingual, and Lagrange was equally at home in French and in Italian: as he once told d'Alembert (*Lag*.XIII.88|1767) his favorite poet was Ariosto.

His education was the traditional one, based on the classics; mathematically he must have been very much on his own. Italy's glorious mathematical tradition was by then a mere memory. Even Fagnano, surviving himself in the small provincial town of Senigallia, could offer Lagrange nothing but formal words of encouragement when the latter, at the age of eighteen, sought to initiate with him a correspondence from which he obviously expected much more (*Fag*.III.179–209|1754). In that same year 1754 Lagrange took it upon himself to address the great Euler, but got no answer.

The next year everything started changing. In September 1755 (cf. *Eu*.IV A-5.378 = *Lag*.XIV.147) he was appointed to a teaching position for mathematics and mechanics at the Artillery School in Turin, where he seems to have had some talented pupils, mostly older than himself. At the same time his study of Euler's writings was bearing fruit. On 12 August 1755 he wrote to Euler as follows:

"Meditanti mihi assidue . . . praeclarissimum librum tuum de methodo maximorum et minimorum ad lineas curvas applicata, factum tandem est, ut . . . inciderim in viam longe breviorem problemata huiuscemodi resolvendi . . . Quanquam enim merito haesitandum fuerat, an mihi, qui obscuri adhuc nominis sum, te tantum virum, omni pene scientiarum genere clarissimum, interpellare liceret; maximus tamen, ac plane singularis affectus meus in te ex operum tuorum studio iam pridem conceptus, effecit, ut opportunam hanc illius tibi quomodocumque testandi occasionem . . . de manibus dimittere nullo modo potuerim" ["Meditating assiduously upon your most splendid book on the method of maxima and minima as it applies to curves, I have at last been led to a much shorter way of solving problems of that kind . . . Well could I hesitate whether such an obscure person as I am still ought to approach a man of your eminence, of such high repute in almost every branch of science; nevertheless the great and indeed singular affection which I have conceived for you since long ago through the study

of your works has made it impossible for me not to seize this occasion of somehow giving vent to those feelings": *Eu*.IV A-5.366 = *Lag*.XIV.138–139]. With this letter went a brief sketch which in effect created the classical calculus of variations.

Euler's response must have exceeded by far Lagrange's expectations. Not only was it prompt and generous; it was enthusiastic. "You seem to have brought the theory of maxima and minima to almost its highest degree of perfection; my admiration for your penetration knows no bounds", Euler wrote to his young correspondent [". . . *theoriam maximorum ac minimorum ad summum fere perfectionis fastigium erexisse videris; eximiam ingenii tui sagacitatem satis admirari non possum*": *Eu*.IV A-5.375 = *Lag*.XIV.144|6 Sept.1755]. The next year, on his insistent recommendation, Lagrange was made a corresponding member of the Berlin Academy, and the plan was formed of finding a suitable position for him there. The Seven Years' War, then Maupertuis' death in 1759, brought these attempts to nought.

There followed for Lagrange ten years of "assiduous meditation", with the sole interruption of a visit to Paris. Invited by a distinguished Neapolitan diplomat, the Marchese Caraccioli, to accompany him to Paris and London, he left Turin in November 1763, but, having fallen seriously ill in Paris, he could not go on to England. He went back to Turin in May 1764, stopping in Geneva and meeting Voltaire in Ferney on the way home. In Paris he met all the French mathematicians, CLAIRAUT, LALANDE, d'ALEMBERT, CONDORCET (cf. *Lag*.XIV.14). Clairaut described him to Daniel Bernoulli as "*un jeune homme très singulier tant par ses talens que par sa modestie; son caractère est doux et mélancolique; il ne connoît d'autre plaisir que l'étude*" ["a young man, no less remarkable for his talents than for his modesty; his temperament is mild and melancholic; he knows no other pleasure than study": *Eu*.IV A-5.330, note [2]|27 Dec. 1763]. With d'Alembert he had already started a correspondence (*Lag*.XIII.3–4|1759), on rather formal terms at first; in Paris their acquaintance soon matured into an intimate friendship which lasted unbroken until d'Alembert's death in 1783.

By the time Euler left Berlin in 1766 for Petersburg, it had become obvious that no one but Lagrange was worthy of filling his place; Euler and d'Alembert concurred on this point, and Frederic II was quickly convinced. Lagrange accepted all the more readily, as nothing except empty promises had occurred in Turin since 1755 to improve his very modest position there (cf. *Lag*.XIII.65|1766). He set off for Berlin in August 1766, visiting his friends d'Alembert in Paris and Marchese Caraccioli in London on his way, and landing in Hamburg in October. The next year, unhappy about his solitary life as a bachelor in unfamiliar surroundings, he sent for a kinswoman of his in Turin and married her in September 1767; to d'Alembert he described this as a pure matter of convenience, praising his wife as "an excellent housewife and without any pretensions whatsoever" (*Lag*.XIII.143|1769). Thereafter his life seems to have been quiet and undisturbed, except for occasional spells of poor health, until his wife fell gravely ill during the latter part of his stay in Berlin. She died in 1783, and d'Alembert, from his deathbed, could still dictate a touching last word of sympathy to his friend (*Lag*.XIII.377). Euler's death occurred in that same year.

For some time Lagrange had been almost without intellectual companionship; LAMBERT, whom Euler had recruited for the Berlin Academy shortly before leaving it, and who had become Lagrange's good friend despite his uncouth manners and appearance, had died prematurely in 1777 (cf. *Lag*.XIII.333–334). In 1786, the king's death deprived the Academy of its founder and protector. In May of the next year Lagrange left Berlin forever to establish himself in Paris. By that time, as he soon confided to his friends and fellow-academicians there, he had lost his taste for mathematical research (cf. his *Eloge* by Delambre, *Lag*.I,p.XXXVII). He had barely passed the age of fifty.

The remainder of his life was on the whole a happy one.

His nearest colleagues became his close friends; in 1792 he married the young daughter of one of them, the astronomer LE MONNIER; her loving care and devotion was to sustain him until his last moments. He watched, at first with sympathetic interest, then with increasing horror, the progress of the revolution, losing to the so-called Terror two of his best friends (Lavoisier and Condorcet). After the worst was over his merits were again recognized. His teaching at the Ecole Normale and at the Ecole Polytechnique was the occasion for him to work out and write up some of his old ideas, going back to his *début* in 1754, on the theory of functions (this making up the main content of his *Théorie des Fonctions Analytiques* of 1797 and of his *Leçons sur le calcul des fonctions* of 1806: *Lag*.IX,X), and to compile his *Traité de la résolution des équations numériques* of 1798 (*Lag*.VIII). He did keep up his interest in the latest mathematical discoveries; in 1804 he was expressing to GAUSS his admiration for the *Disquisitiones Arithmeticae* of 1801:

"*Vos Disquisitiones vous ont mis tout de suite au rang des premiers géomètres . . . J'ai depuis longtemps abandonné ces sortes de recherches, mais elles ont conservé beaucoup d'attrait pour moi, et je me contente maintenant de jouir sur cette matière, comme sur plusieurs autres, du fruit des veilles d'autrui . . .*" ["Your *Disquisitiones* have at once put you on a par with the geometers of the top rank . . . For a long time I have given up this kind of investigations, but they keep much attraction for me, and it is now enough for me, on these matters as on several others, to enjoy the fruits of the labors of others": *Lag*.XIV.299; cf. *Lag*.XIV.300|1808]. Perhaps he felt flattered by the compliment which the young Gauss, staking out his claim as Lagrange's rightful successor, had paid him by adopting for his masterpiece, in latinized form, the same title that Lagrange had modestly given to his great *Recherches d'Arithmétique* of 1775.

Late in life, laden with honors by Napoleon, he undertook a revision of his *Méchanique Analytique* of 1788 (*Lag*.XI–XII); this task, which he was unable to bring to completion, seems to have exhausted him. Mortally ill, on 8 April 1813,

he held a long conversation wtih his friends Monge, Lacépède
and Chaptal — a truly Socratic dialogue, according to Delambre's account (*Lag*.I,p.XLV), where he reminisced about
the past and welcomed his approaching end. Soon after
this he lost consciousness; two days later he was dead.

§II.

   Modest and diffident to a fault, Lagrange offers in many
ways a complete contrast to Euler, whose ebullient enthusiasm for his own discoveries, as well as for those of his
contemporaries, never knew any bounds. As he once wrote
to LAPLACE with obvious sincerity, he "took far more
pleasure in the work of others than in his own, being always
dissatisfied with the latter" ["*je jouis beaucoup plus des travaux
des autres que des miens, dont je suis toujours mécontent*":
Lag.XIV.71|1777]. Similarly, about a memoir of his on ce-
lestial mechanics, he wrote to Condorcet: "Please feel entirely
certain that I attach no value to my feeble efforts and that
no one can have less pretensions than I have in anything"
["*je vous prie d'être intimement convaincu que je n'attache aucun
prix à mes faibles productions, et qu'on ne saurait avoir moins de
prétentions que je n'en ai en quoi que ce soit*": *Lag*.XIV.15|1773].
   Much and perhaps the greater part of Lagrange's best
work is directly inspired by that of Euler, which he had
indeed "meditated assiduously" (*loc.cit.* §I) for many years,
and which, already as a young man, he "knew by heart,
down to the minutest details", as Lalande had written to
Euler in 1764 (*Eu*.IV A-5.330, note [2]). This applies particularly to Lagrange's work on number theory (cf. above,
Chap.III, §VI), a subject which started attracting his active
interest only in 1768 (cf. his letter to d'Alembert,
Lag.XIII.118–119|1768) and held it for less than ten years.
His writings on arithmetical topics consist of the following:

   (A) An early paper on Pell's equation (*Lag*.I.671–731),
composed and sent to Turin in the summer of 1768; it
contained the first proof ever written up for publication of

the fact (possibly already proved by Fermat; cf. Chap.II, §XIII) that Pell's equation has always a solution, and therefore infinitely many. Lagrange was later to criticize it as "very long and very indirect" (*Lag*.VII.159 = *Eu*.I-1.632, Art.84); to his annoyance it appeared only in 1773; by that time it had been superseded by his later work.

(B) Three papers, presented to the Berlin Academy in 1768, 1769 and 1770 (*Lag*.II.377–535,581–652,655–726), dealing chiefly with continued fractions and their application to the solution of diophantine equations. This included a definitive treatment of Pell's equation (*Lag*.II.494–496), essentially equivalent to the one described above in Chap.II, §XIII; a proof for the fact that all quadratic irrationals have periodic continued fractions (*Lag*.II.603–615; cf. above, Chap.III, §XII); the highly original method of solution of equations $z^2 = Ax^2 + By^2$ (*Lag*.II.383–399) which has been summarized in Chap.II, §XIV; and methods of solution for all equations of degree 2 in two unknowns, to be solved in integers (*Lag*.II.655–726). Also included was a method alleged to provide a solution for all equations $F(x, y) = a$ in integers when F is a homogeneous polynomial of arbitrary degree n (*Lag*.II.662–696); as Legendre pointed out in his *Essai sur la Théorie des Nombres* of 1798 (I^e Partie, §XV, Rem.III, no.126), this procedure is not effective for $n > 2$.

(C) The three papers listed under (B) had been written before Lagrange received Euler's *Algebra* (cf. Chap.III, §V*c*). As soon as this was in his hands, some time during the year 1770, he formed the plan of getting it translated into French by his junior colleague in Berlin, Johann III Bernoulli (a grandson of Euler's teacher Johann I) and accompanying this translation with an improved exposition of the main contents of the three memoirs in question. This was promptly carried out, and a manuscript sent to the printer in Lyon in the next year (cf. *Lag*.XIV.4|1771); it came out in two volumes in June 1773. Lagrange's contribution, making up pages 369–658 of volume II (= *Lag*.VII.5–180 = *Eu*.I-1.499–651), bore the title *Additions, De l'Analyse indéterminée*.

(D) Lagrange's proof for Fermat's theorem on sums of

four squares (*Lag*.III.189–201; cf. Chap.III, §XI); it appeared in the Berlin *Memoirs* in 1772.

(E) A proof of Wilson's theorem (*Lag*.III.425–438; cf. Chap.II, §VII); this was read to the Academy in May 1771, published in the Berlin *Memoirs* of 1771, and was commented upon by Euler in a letter to Lagrange (*Eu*.IV A-5.496 = *Lag*.XIV.235–240|1773; cf. Chap.III, §VI).

(F) Lagrange's great work on binary quadratic forms, the *Recherches d'Arithmétique*, published in the Berlin *Memoirs* of 1775 (*Lag*.III.697–758; cf. Chap.III, §IX), with a second part or rather a supplement published there in 1777 (*Lag*.III.759–795).

(G) Lastly, a treatment of the diophantine equation $y^2 = x^3 - 2x$, or equivalently of $X^4 - 2Y^4 = \pm Z^2$, originating from the work of Fermat (cf. Chap.II, §XV and App.V) and lately considered by Euler (cf. Chap.III, §XVI). This is no more, and no less, than a carefully done exercise on Fermat's method of infinite descent, but applied for the first time to an equation of genus 1 and of rank >0 (i.e. with infinitely many solutions). This was read to the Academy in 1777 and appeared there in the same year (*Lag*.IV.377–398).

§III.

As appears from this list, the greater part of Lagrange's contributions to arithmetic dealt with questions already considered by Fermat and Euler; these matters have been discussed in some detail in Chapters II and III and need not be taken up again here. We also refrain from describing Lagrange's contribution to the theory of continued fractions as such, and his application of that theory to the numerical solution of algebraic equations and to indeterminte equations of degree >2, as these belong to so-called "diophantine approximation" rather than to number theory proper. On the other hand, Lagrange's work on binary quadratic forms deserves a more thorough description than could be given in Chap.III, §IX, where it was viewed only through the failing eyes of the aging Euler.

In 1768, announcing to d'Alembert his first big memoir on indeterminate equations (the first one of those listed under (B) in §II), he had written that he had thoroughly exhausted the subject of equations of degree 2 in two unknowns, to be solved either in fractions or in integers ["*Je crois avoir entièrement épuisé cette matière, sur laquelle M. Euler paraît s'être vainement exercé*": Lag.XIII.125|6 Dec.1768, and again a few weeks later: "*Je crois n'avoir presque rien laissé à désirer sur ce sujet*": Lag.XIII.128|28 Feb.1769]. Of course he was right in pointing out that Euler, in dealing with such equations, had always been compelled to assume that at least one solution was known to him beforehand (*Lag*.II.378–379|1768 and *Lag*.XIII.128|1769; cf. Chap.III, §XIII), whereas his own methods were fully effective without any such limitation. Nevertheless he was soon to discover that he had not yet said the last word on this subject; already in 1770 he was publishing a variant of his own solution under the title "*Nouvelle méthode pour résoudre les problèmes indéterminés en nombres entiers*" (*Lag*.II.655–726). The next year, working out still another variant for insertion into his *Additions* to Euler's *Algebra*, and inspired (as he says) by the method of descent which he had used to solve the equation $z^2 = Ax^2 + By^2$ in rational numbers, he introduced a new idea of such far-reaching importance that it has dominated the whole subject of quadratic forms ever since; in modern language it consists in an equivalence concept with respect to which forms of given discriminant fall into classes, coupled with a process of reduction which shows the number of classes to be finite (*Lag*.VII.125–126 = *Eu*.I-1.603, Art.70).

In the *Additions* this idea had been presented only incidentally, with no stress laid upon it. A few years later, having better realized its scope, Lagrange devoted to it the whole of his *Recherches* of 1775. As we have seen, Euler greeted this with enthusiasm and lost no time in summarizing it in his own way (cf. Chap.III, §IX).

§IV.

Understandably, what impressed Euler most in the *Re-cherches* was that it contained proofs for some of the facts he had discovered empirically in his younger days (cf. Chap.III, §IX); in this he had the excuse that this was the aspect chiefly emphasized by Lagrange himself in discussing the consequences of his main idea. To do full justice to Lagrange, one must view his memoir in the light of the later developments which it inspired, and more specifically of the work of Gauss and his successors.

Fermat, the creator of the theory of binary quadratic forms, had never ventured beyond a few specific cases of the type $X^2 \pm AY^2$. Euler had systematically considered forms of the type $mX^2 + nY^2$ but no others, except that he could not avoid the form $X^2 + XY + Y^2$ in dealing with $X^2 + 3Y^2$ (cf. Chap.III, §VIII). Lagrange's initial step, and in itself a momentous one, was to realize that a coherent general theory required the simultaneous consideration of all forms $aX^2 + bXY + cY^2$.

No less obvious perhaps to our eyes, but no less essential, was the recognition of the fact that two such forms f, F "represent" the same integers (i.e., that the equations $f(x, y) = m$, $F(x, y) = m$ have solutions in integers for the same values of m) whenever there is a relation

(1) $$F(X, Y) = f(\alpha X + \beta Y, \gamma X + \delta Y)$$

where $\alpha, \beta, \gamma, \delta$ are integers satisfying $\alpha\delta - \beta\gamma = \pm 1$. In modern terms, this is accounted for by the fact that the substitutions

(2) $$(X, Y) \to (\alpha X + \beta Y, \gamma X + \delta Y)$$

with $\alpha, \beta, \gamma, \delta$ as above, are invertible, the inverse of (2) being the substitution

$$(X, Y) \to (\pm(\delta X - \beta Y), \pm(-\gamma X + \alpha Y)).$$

For forms f, F related as above, Lagrange, disinclined as always from bringing in new concepts and new terminology, has the rather inadequate phrase "forms which can be transformed into one another"; Gauss introduced for them the term *"equivalent forms"* which has been in use ever since. Lagrange shows that such forms have the same discriminant $b^2 - 4ac$; he is also well aware of the fact that the relation between them is truly "an equivalence relation" in the modern sense, i.e. that it is symmetric and transitive; in modern language this amounts to saying that the substitutions (2) make up a group, viz., the group $GL(2, \mathbf{Z})$ of automorphisms of the additive group of vectors (x, y) with integral x and y. For brevity's sake, if S is the substitution (2), we shall write det S for its determinant $\alpha\delta - \beta\gamma$, $F = f \circ S$ for the relation (1), and $\mathscr{G}$ for the group $GL(2, \mathbf{Z})$. The distinction, made later by Gauss, and so essential to his purposes, between "proper" and "improper" equivalence is not made by Lagrange, nor yet, some time later, by Legendre; it depends upon the fact that the substitutions S in $\mathscr{G}$ for which det $S = 1$ make up a subgroup $SL(2, \mathbf{Z})$ of $\mathscr{G}$ of index 2, two equivalent forms f, F being called "properly equivalent" if there is S in that subgroup for which $F = f \circ S$, and "improperly equivalent" otherwise.

Once this equivalence concept has been clarified, forms with a given discriminant can be distributed into *classes*, two forms being put into the same "Lagrangian" (resp. "Gaussian") class if and only if they are equivalent (resp. properly equivalent); here again the first use of the word "class" in this sense belongs to Gauss. A "Lagrangian" class is either a "Gaussian" class, or it consists of two such classes. Here we shall use the word "class" exclusively in the Lagrangian sense.

Still another distinction has to be pointed out between the usage of Lagrange and that of Gauss. For Gauss the binary quadratic form (a, b, c) is defined as the form $aX^2 + 2bXY + cY^2$, with an even middle coefficient $2b$; Gauss always insisted that this was the most appropriate notation, classifying forms according to their "determinant" $b^2 - ac$. On the other hand, Lagrange took as his starting point the forms $aX^2 + bXY + cY^2$, where b may be even or odd, while sometimes giving precedence to those with even middle coefficient. Here we will write (a, b, c) for the "Lagrangian" form $aX^2 + bXY + cY^2$, calling $b^2 - 4ac$ its discriminant (which has the same parity as b). Such forms, after disappearing from the literature (under the influence of Gauss) during much of the nineteenth century, became prominent again after Dedekind re-interpreted binary quadratic forms in terms of the ideal-theory of quadratic fields. Of course the distinction between "Lagrangian" and "Gaussian" forms is merely one of notation as long as fields (or rings) of characteristic 2 are not considered, since a form $f = aX^2 + bXY + cY^2$ with b odd can always be replaced by the form $2f$ if one wishes to consider it within the framework of the Gaussian theory. For brevity we will exclude the forms (a, b, c) whose discriminant $b^2 - 4ac$ is 0 or a square; these are the forms which can be written as products of two linear forms with integral coefficients.

Another observation of some moment, particularly in view of the problems previously studied by Euler, concerns the integers "properly represented" by a form f, i.e. those which can be written as $f(x, y)$ with x prime to y; these are the same for two equivalent forms, since a substitution in $GL(2, \mathbf{Z})$ maps every pair (x, y) with x prime to y onto a similar pair. For $f = (a, b, c)$, both a and c are properly represented by f, since we have $a = f(1, 0)$, $c = f(0, 1)$; therefore, if $F = (A, B, C)$ is equivalent to f, both A and C are properly represented by f. Conversely, let m be an integer properly represented by f, so that we can write $m = f(\alpha, \gamma)$ with α prime to γ; then there are integers β, δ such that $\alpha\delta - \beta\gamma = 1$, and m is the coefficient of X^2 in the form F defined by (1). Now let d be a "divisor" of the form f, by which we mean, with Euler, any divisor of an integer m properly represented by f, provided it is prime to the discriminant D of f. Then there is F, equivalent to f, whose first coefficient is m, so that we can write $F = (m, n, p)$; as F has the same discriminant as f, we have

$$D = n^2 - 4mp = n^2 - 4d(m/d)p;$$

therefore d is the first coefficient of the form $(d, n, (m/d)p)$

with the discriminant D, and is properly represented by it (cf. Chap.III, §IX).

As Euler rightly observed (cf. Chap.III, §IX), Lagrange's most decisive step consists then in showing that every form (a, b, c) can be transformed, by successive substitutions

$$(X, Y) \rightarrow (X - rY, Y)$$

and

$$(X, Y) \rightarrow (X, Y - sX)$$

into an equivalent form (A, B, C) for which $|B|$ is $\leq |A|$ and $\leq |C|$ $(Lag.VII.125-126 = Eu.I-1.603-605, Art.70; Lag.III. 697-700)$. His proof is the one described above in Chap.III, §IX.

Lagrange had no word to designate the forms (A, B, C) satisfying the above conditions; we shall call them *"reduced in the Lagrangian sense"*, or simply *reduced* since no confusion is to be feared in the present context (Gauss, who introduced the word, used it in a somewhat different sense). For such a form, as we have seen (Chap.III, §IX), A, B and C are bounded in terms of $D = B^2 - 4AC$, so that it is easy to list all such forms with a given discriminant; Lagrange's proof shows that there is at least one reduced form in each class. The process, described in Chap.III, §IX, by which a given form is transformed into a reduced one may be called *"Lagrangian reduction"*.

Thus far Euler had been able to follow Lagrange. Perhaps he attached less value to the last problem solved in the *Recherches*, or else he was deterred at this point by the complicated calculations which this involved. In the eyes of Lagrange and of his successors, however, this was the indispensable complement of the process of Lagrangian reduction, the question being to obtain a set of representatives (in the strict sense) for the classes of forms with a given discriminant, i.e. a set of forms containing one and only one form from each class. Lagrange solved it by inventing a procedure for obtaining all the reduced forms equivalent to a given one without having recourse to trial and error (*"sans aucun tâtonnement"*: *Lag.III.737*). Clearly, by applying such a procedure successively to each form in a list of all the reduced forms with a given discriminant, and eliminating the superfluous ones accordingly, one will obtain a set of representatives as desired.

At first Lagrange seems to have sought a unified treatment for all classes of forms, using for this a kind of canonical decomposition for substitutions in $GL(2, \mathbf{Z})$; he soon discovered, however, that in this respect forms behave quite differently according to the sign of their discriminant. As he knew, a form f has a discriminant $D < 0$ if and only if all the values $f(x, y)$ it takes, for integers x, y not both 0, have the same sign; since Gauss such forms are called *definite*, while forms with $D > 0$ are called *indefinite*. Clearly, among definite forms, it is enough to consider those with only positive values; for such a form (a, b, c), a and c must be >0. Lagrange found that, if $F = (A, B, C)$ is definite and reduced (in the "Lagrangian sense"), then there are no reduced forms equivalent to F except its transforms $(A, \pm B, C)$, $(C, \pm B, A)$ by the substitutions $(X, Y) \to (X, \pm Y)$, $(X, Y) \to (Y, \pm X)$. His proof for this was later superseded by one of Legendre's (in his *Essai* of 1798, I^e Partie, §VIII) which is both simpler and more informative; it will be described in Appendix II.

For indefinite forms the question is more difficult, and in a sense more interesting; Lagrange's method for dealing with it is manifestly inspired by his own solution of Pell's equation, which indeed may be regarded as the special case in which the given reduced form is $X^2 - NY^2$. In his *Essai* of 1798 (I^e Partie, §XIII), Legendre gave an improved presentation of Lagrange's solution by making more explicit use of the theory of continued fractions; this was then adapted by Gauss to his own concept of "proper equivalence" (*Disq.* Art.184–193; cf. Dirichlet's definitive presentation in the *Vorlesungen über Zahlentheorie*, ed. R. Dedekind, Braunschweig 1894, §§72–82). Lagrange's original treatment of the problem will be described in Appendix III.

§V.

Adrien-Marie LE GENDRE (or LEGENDRE) was born in Paris in 1752; he died there in 1833. "The illness that put an end to his life was a long and painful one", said

ESSAI

SUR LA THÉORIE

DES NOMBRES;

Par A. M. LE GENDRE, de l'Institut national.

———————————

A PARIS,

Chez DUPRAT, Libraire pour les Mathématiques, quai
des Augustins.

AN VI.

POISSON at his funeral; "it was our colleague's often-repeated wish that one should mention only his works when speaking of him, and indeed they are all his life". The year before he died, however, writing to the young JACOBI to congratulate him on his marriage, he indulged briefly in some reflections about his own personal life: "I married at a much later age than you, after a bloody revolution which had destroyed what little fortune I possessed; we had serious embarrassments and some really difficult moments, but my wife helped me mightily to put my affairs progressively in order . . . so as to secure . . . a small fortune, the remains of which, after great losses incurred in subsequent revolutions, will still be adequate for taking care of my needs in my old age, and of those of my beloved wife when I shall be no more. But I have been talking about myself too much . . ." ["*Je me suis marié beaucoup plus tard que vous et à la suite d'une révolution sanglante qui avait détruit ma petite fortune; nous avons eu de grands embarras et des moments bien difficiles à passer mais ma femme m'a aidé puissamment à restaurer progressivement mes affaires . . . de manière à me procurer bientôt une existence honorable et une petite fortune dont les débris, après de nouvelles révolutions qui m'ont causé de grandes pertes, suffiront encore pour pourvoir aux besoins de ma vieillesse et suffiront pour pourvoir à ceux de ma femme bien-aimée quand je n'y serai plus. Mais c'est trop parler de moi . . .*": Jac.I.460|30 June 1832]. In 1792, indeed, he had married a girl, not quite half his age at the time, who was to survive him for 23 years.

He had early attracted the attention of the senior mathematicians in Paris. Born into a well-to-do family, he nevertheless accepted a teaching position at the Ecole Militaire and held it from 1775 to 1780. In 1782 a prize-winning essay on ballistics, which he had sent to Berlin, caused Lagrange (*Lag*.XIV.116|1782) to inquire about him from Laplace, who answered with high praise (*Lag*.XIV.121|1783); in the same year Legendre was appointed to succeed Laplace on the latter's promotion to associate membership, as "adjunct" ("*adjoint mécanicien*") with the Paris Academy; he was himself promoted in 1785. The Academy was abolished in 1793, and eventually restored as part of the Institut de France; as he was to write to Jacobi, he experienced some difficulties in supporting himself during the troubled times of the French revolution, occupying various positions which he seems to have owed largely to his acknowledged skill in numerical computation. Already in 1787 he had been a member of an Academy committee on geodesy, whose business brought him to London where he was made a Fellow of the Royal Society; in 1791, and again from 1794 onwards, he was a member of committees which eventually established the metric system. From 1799 to 1815 he was examiner for the Ecole Polytechnique—an important position in those days, but hardly a prominent one. In 1813 he succeeded Lagrange on the Bureau des Longitudes, and held that position until his death.

One utterance of Poisson's, in his speech delivered at Legendre's funeral, might not have won his late colleague's approval. "*Les questions relatives aux propriétés des nombres*", he said, "*isolées de toute application, n'ont qu'un seul attrait, à la vérité bien puissant sur les mathématiciens: l'extrême difficulté qu'elles présentent*" ["Questions pertaining to the properties of numbers, remote from any application, have only one attraction, but a potent one, it must be acknowledged, for mathematicians: they are exceedingly difficult"]. This can be contrasted with Legendre's own statement, in the preface of his *Essai* of 1798 on number theory: "*Il est à croire . . . qu'Euler avoit un goût particulier pour ce genre de recherches, et qu'il s'y livroit avec une sorte de passion, comme il arrive à presque tous ceux qui s'en occupent*" ["It appears . . . that Euler had a special inclination towards such investigations and that he took them up with a kind of passionate addiction, as happens to nearly all those who concern themselves with them": *Essai sur la Théorie des Nombres*, Paris An VI (= 1798), p.vi]. Surely, when writing these lines, Legendre must have been giving vent to his own feelings about what remained always, along with elliptic functions, his favorite subject.

§VI.

Legendre's first venture into number theory was a lengthy essay submitted in 1785 to the Paris Academy and published in 1788 under the title *"Recherches d'Analyse Indéterminée"* in the *Mémoires de Mathématique et de Physique . . . de l'Académie Royale des Sciences*, Année 1785, pp.465–559: "an excellent memoir", as the young Gauss wrote to his teacher Zimmermann in May 1796 after discovering it in the Göttingen University library [*"eine vortreffliche Abhandlung"*: cf. L. Schlesinger, *Ueber Gauss' Arbeiten zur Funktionentheorie*, p.21, in *Gau.X-2*). By the time Legendre composed this essay, Euler was dead and Lagrange had ceased to be active in that field. Obviously Legendre had been an eager student of their work, including at least the first volume of Euler's *Opuscula Analytica* published in 1783; this contained on page 84 ($= Eu.$I-3.512 in E 552; cf. Chap.III, §VIII and Appendix I) a statement so close to the law of quadratic reciprocity as to be almost undistinguishable from it.

As years went by, Legendre launched a more ambitious project; this took the form of a full-sized volume, where he sought to give a comprehensive account of number theory as he saw it at the time, including, beside his own research, all the main discoveries of Euler and Lagrange, as well as numerical evidence (in the form of extensive tables) for many results whose proofs he felt to be shaky. It was published in Paris in 1798 under the title *Essai sur la théorie des nombres*: "I am not offering it as a complete treatise", the author said in his preface, "but merely as an essay purporting to show roughly the present state of the theory" [*"Je le donne non comme un traité complet, mais simplement comme un essai qui fera connoître à-peu-près l'état actuel de la science"*: pp.ix–x].

Following the appearance of Gauss's *Disquisitiones* in 1801, the work underwent an extensive revision, which however did not affect its main features; this led to a second edition

in 1808. Supplements, first printed in the *Memoirs* of the
Paris Academy, were added to this subsequently. Finally
a definitive edition appeared in 1830 in two large volumes;
feeling that he had now done his utmost for the improve-
ment of his treatise [*"L'ouvrage ayant ainsi reçu tous les
perfectionnements que l'auteur a pu lui procurer"*: p.v], Legendre
let it appear with the title *Théorie des Nombres*. By then, as
his younger contemporaries well knew, Gauss's *Disquisitiones*
had made it almost wholly obsolete.

It is said that Gauss's principle, in publishing his work,
was to remove the scaffolding before letting the public
inspect the building. Of Legendre's number theory it may
fairly be said that there is often more scaffolding in it
than solid masonry; even the foundations are at times so
shaky that the visitor feels seldom quite safe, in spite of
the warning signs which Legendre does insert from time
to time. In discussing its contents it will be well to begin
with the *Recherches* of 1785, where matters are more clearcut.
This memoir is divided into four parts, the first two of
which deserve only the briefest mention. Part I, on congruences of higher degree modulo a prime (pp.465–490)
brings only a number of easy consequences of Euler's
results described above (Chap.III, §VI). Part II (pp.490–
507) gives procedures for splitting a polynomial with integral
coefficients into factors of lesser degree, over $\mathbf{Q}$ or over
quadratic extensions of $\mathbf{Q}$.

Part III of the *Recherches* (pp.507–513) is devoted firstly
(pp.507–509) to Lagrange's solution of the equation
$ax^2 + by^2 + cz^2 = 0$, described above (Chap.II, §XIV); this
is supplemented (pp.509–513) by the proof of the following
important criterion, which is one of Legendre's main claims
to fame:

*Theorem. Let a, b, c be three integers, not all of the same sign
and such that abc is square-free. Then the equation*

$$ax^2 + by^2 + cz^2 = 0$$

has a solution in integers x, y, z, not all 0, if and only if $-bc$,

$-ca$, $-ab$ *are quadratic residues modulo* $|a|$, *modulo* $|b|$ *and modulo* $|c|$, *respectively.*

While Legendre's proof for this is straightforward, it is somewhat artificial; for another proof, cf. Appendix I (cf. also the fuller treatment given to this equation in Dirichlet's *Vorlesungen über Zahlentheorie*, ed. R. Dedekind, Braunschweig 1894, §157).

In Part IV (pp.513–552) the above criterion is applied (pp.513–523) to an attempted proof for the theorem which Legendre was later to call "a law of reciprocity between primes" and which has remained famous as "the law of (quadratic) reciprocity". This establishes the relation between the quadratic residue character of an odd prime p modulo another one q, and the character of q modulo p. In order to express this, Legendre introduced in 1785 the notation $N^{(p-1)/2}$, where p is any prime and N an integer prime to p, explaining that this was not to be understood literally but meant the smallest residue (in absolute value) of $N^{(p-1)/2}$ modulo p, which is $+1$ or -1 according as N is or is not a quadratic residue modulo p. In 1798 he replaced this cumbersome and inadequate notation by the celebrated "Legendre symbol"

$$\left(\frac{N}{p}\right)$$

defined as having that same meaning[1] (cf. Chap.III, §VI).

The reasons for the partial success of Legendre's idea will be explained in Appendix I. What Legendre does is to distinguish eight cases according to the values of p and of q modulo 4 and to the value of (p/q); this same arrangement was observed by Gauss in his first proof of the law in question (*Disq.*Art.136). In each of these cases, Legendre introduces an appropriate equation of the form

$$ax^2 + by^2 + cz^2 = 0$$

with $a \equiv b \equiv c \equiv 1 \pmod{4}$; this can have no non-trivial

---

[1] For reasons of typographical convenience, this will be printed here as (N/p), except in displayed formulas.

solution, since the congruence

$$ax^2 + by^2 + cz^2 \equiv 0 \ (\text{mod } 4)$$

has none. Therefore, by Legendre's criterion, $-bc$, $-ca$ and $-ab$ cannot simultaneously be quadratic residues modulo $|a|$, modulo $|b|$ and modulo $|c|$, respectively. In each case Legendre tries to choose a, b, c so that this will lead to the desired conclusion.

Take firstly $p \equiv 1, q \equiv -1 \ (\text{mod } 4)$, $(p/q) = -1$, and the equation $x^2 + py^2 - qz^2 = 0$; as $(-1/q) = -1$, we have $(-p/q) = 1$; therefore (q/p) cannot be 1; it must be -1, as required by the reciprocity law. The same reasoning applies to the case $q \equiv q' \equiv -1 \ (\text{mod } 4)$, $(q/q') = 1$, and to the equation $x^2 - qy^2 - q'z^2 = 0$.

Now take $q \equiv q' \equiv -1 \ (\text{mod } 4)$, $(q/q') = -1$. In that case Legendre introduces the equation $px^2 - qy^2 - q'z^2 = 0$, where p is an auxiliary prime subject to the conditions

$$p \equiv 1 \ (\text{mod } 4), \quad \left(\frac{p}{q}\right) = -1, \quad \left(\frac{p}{q'}\right) = -1.$$

Assuming that there is such a prime, and using what has been proved before, the conclusion $(q'/q) = 1$ follows. For this to be valid, there has to be a prime in one of the arithmetic progressions

$$\{4qq'x + m | x = 0, 1, 2, ...\}$$

where one takes for m all the integers >0 and $<4qq'$ which are $\equiv 1 \ (\text{mod } 4)$ and non-residues modulo q and modulo q'.

Legendre was rightly convinced that every arithmetic progression $\{ax + b\}$, with a prime to b, contains infinitely many primes. "Perhaps it would be necessary to prove this rigorously", he had written in 1785 (*Recherches*, p.552). "We need have no doubt about it", he wrote in 1798 (*Essai*, p.220) this time attempting to support his statement by an altogether unconvincing argument (*Essai*, pp.12–16) which he later expanded, even more disastrously, into a whole chapter of his *Théorie des Nombres* (t.II, §XI, pp.86–104). It was left for DIRICHLET to prove the theorem of the arithmetic progression, as he did in 1837 by a wholly original method

(*Dir*.I.315–342) which remains as one of his major achievements; by the same method he also proved (*Dir*.I.499–502) that every quadratic form $ax^2 + bxy + cy^2$ represents infinitely many primes provided a, b, c have no common divisor, as had been announced by Legendre (*Théorie des Nombres*, t.II, pp.102–103).

In the case we have just discussed, the theorem of the arithmetic progression, taken by Legendre as a kind of axiom, gave at least some semblance of justification to his argument; but there is worse to come. Take the case of two primes p, p', both $\equiv 1$ (mod 4). Legendre seeks to treat it by means of the equation $px^2 + p'y^2 - qz^2 = 0$, where q is an auxiliary prime satisfying

$$q \equiv -1 \ (\text{mod } 4), \ \left(\frac{q}{p'}\right) = 1, \ \left(\frac{p}{q}\right) = -1,$$

or alternatively by means of $x^2 + py^2 - p'qz^2 = 0$, with q satisfying

$$q \equiv -1 \ (\text{mod } 4), \ \left(\frac{p}{q}\right) = -1$$

(*Recherches*, pp.519–520; *Essai*, pp.216–217 and 220–221; *Théorie des Nombres*, t.I, pp.233–234). Is there such a prime? Clearly its existence would follow from Dirichlet's theorem *and* the reciprocity law; but it is doubtful (as Gauss pointed out: *Disq*.Art.297) that it could be proved otherwise. Thus Legendre's method of proof ends up unavoidably with a piece of circular reasoning out of which there seems to be no escape. This did not prevent him, in a letter of 1827 to Jacobi, from complaining, as bitterly as unjustly, that Gauss had "claimed for himself" the discovery of the reciprocity law (*Jac*.I.398). He must have realized, however, that Gauss's criticism of his proof had its validity; in his *Théorie des Nombres* of 1830 he chose to be on the safe side by inserting (t.II, pp.57–64) Gauss's third proof (cf. *Gau*.II.3–8|1808), and, for good measure (t.II, pp.391–393) the "cyclotomic" proof communicated to him by Jacobi at the beginning of their correspondence (*Jac*.I.394|1827).

The remainder of Part IV of the *Recherches* of 1785, or at least its most valuable portion (pp.531–548) is devoted to a series of observations ("*quelques remarques assez singulières*", as he calls them, p.531; "*l'ébauche d'une théorie entièrement nouvelle*", he was to call this in his *Essai* of 1798, p.ix) concerning the representations, not only of integers, but also of binary quadratic forms as sums of three squares. These were based on the numerical evidence contained in the tables at the end of the memoir (pp.553–559); "*on peut les regarder comme autant de théorèmes*", he wrote, "*car quoique je n'en donne pas la démonstration complète [sic!], ils sont fondés au moins sur une induction très-étendue*" ["one may regard them as so many theorems, for, while I am not giving the full proof for them, they are at least based on wide numerical evidence": p.531]. Even Euler might not have been quite so casual about such matters.

This sketch was later expanded into the whole third section of the *Essai* of 1798 (pp.321–400) and again rewritten for the 1808 edition; by that time the same ground had been covered, far more concisely, more solidly and more fully, in Gauss's *Disquisitiones* (*Disq*.Art.266–293). Nevertheless, was it a mere coincidence that the young Gauss initiated his study of sums of three squares in July 1796 (cf. nos. 17 and 18 of his *Tagebuch*, *Gau*.X-1.496–497) a few weeks after discovering Legendre's *Recherches* in the Göttingen library? At any rate Gauss's first step (*ibid.*, no.17) was to verify the relationship, noted by Legendre in 1785, between the representations of an integer N as a sum of three squares and certain binary quadratic forms of determinant $-N$, also decomposable into sums of three squares. From this one can derive relations between the number of representations of N by sums of three squares and suitable class-numbers for forms of determinant $-N$; having discovered such relations empirically in 1785, Legendre sought to establish them in the *Essai* of 1798 (pp.366–400) and again in his later publications; this included in particular Fermat's celebrated assertion that every integer is a sum of three "triangular numbers" (*Essai*, p.399; cf. above, Chap.II, §XIV).

As Gauss rightly observed, these attempts involved Legendre in serious complications from which it might not have been easy for him to extricate himself, and which left his main conclusions still open to doubt ["*compluribus difficultatibus implicatus est, quae effecerunt ut theoremata palmaria demonstratione rigorosa munire non licuerit*": *Disq.*, Addit ad Art.288–293, *Gau.*I.466]. Gauss, of course, gave complete proofs for such results, and for many more of similar import, in the *Disquisitiones*.

On another matter of great importance Legendre also appears as a forerunner of Gauss, and more so perhaps than Gauss was willing to concede. Lagrange had shown, as we have seen (cf. above, §IV), that, if d is any "divisor" of a quadratic form, then it can be properly represented by some form with the same discriminant. In particular, let d be a "divisor" of $X^2 + NY^2$; i.e. such that $-N$ is a quadratic residue modulo d (cf. Chap.III, §VIII); then d can be properly represented by some form $f = aX^2 + 2bXY + cY^2$ of determinant $b^2 - ac = -N$. Let d' be another such divisor, similarly represented by a form f'; assume that d' is prime to d. As $-N$ is a quadratic residue modulo d and modulo d', it is such also modulo dd'; therefore dd' can be represented by some form F, also with the determinant $-N$. It was Legendre's great merit to perceive that F depends only upon f and f'; in other words, there is a form F such that every product of two mutually prime integers, properly represented by f and by f' respectively, is properly represented by F.

In order to do this, Legendre takes two forms

$$f(X, Y) = aX^2 + 2bXY + cY^2,$$
$$f'(X', Y') = a'X'^2 + 2b'X'Y' + c'Y'^2$$

with the same determinant $\delta = b^2 - ac = b'^2 - a'c'$. Putting $Z = aX + bY, Z' = a'X' + b'Y'$, he gets:

$$af(X, Y) = Z^2 - \delta Y^2, \qquad a'f'(X', Y') = Z'^2 - \delta Y'^2.$$

To this he applied "Brahmagupta's identity" (cf. above,

Chap.III, §VIII); this gives:

$$f(X, Y)f'(X', Y') = \frac{1}{aa'}[(ZZ' \pm \delta YY')^2 - \delta(ZY' \pm YZ')^2].$$

The right-hand side can be written as $AU^2 + 2BUV + CV^2$
by putting $A = aa'$, $C = (B^2 - \delta)/A$, and further

$$V = ZY' \pm YZ', \qquad AU + BV = ZZ' \pm \delta YY',$$

where B has still to be determined suitably. An easy calculation gives $U = XX' + mYX' + m'XY' + nYY'$, with m,
m', n given by

$$m = \frac{b \mp B}{a}, \qquad m' = \frac{b' - B}{a'}, \qquad n = mm' \mp C$$

(*Essai*, IVe Partie, §III, nos.362–369).

Assume now that a and a' are mutually prime; if not, as
Legendre observes, one can replace f' by an equivalent form
so as to fulfil that condition, provided of course a', $2b'$ and
c' have no common divisor. Take then $B \equiv \pm b$ (mod a),
$B \equiv b'$ (mod a'); then we have

$$B^2 \equiv b^2 \equiv \delta \text{ (mod } a), B^2 \equiv b'^2 \equiv \delta \text{ (mod } a'),$$

so that all the coefficients in the above formulas are integers.
In view of the double sign, Legendre had thus obtained
two forms

$$F(U, V) = AU^2 + 2BUV + CV^2$$

with the determinant $B^2 - AC = \delta$ and the property that
there is an identity

$$f(X, Y)f'(X', Y') = F[B(X, Y; X', Y'), B'(X, Y; X', Y')]$$

where B, B' are two bilinear forms in X, Y and X', Y'. Brahmagupta's identity appears here as a special case; at least
another one had been noticed by Euler (cf. his *Algebra*, *Eu.*I-
1.424, Art.178, and above, Chap.III, §XIV).

Anyone familiar with the Gaussian concept of composition
for binary quadratic forms will immediately recognize it in
the above elementary construction; the form F correspond-

ing to the upper signs in those formulas (which clearly depends symmetrically upon f and f') is the one derived from f and f' by "composition". No doubt the Gaussian theory, as Gauss chose to describe it (Disq.Art.234–260), is far more elaborate; so much so, indeed, that it remained a stumbling-block for all readers of the *Disquisitiones* until Dirichlet restored its simplicity by going back very nearly to Legendre's original construction (*Dir*.II.107–114|1851). Was Gauss inspired by Legendre's *Essai* to develop his own theory? This seems at least plausible, notwithstanding Gauss's statement that he did not see Legendre's book until "the greater part" ("*maxima pars*": *Gau*.I.7) of the *Disquisitiones* was in print; actually he started work on "composition" in the Fall of 1798 (cf. *Gau*.I.476, Zu Art.234) about the time when he visited PFAFF in Helmstedt (cf. his letter to BOLYAI of 29 Nov. 1798); perhaps he could have seen the *Essai* there[2].

Legendre seems to take it for granted that the classes of the two forms F which (because of the double sign) he derives from f and f' depend only upon the classes of f and f'; this is not quite obvious, and would deserve proof. But there is at the same time an essential difference between his treatment and that of Gauss. Legendre never made the distinction between proper and improper equivalence; his classes of quadratic forms are the Lagrangian classes (cf. above, §IV). Thus it was left to Gauss to discover that the "Gaussian" classes, under the operation of composition, make up a finite commutative group, and to draw the consequences of this fact. If C is a "Gaussian" class, then the corresponding "Lagrangian" class is either C if $C = C^{-1}$ in the sense of the group operation, or else it is the union of C and C^{-1}; if C' is another "Gaussian" class, then composition in the

---

[2] The relevant Section V of the *Disquisitiones* was not in print until much later; it contains the theory of ternary quadratic forms, which Gauss attacked only in February 1799 (cf. no.96 of the *Tagebuch*, *Gau*.X-1.539) and did not bring to completion before February 1800 (cf. *ibid*.,no.103, *Gau*.X-1.545).

sense of Gauss produces a class CC', while Legendre's operation produces the Lagrangian classes corresponding both to CC' and to CC'^{-1} and is thus a two-valued operation, which accounts for the double sign in his formulas; so it appears also in the tables (pp.432–434 of the *Essai*) where he lists the result of his operation, e.g. for the determinant -89. In this matter, too, Legendre had missed perhaps the most valuable feature of his discovery.

One last satisfaction was granted Legendre in his old age. Since the work of Euler, Fermat's so-called "last theorem", for exponents greater than 4, had remained as a challenge to all arithmeticians. In 1798 and again in 1808 Legendre had been unable to do more than reproduce in the *Essai* Fermat's proof for exponent 4 and Euler's for exponent 3 (cf. above, Chap.II, §§X and XII, and Chap.III, §XIV). Interest in the problem was revived in Paris in the following decades, particularly after the Paris Academy, in 1816, made it the subject of their annual prize-competition for 1818; Olbers brought this to Gauss's attention, and Gauss, with characteristic caution, answered that the problem had little interest for him as such, but that, with luck, its solution might perhaps turn up as a by-product of a wide extension of the higher arithmetic ("*einer grossen Erweiterung der höheren Arithmetik*: *Gau*.X-1.75|1816) which he was contemplating.

In the meanwhile Sophie GERMAIN, whose talent had early attracted the notice of Lagrange, Legendre and Gauss, had started working on Fermat's theorem, obtaining some valuable results based on ingenious congruence arguments (cf. e.g. H. M. Edwards, *Fermat's last theorem*, Springer 1977, p.64, or P. Ribenboim, *Thirteen lectures on Fermat's last theorem*, Springer 1979, pp.53–55). These need not be discussed here; what concerns us is rather the application of the infinite descent by Dirichlet and by Legendre, in the year 1825, to Fermat's equation for exponent 5.

Infinite descent *à la Fermat* depends ordinarily upon no more than the following simple observation: if the product $\alpha\beta$ of two ordinary integers (resp. two integers in an algebraic number-field) is equal to an m-th power, and if the g.c.d.

of α and β can take its values only in a given finite set of integers (resp. of ideals), then both α and β are m-th powers, up to factors which can take their values only in some assignable finite set. For ordinary integers this is obvious; it is so for algebraic number-fields provided one takes for granted the finiteness of the number of ideal-classes and Dirichlet's theorem about units. In the case of a quadratic number-field $\mathbf{Q}(\sqrt{N})$, this can be replaced by equivalent statements about binary quadratic forms of discriminant N. A typical case is given by Euler's proofs for Fermat's theorem (cf. Chap.III, §XIV); there the initial step consists in writing the equation in the form

$$(x-y)(x-jy)(x-j^2y) = z^3,$$

where $j = (-1+\sqrt{-3})/2$ is a cubic root of unity, and applying the above principle to the factors in the left-hand side. As shown in Chap.II, Appendix I, and as Euler knew (*loc.cit.*), this can easily be replaced by an argument based on the theory of the form $X^2 + 3Y^2$.

In seeking to deal with Fermat's equation $x^p - y^p = z^p$, where p is an odd prime >3, it is natural to split $x^p - y^p$ into linear factors in the "cyclotomic field" $\mathbf{Q}(\varepsilon)$, where ε is a p-th root of unity; this was all the more tempting after 1801, since Gauss had developed the algebraic side of the theory of $\mathbf{Q}(\varepsilon)$ in section VII of the *Disquisitiones*; it is clearly what Gauss had in mind in writing to Olbers in 1816 (*loc.cit.*). But ideal-theory had to be created first.

Gauss himself, however, had shown how $x^p - y^p$ can be split into factors in the quadratic number-field $k = \mathbf{Q}(\sqrt{\pm p})$, where the sign is the one for which $\pm p \equiv 1 \pmod 4$; this field, as he had shown, is contained in $\mathbf{Q}(\varepsilon)$, from which fact it follows that the polynomial

$$F = \frac{x^p - y^p}{x - y} = x^{p-1} + x^{p-2}y + \dots + y^{p-1},$$

which (as Gauss had shown) is irreducible over $\mathbf{Q}$, splits into two factors $P + Q\sqrt{\pm p}, P - Q\sqrt{\pm p}$, where P, Q are two polynomials of degree $(p-1)/2$ in x, y, with half-integer coefficients; this gives an identity

$$F = P^2 \mp pQ^2,$$

so that the principle described above can be applied to Fermat's equation when it is written in the form

$$(x-y)(P+Q\sqrt{\pm p})(P-Q\sqrt{\pm p}) = z^p.$$

Moreover, instead of the theory of the quadratic field k, which was still to be worked out, one could, even in Legendre's lifetime, apply the equivalent theory of the binary quadratic forms with the discriminant $\pm p$, as it was known from the work of Lagrange and that of Gauss.

This is precisely what was attempted in 1825 by a young student in Paris, Lejeune Dirichlet, then barely 20 years old; he had gone there to study mathematics, since this could hardly be done in Germany at the time. Naturally he tried his hand at the first outstanding case $p = 5$; he could do so all the more readily, since the required identity, viz.:

$$x^4 + x^3y + x^2y^2 + xy^3 + y^4 = (x^2+\tfrac{1}{2}xy + y^2)^2 - 5(\tfrac{1}{2}xy)^2$$

had been known since the days of Euler (Eu.I-3.280 in E 449|1772) who had derived from it some partial results concerning the form $X^2 - 5Y^2$ (cf. above, Chap.III, §IX).

In this first attempt Dirichlet was only partly successful. If the equation is written symmetrically as $x^5 + y^5 + z^5 = 0$, where x, y, z have no common divisor, it is obvious that one of them must be even while the other two are odd. By considering the congruence

$$x^5 + y^5 + z^5 \equiv 0 \pmod{25}$$

one sees also that one of the unknowns must be $\equiv 0$ (mod 5) while the other two are not; the same conclusion can be derived, somewhat more elaborately, from Sophie Germain's theorems mentioned above. At first Dirichlet's treatment (by infinite descent) could be applied only to the case where one and the same of the three unknowns is a multiple of 2 and of 5.

This is where Legendre, then well over 70 years old,

stepped in. After presenting Dirichlet's paper to the Academy
in July 1825 (cf. *Dir*.I.3–13), it took him only a few weeks
to deal with the remaining case. The technique he used did
not differ much from Dirichlet's.

This had perhaps been a modest Everest to climb, and
Dirichlet had guided him almost to the top. But Legendre
got there first. He described the whole proof in a paper
presented to the Academy in September 1825 and printed
in its Memoirs as a "Second Supplement" to his *Essai*; in it
he duly quoted "Mlle Sophie Germain" and also, strangely
enough, one "Lejeune Dieterich". The next year Dirichlet
was back in Germany, preparing his proof for publication
in Crelle's newly founded journal where it appeared in 1828
(*Dir*.I.21–46); in it he included the case he had had to omit
previously. Legendre inserted his own version of the proof,
with no reference to Dirichlet, in volume II of his *Théorie
des Nombres* of 1830 (6e Partie, §IV, pp.361–368). Had he
by that time persuaded himself that his alone was the glory?
If so, the old man deserves forgiveness. Towards Jacobi
and Abel he had behaved with great generosity, unstintingly
lavishing praise on their discoveries in his other topic of
predilection, the theory of elliptic functions. As to Dirichlet,
he was soon to take his flight and soar to heights undreamt
of by Legendre.

Appendix I

Hasse's Principle for Ternary Quadratic Forms

By a *form* we shall understand here a homogeneous polynomial F of any degree δ in some indeterminates $X_1, X_2, ..., X_n$, with coefficients in $\mathbf{Z}$. Clearly the equation $F = 0$ has a non-trivial solution in $\mathbf{Q}$, i.e. a solution in rational numbers, not all 0, if and only if it has a "proper" solution in $\mathbf{Z}$, i.e. a solution in integers without a common factor. On the other hand, we will say that a solution $(x_1, x_2, ..., x_n)$ of the congruence $F \equiv 0 \pmod{m}$ is *proper* if $x_1, x_2, ..., x_n$ are integers whose g.c.d. is prime to m. Obviously, if m' is prime to m, $F \equiv 0$ has a proper solution modulo mm' if and only if it has such solutions modulo m and modulo m'.

Let G be another such form, with indeterminates $Y_1, Y_2, ..., Y_n$; we shall say that the equations $F(X) = 0$, $G(Y) = 0$ are *equivalent* if there is a linear substitution S with coefficients in $\mathbf{Z}$, of determinant $D \neq 0$, transforming F into aG with $a \neq 0$; this means that we have

$$F[S(Y)] = aG(Y).$$

Then the substitution $T = aD.S^{-1}$ has also integral coefficients, and we have

$$G[T(X)] = a^{\delta-1}D^{\delta}F(X),$$

which shows that the relation between F and G is symmetric; as it is obviously reflexive and transitive, it is an equivalence relation in the usual sense.

Lemma 1. Let the forms F, G be such that the equations $F = 0$,

$G = 0$ *are equivalent; let p be a prime. Then the congruence* $G \equiv 0 \pmod{p^\mu}$ *has a proper solution for all μ if and only if $F \equiv 0$ has the same property.*

Notations being as above, let p^α be the highest power of p dividing a; let $x = (x_1, \ldots, x_n)$ be a proper solution of $F \equiv 0$ $\pmod{p^{\mu+\alpha}}$; put $y = T(x)$, and call p^λ the highest power of p dividing all the y_i. Put $y' = p^{-\lambda}y$; as we have

$$S(y') = p^{-\lambda}S(y) = p^{-\lambda}aDx,$$

and as x is proper, p^λ must divide aD. Now we have

$$aG(y') = F[S(y')] = (p^{-\lambda}aD)^\delta F(x);$$

as the right-hand side is $\equiv 0 \pmod{p^{\mu+\alpha}}$, $G(y')$ must therefore be $\equiv 0 \pmod{p^\mu}$, so that y' is a proper solution of that congruence.

Using elementary properties of the ring of p-adic integers, it is easy to show that the assumption (resp. the conclusion) in lemma 1 amounts to saying that $F = 0$ (resp. $G = 0$) has a non-trivial solution in the p-adic field $\mathbf{Q}_p$. After this remark, the lemma becomes obvious.

Here we shall be concerned exclusively with non-degenerate quadratic forms; these are the forms of degree 2

$$F(X) = \sum_{i,j=1}^{n} a_{ij}X_iX_j$$

for which the determinant $\det(a_{ij})$ is not 0. The so-called Hasse-Minkowski principle says that for such a form the equation $F = 0$ has a non-trivial solution in $\mathbf{Q}$ (and therefore a proper solution in integers) if and only if it has one in the field $\mathbf{R}$ of real numbers and the congruence $F \equiv 0$ $\pmod{m}$ has a proper solution for every integer m. The latter condition is tantamount to saying that the equation $F = 0$ has, for every prime p, a non-trivial solution in p-adic numbers[1]. The value of this principle for modern

---

[1] For a proof, based on an elementary treatment of quadratic forms from the point of view of p-adic fields, cf. J.-P. Serre, *Cours d'Arithmétique*, P.U.F., Paris 1970, Chap.I–IV.

number theory is greatly enhanced by the fact that it can
be extended to all algebraic number-fields[2]. As was discovered in substance by Legendre in 1785, a somewhat stronger
result holds for ternary forms:

*Theorem 1. Let $F(X, Y, Z)$ be a non-degenerate ternary quadratic
form. Then the equation $F = 0$ has a non-trivial solution in $\mathbf{Q}$ if
and only if it has one in $\mathbf{R}$ and the congruence $F \equiv 0 \pmod{p^{\mu}}$
has a proper solution for every power p^{μ} of every odd prime p.*

Of course the condition is necessary; only the converse
need be proved. By an elementary substitution, the equation
$F = 0$ can be transformed into an equivalent one of the
form

$$F(X, Y, Z) = aX^2 + bY^2 + cZ^2 = 0;$$

here, as F is non-degenerate, a, b and c are not 0; as $F = 0$
has solutions in $\mathbf{R}$, a, b and c are not all of the same sign.
Assume that F satisfies the conditions in our theorem; in
view of lemma 1, this implies that every equation $G = 0$
equivalent to $F = 0$ satisfies the same conditions. Now consider
Lagrange's method of solution for an equation $F = 0$,
as it has been described in Chap.II, §XIV; in it one replaces
$F = 0$, firstly by an equivalent equation $Z^2 = AX^2 + BY^2$
with square-free A and B, not both negative, then successively
by similar equations $Z^2 = MX^2 + NY^2$, each one equivalent
to the next one, until one reaches an equation for which M
is either 1 or does not satisfy any congruence $M \equiv m^2$
$\pmod{|N|}$. In the former case this last equation, and therefore
also the original one, have non-trivial solutions. We shall
prove our theorem by showing that the latter case cannot
occur.

In fact, if M does not satisfy any congruence $M \equiv m^2$
$\pmod{|N|}$, then, since N is square-free, it must have a prime
divisor p such that M does not satisfy any congruence $M \equiv
m^2 \pmod{p}$; clearly p must be odd. Consider the congruence

$$Z^2 \equiv MX^2 + NY^2 \pmod{p^2};$$

---

[2] As to the history of Hasse's discovery of his "principle", cf. H. Hasse,
Crelles J. 209 (1962), pp.3–4.

as we have seen, it follows from our assumptions about F that this has a proper solution (x, y, z). Here p cannot divide x; for, if it did, it would divide z; p^2 would divide Ny^2; as N is square-free, p would divide y, and (x, y, z) would not be proper. Taking now x' such that $xx' \equiv 1 \pmod{p}$, we have $M \equiv (x'z)^2 \pmod{p}$, which contradicts the assumption about M.

In order to derive from this Legendre's original formulation of his theorem, we need some lemmas.

Lemma 2. Let p be a prime. If a, b and c are prime to p, the congruence $aX^2 + bY^2 + cZ^2 \equiv 0 \pmod{p}$ has a proper solution.

As we have seen, this is a theorem of Euler's; for the proof, cf. above, Chap.III, §XI.

Lemma 3. Let p be a prime; put $F(X) = \Sigma_{i=1}^{n} a_i X_i^2$, where all the a_i are prime to p. Assume that the congruence $F(X) \equiv 0 \pmod{p^\mu}$ has a proper solution, with $\mu \geq 1$ if $p \neq 2$ and $\mu \geq 3$ if $p = 2$. Then the congruence $F(X) \equiv 0 \pmod{p^{\mu+1}}$ has a proper solution.

Let $(x_1, ..., x_n)$ be a proper solution of $F(X) \equiv 0 \pmod{p^\mu}$; then we can write $F(x) = p^\mu r$, and one of the x_i, say x_1, is prime to p. Put

$$y_1 = x_1 + p^\mu t \quad \text{resp.} \quad y_1 = x_1 + 2^{\mu-1}t$$

according as $p \neq 2$ or $p = 2$; put $y_i = x_i$ for $i = 2, ..., n$. We have

$$F(y) \equiv p^\mu(r + \delta a_1 x_1 t) \pmod{p^{\mu+1}}$$

with $\delta = 2$ if $p \neq 2$, $\delta = 1$ if $p = 2$. Taking t such that

$$r + \delta a_1 x_1 t \equiv 0 \pmod{p},$$

we see that $(y_1, ..., y_n)$ is a proper solution of $F(X) \equiv 0 \pmod{p^{\mu+1}}$.

Now take again an equation $F = 0$, with F of the form

$$F(X, Y, Z) = aX^2 + bY^2 + cZ^2,$$

where a, b, c are not 0 and are not all of the same sign. If abc is not square-free, and e.g. $a = a'm^2$ with $m > 1$, then, replacing Y, Z by mY, mZ, we can replace $F = 0$ by an equivalent equation with the coefficients a', b, c. If e.g. a and b have a common divisor $d > 1$, then, writing $a = a'd$, $b = b'd$ and replacing Z by dZ, we can replace $F = 0$ by an equivalent equation with the coefficients a', b', cd. As $|abc|$ decreases in both procedures, we can, by repeating them if necessary, obtain an equation of the same type, equivalent to $F = 0$ and for which abc is square-free, or, what amounts to the same, for which a, b and c are square-free and pairwise mutually prime. For such equations, we have Legendre's theorem (cf. above, §VI):

Theorem 2. Let a, b, c be integers, not all of the same sign, such that abc is square-free and not 0. Then the equation

$$F(X, Y, Z) = aX^2 + bY^2 + cZ^2 = 0$$

has a non-trivial solution in $\mathbf{Q}$ if and only if $-bc$, $-ca$ and $-ab$ are quadratic residues modulo $|a|$, modulo $|b|$ and modulo $|c|$, respectively.

If $F = 0$ has a non-trivial solution in $\mathbf{Q}$, it has a proper solution (x, y, z) in $\mathbf{Z}$. Here z must be prime to ab; for, if e.g. p is a prime dividing z and a, it must divide by^2, hence y since b is prime to a; then p^2 divides ax^2 and p divides x since a is square-free, so that (x, y, z) is not proper. Take now z' such that $zz' \equiv 1 \pmod{|ab|}$; then we have $-bc \equiv (byz')^2 \pmod{|a|}$ and $-ca \equiv (axz')^2 \pmod{|b|}$; the proof for $-ab$ is of course similar. Therefore the conditions in the theorem are necessary.

Assuming now that they are satisfied, we shall verify that those in theorem 1 are also satisfied. Let p by any odd prime. If it does not divide abc, then, by lemma 2, the congruence $F \equiv 0 \pmod{p}$ has a proper solution; in view of lemma 3, this implies, by induction on μ, that it has such solutions modulo p^μ for all μ. Now let p be a divisor of abc, e.g. of c and therefore not of ab. As $-ab$ is a quadratic residue modulo $|c|$, it is such modulo p, so that we can write $-ab \equiv m^2 \pmod{p}$. Then (m, a) is a proper solution of the congruence $aX^2 + bY^2 \equiv 0 \pmod{p}$; applying lemma 3 and induction on μ, we see that, for every μ, the same congruence has a proper solution (x, y) modulo p^μ, so that $(x, y, 0)$ is a proper solution of $F \equiv 0 \pmod{p^\mu}$. This completes the proof.

Corollary. Let N be a square-free integer, other than 0 or -1; let p be an odd prime, not a divisor of N. If p is a quadratic residue modulo $4|N|$, it can be written as $a^2 + Nb^2$ with rational a and b.

Apply Legendre's theorem to the equation $X^2 + NY^2 = pZ^2$. This will have a non-trivial solution in $\mathbf{Q}$ if p is a quadratic residue modulo $|N|$ and $-N$ is a quadratic residue modulo p. In view of the results described in Chap.III, Appendix I, both conditions will be fulfilled if p is a quadratic residue modulo $4|N|$ (and also if p is a quadratic residue modulo $|N|$ and $N \equiv -1 \pmod 4$). Then, if (x, y, z) is a non-trivial solution of $X^2 + NY^2 = pZ^2$, z cannot be 0, so that $p = a^2 + Nb^2$ with $a = x/z$, $b = y/z$. This had been predicted by Euler for the case $N > 0$, $p \equiv 1 \pmod{4N}$ (cf. above, Chap.III, §IX, and *Corr.*I.605–606|1753).

Finally, since Legendre applied this theorem to a proof (only partly successful) of the law of quadratic reciprocity, it seems worthwhile to comment briefly upon the relation between the two topics; for this purpose it is convenient to introduce a symbol $[F]_v$, where F is a ternary form over $\mathbf{Z}$ and v is either a prime or the symbol ∞. If p is a prime, we put $[F]_p = +1$ if the equation $F = 0$ has a non-trivial solution in the field $\mathbf{Q}_p$ of p-adic numbers, or, what amounts to the same, if the congruence $F \equiv 0 \pmod{p^\mu}$ has a proper solution for all μ; otherwise we put $[F]_p = -1$. We put $[F]_\infty = +1$ or -1 according as $F = 0$ has non-trivial solutions in $\mathbf{R}$ or not. For $F = Z^2 - aX^2 - bY^2$, $[F]_v$ is nothing else than the so-called Hilbert symbol $(a, b)_v$. As every equation $F = 0$ is equivalent to one of the latter type, it follows from lemmas 1, 2 and 3 that, for any F, we have $[F]_p = +1$ for all except finitely many primes p.

The Hasse-Minkowski principle in its most general form, applied to a ternary form F, says that $F = 0$ has a non-trivial solution in $\mathbf{Q}$ if and only if $[F]_\infty = +1$ and $[F]_p = +1$ for all primes p. On the other hand, theorem 1 (or, what amounts to the same, Legendre's theorem) says that it is enough, for $F = 0$ to have a non-trivial solution in $\mathbf{Q}$, that F should satisfy $[F]_\infty = +1$ and $[F]_p = +1$ for all *odd* primes p; thus these conditions imply $[F]_2 = +1$.

This last result is actually a special case of Hilbert's product-formula, valid for every ternary form F over $\mathbf{Z}$:

$$[F]_\infty \cdot \prod_p [F]_p = 1$$

where the product is taken over all primes. This says in effect that the number of factors equal to -1 in the left-hand side must be even. In particular, if all those factors except possibly $[F]_2$ are $+1$, then also $[F]_2$ must be $+1$; this, combined with the Hasse-Minkowski principle, gives Legendre's result.

Hilbert's formula contains the quadratic reciprocity law, from which in its turn it can be deduced by elementary considerations (cf. J.–P. Serre, *loc.cit.*). Apply it for instance to the form $F = pX^2 \pm qY^2 - Z^2$, where p and q are odd primes and the sign is determined by $\pm q \equiv 1 \pmod 4$. Here we have $[F]_\infty = +1$. It is easily seen that either $(0, 1, 1)$ or $(2, 1, 1)$ is a proper solution of $F \equiv 0 \pmod 8$; in view of lemma 3, this gives $[F]_2 = +1$. Lemmas 2 and 3 show that $[F]_r = +1$ for every prime r other than p and q. Finally, reasoning just as above in the proof of theorem 2, one gets

$$[F]_p = \left(\frac{\pm q}{p} \right), \, [F]_q = \left(\frac{p}{q} \right),$$

so that these last symbols must be both $+1$ or both -1. As we have observed in Chap. III, Appendix I, this is nothing else than the quadratic reciprocity law. The fact that Legendre's theorem implies at any rate one special case of the product-formula accounts for his partial success in applying it to the proof of the law in question.

Appendix II

A Proof of Legendre's on Positive Binary Quadratic Forms

In his *Additions* to Euler's *Algebra*, Lagrange had raised the question of finding the minimum of $|f(x, y)|$ for a given binary form f with integral coefficients when x, y are integers, not both 0; he had solved it for quadratic forms, both definite and indefinite (*Lag*.VII.61–74, Art.31–36 = *Eu*.I-1.552–562), by applying to it the theory of continued fractions.

In those same *Additions*, he had introduced the process of "Lagrangian reduction" (cf. above, §III), later devoting to it the whole of his *Recherches* of 1775. Actually there is a rather close connection between this topic and the minimum problem mentioned above, although there is nothing to show whether Lagrange was aware of this connection, which, in the case of definite forms, seems to have been first pointed out by Legendre in his *Essai* of 1798. It depends upon the following observations.

For brevity's sake, if $f = (a, b, c)$ is any binary quadratic form (in the notation explained above in §IV), we will write $\mu(f)$ for the minimum of $|f(x, y)|$ when x, y are given all integral values, not both 0. The forms for which $\mu(f) = 0$ are those whose discriminant $b^2 - 4ac$ is 0 or a square; as we have excluded these once for all (cf. above, §IV), $\mu(f)$ is an integer >0.

Lemma 1. Let f be a binary quadratic form (whose discriminant is neither 0 nor a square). Then there is a form $F = (A, B, C)$ equivalent to f, such that $A = \pm\mu(f)$ and $|B| \leq |A| \leq |C|$.

Take (a, b) such that $|f(a, b)| = \mu(f)$. If we had $a = da'$, $b = db'$ with $d > 1$, we would have

$$|f(a', b')| = |d^{-2}f(a, b)| < \mu(f).$$

Therefore a and b are mutually prime, so that $f(a, b)$ is properly represented by f. As shown above in §IV, there is then a form $f' = (m, n, p)$ equivalent to f with $m = f(a, b) = \pm\mu(f)$. As in Lagrangian reduction, take ν such that $-|m| \leq n - 2\nu m \leq |m|$; then the substitution $(X, Y) \to (X - \nu Y, Y)$ transforms f' into a form $F = (A, B, C)$ with $A = m = \pm\mu(f)$, $|B| \leq |A|$. As C is properly represented by F, it is properly represented by f, so that we have $|C| \geq \mu(f)$, which proves the lemma.

As Legendre discovered (cf. his *Essai* of 1798, I^e Partie, §VIII), there is, for definite forms, a kind of converse to lemma 2. Stated in somewhat more precise form than in Legendre, it is as follows:

Lemma 2. Let $F = (A, B, C)$ be a binary quadratic form with $A > 0$, $0 \leq B \leq A \leq C$. Then, if x, y are integers, we have $F(x, y) > C$ except for $x = 0$, $y = \pm 1$ or for $A = B$, $x = -y = \pm 1$, in which cases we have $F(x, y) = C$, or possibly for $y = 0$, in which case we have $F(x, 0) = Ax^2$ (and $F(x, 0) > A$ unless $x = 0$ or ± 1, $F(x, 0) \geq A$ unless $x = 0$).

Let $D = B^2 - 4AC$ be the discriminant of F; it is < 0. If $F(x, y) \leq C$, we have

$$4AF(x, y) = (2Ax + By)^2 + |D|y^2 \leq 4AC$$

and therefore $y^2 \leq 4AC|D|^{-1}$. As $0 \leq B \leq A \leq C$, we have $|D| \geq 3AC$; this gives $y^2 \leq 4/3$, $y = 0$ or ± 1. If $y = \pm 1$, the above inequality can be rewritten as

$$(2Ax \pm B)^2 + |D| - 4AC = 4A(Ax^2 \pm Bx) \leq 0$$

i.e. $Ax^2 \pm Bx \leq 0$; this cannot be so unless either $x = 0$ or

$A = B, x = \mp 1$, in which cases we have $F(x, y) = C$. As to our statements for the case $y = 0$, they are obvious.

Lemma 3. Let $F = (A, B, C)$ be a form satisfying the conditions in lemma 2. Then $\mu(F) = A$; moreover, if $C > A$, we have $F(a, b) = A$ if and only if $(a, b) = (\pm 1, 0)$; if $C = A > B$, we have $F(a, b) = A$ if and only if $(a, b) = (\pm 1, 0)$ or $(0, \pm 1)$; if $C = A = B$, we have $F(a, b) = A$ if and only if $(a, b) = (\pm 1, 0), (0, \pm 1)$ or $(\pm 1, \mp 1)$.

This follows at once from lemma 2.

Theorem. Let $F = (A, B, C), F' = (A', B', C')$ be two equivalent forms, both satisfying the conditions in lemma 2. Then $F = F'$.

As F and F' are equivalent, we can write

$$F'(X, Y) = F(\alpha X + \beta Y, \gamma X + \delta Y), \qquad \alpha\delta - \beta\gamma = \pm 1;$$

also, we have $\mu(F) = \mu(F')$ and therefore $A = A'$; this gives $F(\alpha, \gamma) = A$, and therefore, by lemma 3, $\gamma = 0$, $\alpha = \pm 1$ unless $C = A$. If $C = A$, there are, by lemma 3, four or six values of (a, b) for which $F(a, b) = A$; therefore the same must be true of F', so that, by the same lemma, we have $C' = A' = A$; as F and F' must have the same discriminant, and as B and B' are $\geqslant 0$, this gives $B = B'$, $F = F'$. Now take the case $C > A$, $\gamma = 0$; then we must have $\alpha = \pm 1, \delta = \pm 1$. Here we have

$$B' = 2A\alpha\beta + B\alpha\delta = \pm 2A\beta \pm B,$$

and hence $2A\beta = \pm B \pm B'$. As B and B' are $\geqslant 0$ and $\leqslant A$, this implies $B = B'$, and therefore, as above, $C = C'$, $F = F'$.

Now, for every form f, the process of Lagrangian reduction produces an equivalent form $F = (A, B, C)$ with $|B| \leqslant |A|$ and $|B| \leqslant |C|$; if f is positive-definite, both A and C must be > 0, while if f is indefinite they must be of opposite signs since $B^2 - 4AC$ is then > 0 and $|AC|$ is $\geqslant B^2$. As it is obvious that the forms $(A, \pm B, C)$ and $(C, \pm B, A)$ are all equivalent (cf. above, §IV), there must be, in each Lagrangian class of positive forms, at least one satisfying the conditions $A > 0$,

$0 \leqslant B \leqslant A \leqslant C$ of lemma 2; as Lagrange found, this is unique, and the above lemmas, in substance, make up Legendre's proof (obtained, as he says, "*par une méthode particulière*": *Essai*, p.xij) for this result.

As to the determination of $\mu(f)$ for an indefinite form f, lemma 1 says that there is a form $F = (A, B, C)$ equivalent to f and reduced "in the Lagrangian sense", for which $\mu(f) = |A|$; thus the problem can be solved by constructing all such forms. Lagrange's method for doing this will now be described in Appendix III.

Appendix III

A Proof of Lagrange's on Indefinite Binary Quadratic Forms

Here, for convenience, we shall use matrix notation in addition to the notations introduced above in §IV. Thus we write

$$S = \begin{pmatrix} \alpha & \beta \\ \gamma & \delta \end{pmatrix}$$

for the substitution given by (2), §IV. We write again $f \circ S$ for the form F given by (1), §IV, i.e. for the transform of the form f by the substitution S, and $\mathcal{G}$ for the group $GL(2, \mathbf{Z})$ of the substitutions S with integral coefficients α, β, γ, δ such that $\det S = \pm 1$. We write the group $\mathcal{G}$ multiplicatively; thus, for S and T in $\mathcal{G}$, ST will denote the substitution

$$(x, y) \rightarrow ST(x, y) = S[T(x, y)];$$

then, for a form f, we have $f \circ ST = (f \circ S) \circ T$. Of course those notations are foreign to Lagrange, having been developed only in the nineteenth and twentieth century.

Expressed in this language, Lagrange's first observation is as follows:

Lemma 1. Every substitution $\begin{pmatrix} \alpha & \beta \\ \gamma & \delta \end{pmatrix}$ in $\mathcal{G}$ can be changed, by multiplication to the right and to the left by substitutions of the form $\begin{pmatrix} \pm 1 & 0 \\ 0 & \pm 1 \end{pmatrix}$, into a similar one for which α, β, γ, δ are all ≥ 0.

Multiplying it to the right by one of these substitutions, we can change it into one for which $\alpha > 0$, $\beta > 0$ (if neither of them is 0) or else into one for which α and β are $\geqslant 0$ and γ, δ are not of opposite signs (if α or β is 0). In the former case the relation $\alpha\delta - \beta\gamma = \pm 1$ implies that γ, δ are not of opposite signs. Multiplying then to the left with one of the substitutions $\begin{pmatrix} 1 & 0 \\ 0 & \pm 1 \end{pmatrix}$ will do what was required.

We will write $\mathcal{G}_+$ for the set of those substitutions $\begin{pmatrix} \alpha & \beta \\ \gamma & \delta \end{pmatrix}$ in $\mathcal{G}$ for which $\alpha, \beta, \gamma, \delta$ are all $\geqslant 0$; this is a semigroup, i.e., the product of any substitutions in $\mathcal{G}_+$ is itself in $\mathcal{G}_+$. As to these, Lagrange showed in substance that they can all be obtained by a succession of substitutions $(x, y) \rightarrow (y, x)$ and $(x, y) \rightarrow (x, y + \mu x)$ with $\mu > 0$. If we put, in matrix notation:

$$J = \begin{pmatrix} 0 & 1 \\ 1 & 0 \end{pmatrix}, \qquad T = \begin{pmatrix} 1 & 0 \\ 1 & 1 \end{pmatrix},$$

then these are the substitutions J, T^μ; we have in fact

$$T^\mu = \begin{pmatrix} 1 & 0 \\ \mu & 1 \end{pmatrix}, \qquad JT^\mu = \begin{pmatrix} \mu & 1 \\ 1 & 0 \end{pmatrix}, \qquad J^2 = \begin{pmatrix} 1 & 0 \\ 0 & 1 \end{pmatrix}.$$

With these notations, Lagrange's result can be stated as follows:

Theorem 1. Every substitution in $\mathcal{G}_+$ can be expressed as a product of substitutions J and T^μ with $\mu > 0$.

Take $S = \begin{pmatrix} \alpha & \beta \\ \gamma & \delta \end{pmatrix}$ in $\mathcal{G}_+$; after replacing S by SJ if $\alpha < \beta$, we may assume that $\alpha \geqslant \beta$. In substance, Lagrange proceeds then by induction on β; following his procedure, we first consider the case $\beta > 1$. As $\alpha \geqslant \beta$, we can write $\alpha = \mu\beta + \rho$ with $0 \leqslant \rho < \beta$, just as in the euclidean process for finding the g.c.d. of α and β, or for constructing the continued fraction for α/β. Here, as $\alpha\delta - \beta\gamma = \pm 1$, α is prime to β, so that $\rho > 0$. Then, putting

$$\sigma = \gamma - \mu\delta, \ S' = \begin{pmatrix} \beta & \rho \\ \delta & \sigma \end{pmatrix},$$

we have $S = S'JT^\mu$. As S, J and T^μ are in $\mathcal{G}$, so is S'; therefore we have

$$\det S' = \beta\sigma - \rho\delta = \pm 1.$$

This implies that σ is $\geqslant 0$, since otherwise we would have:

$$\beta\sigma - \rho\delta \leqslant \beta\sigma \leqslant -\beta < -1.$$

Therefore S' is in $\mathcal{G}_+$. We can then, as Lagrange does, operate on S' in the same manner provided $\rho > 1$, or, what amounts to the same, apply the induction assumption to S' provided the cases $\beta = 1$ and $\beta = 0$ have been dealt with first of all.

In the case $\beta = 1$, we put $\alpha = \mu\beta + \rho$ with $\mu = \alpha$ and $\rho = 0$ if $\alpha\delta - \beta\gamma = -1$, resp. $\mu = \alpha - 1$ and $\rho = 1$ if $\alpha\delta - \beta\gamma = 1$ (in which case α and δ must be $\geqslant 1$; Lagrange, otherwise so scrupulously careful in such matters, thought that this case cannot occur, even giving a fallacious argument to prove it: *Lag.*III.733, lines 8–9).In either case, take again $\sigma = \gamma - \mu\delta$, i.e. $\sigma = \gamma - \alpha\delta = 1$ resp. $\sigma = \gamma - (\alpha-1)\delta = \delta - 1$. Then we get, as before, $S = S'JT^\mu$, with

$$S' = T^\delta \text{ resp. } S' = \begin{pmatrix} 1 & 1 \\ \delta & \delta - 1 \end{pmatrix};$$

repeating the same procedure on S' in the latter case, we find $S' = T^{\delta-1}JT$, which completes the proof for $\beta = 1$. If $\beta = 0$, we must have $\alpha = \delta = 1$ and $S = T^\gamma$.

For every S in $\mathcal{G}_+$, the above procedure gives an expression

$$S = T^{\mu_0}JT^{\mu_1}JT^{\mu_2} \dots JT^{\mu_m}$$

where all the μ_i are $\geqslant 0$, all are > 0 except possibly the first and last one, and where the μ_i, for $m > 0$, determine a continued fraction for α/β:

$$\frac{\alpha}{\beta} = \mu_m + 1/(\mu_{m-1} + 1/ \cdots + 1/(\mu_2 + 1/\mu_1) \cdots).$$

It can be shown easily that the above expression for S is unique, but this point is not considered by Lagrange.

We shall now describe Lagrange's application of theorem 1 to the theory of indefinite binary quadratic forms; we recall that we exclude the forms whose discriminant is 0 or a square.

The process of Lagrangian reduction (Chap.IV, §IV; cf. Chap.III, §IX) applied to a given form leads to a form (A, B, C) reduced "in the Lagrangian sense", i.e. for which $|B|$ is $\leq|A|$ and $\leq|C|$, so that $B^2 \leq |AC|$. Then $B^2 - 4AC$ has the sign of $-4AC$, so that, if it is >0 as we are assuming here, A and C must be of opposite signs, i.e. $AC < 0$. It will be convenient to say that a form $f = (a, b, c)$ is *weakly reduced* if $ac < 0$; this is the same as to say that the polynomial $f(X, 1) = aX^2 + bX + c$ has two real roots of opposite signs. Clearly there are only finitely many weakly reduced forms with a given discriminant. If $f = (a, b, c)$ is weakly reduced, so is $f \circ J = (c, b, a)$.

To bring clarity into Lagrange's treatment of his problem, it is also necessary to give a name to those forms f for which $f(X, 1)$ has a positive root >1 and a negative root > -1; we shall call them *strongly reduced*. This concept is indeed implicit in Lagrange's calculations, as well as in his solution of Pell's equation and in his treatment of the continued fractions for quadratic irrationalities (cf. above, §II(B)), two topics closely related to the present one. For the same reason it has already made its appearance (under the name of "reduced forms") in our discussion of Pell's equation in Chap.II, §XIII. In the *Disquisitiones* Gauss introduced a closely related concept, also under the name of "reduced forms" (*Disq.* Art.183); perhaps he had merely disengaged it from Lagrange's calculations and adapted it to his own purposes.

Lagrange's problem was to decide when two indefinite forms, reduced in his sense, are equivalent. His solution depends upon the facts expressed in the following lemmas. As usual, if ξ is any real number, we write $[\xi]$ for the integer μ determined by $\mu \leq \xi < \mu + 1$, i.e. for the largest integer $\leq\xi$.

Lemma 2. Let f be a weakly reduced form; let ξ be the positive root of $f(X, 1)$. Then, for $\mu > 0$, the form $f' = f \circ JT^\mu$ is weakly reduced if and only if $\mu \leq \xi$, strongly reduced if and only if $\mu = [\xi]$; and the form $f'' = f \circ JT^\mu J$ is never strongly reduced.

Put $f(1, 0) = a$, and call $-\eta$ the negative root of $f(X, 1)$. Then:

$$f'(X,Y) = f(\mu X + Y, X) = a[Y + (\eta + \mu)X] \cdot [Y - (\xi - \mu)X].$$

This is weakly reduced if and only if $\xi - \mu > 0$. The roots of $f'(X, 1)$ are

$$\xi' = \frac{1}{\xi - \mu}, \qquad -\eta' = \frac{-1}{\eta + \mu}.$$

If $\mu > 0$, η' is >0 and <1; as to ξ', it is >1 if and only if $0 < \xi - \mu < 1$, i.e. $\mu = [\xi]$. Finally, the roots of $f''(X, 1)$ are $\xi - \mu$ and $-\eta - \mu$, and the latter is < -1.

From this it follows that every weakly reduced form is equivalent to a strongly reduced one, viz., $f \circ JT^\mu$ for $\mu = [\xi]$ if the positive root ξ of $f(X, 1)$ is >1, and $(f \circ J) \circ JT^\mu = f \circ T^\mu$ for $\mu = [1/\xi]$ if $\xi < 1$ since in the latter case we can apply lemma 2 to $f \circ J$. As every form is equivalent to one reduced in the Lagrangian sense, and as every indefinite form, reduced in that sense, is weakly reduced, this shows that there is at least one strongly reduced form in every Lagrangian class of indefinite forms. In view of this and of lemma 2, Lagrange's problem amounts to finding all the strongly reduced forms equivalent to a given one. The key to this is contained in the following lemma (cf. *Lag.*III.730–732):

Lemma 3. Let S' be a substitution in $\mathcal{G}_+$; put $S = S'JT^\mu$ with some $\mu > 0$. Let f be a weakly reduced form; let $F' = f \circ S'$ and $F = f \circ S$ be its transforms by S' and by S, respectively. Then, if F is weakly reduced, so is F'.

Put $f(1, 0) = a$; let ξ, $-\eta$ be the positive and the negative root of $f(X, 1)$, respectively; we have:

$$f(X, Y) = a(X + \eta Y)(X - \xi Y).$$

As in the proof of theorem 1, let S' be given by $S' = \begin{pmatrix} \beta & \rho \\ \delta & \sigma \end{pmatrix}$, so that we have $S = \begin{pmatrix} \alpha & \beta \\ \gamma & \delta \end{pmatrix}$ with $\alpha = \mu\beta + \rho$, $\gamma = \mu\delta + \sigma$. Put $F = (A, B, C)$, $F' = (A', B', C')$. We have

$$A = f(\alpha, \gamma) = a(\alpha + \eta\gamma)(\alpha - \xi\gamma),$$
$$C = A' = f(\beta, \delta) = a(\beta + \eta\delta)(\beta - \xi\delta),$$
$$C' = f(\rho, \sigma) = a(\rho + \eta\sigma)(\rho - \xi\sigma).$$

Put:

$$\theta = \alpha - \xi\gamma, \qquad \zeta = \beta - \xi\delta, \qquad \omega = \rho - \xi\sigma.$$

To say that $AC < 0$ is the same as to say that θ and ζ have opposite signs; we have to show that, when that is so, also ζ and ω have opposite signs, or, what amounts to the same, that ω has the sign of θ. This follows at once from the fact that $\omega = \theta - \mu\zeta$.

Lemma 4. Let f be a weakly reduced form; let ξ be the positive root of $f(X, 1)$; let μ, ν be integers, both >0; put $f' = f \circ JT^\mu$, $f'' = f' \circ JT^\nu$. Then, if f'' is weakly reduced, f' is strongly reduced, ξ is >1, and $\mu = [\xi]$.

Applying lemma 3 to $S' = JT^\mu$, $S = S'JT^\nu$ and to f, we see that f' must be weakly reduced. As in the proof of lemma 2, the positive root of $f'(X, 1)$ is $\xi' = 1/(\xi - \mu)$. Now apply lemma 2 to f' and f''; it shows that we have $\xi' > \nu$; as ν is ≥ 1, this gives $0 < \xi - \mu < 1$, i.e. $\mu = [\xi]$; by lemma 2, this implies that f' is strongly reduced.

Lemma 5. Let μ_1, μ_2, ..., μ_m be integers, all >0; for $1 \leq i \leq m$, put

$$S_i = JT^{\mu_1}JT^{\mu_2} \ldots JT^{\mu_i}.$$

Let f_0 be a weakly reduced form; for $1 \leq i \leq m$, put $f_i = f_0 \circ S_i$, and assume that f_m is weakly reduced. Then $f_1, f_2, ..., f_{m-1}$ are strongly reduced; moreover, if ξ_i is the positive root of $f_i(X, 1)$, we have, for $1 \leq i \leq m$, $[\xi_{i-1}] = \mu_i$, and ξ_0 is given by the continued fraction:

$$\xi_0 = \mu_1 + 1/(\mu_2 + 1/ \cdots + 1/(\mu_i + 1/\xi_i) \cdots).$$

From lemma 3 it follows that f_{m-1} is weakly reduced, then successively, by induction on j, that f_{m-j} is weakly reduced for $j = 2, 3, ..., m-1$. The assertions in the lemma follow then from lemma 4 applied to f_{i-1}, f_i, f_{i+1} for $1 \leqslant i \leqslant m-1$.

Let f be a weakly reduced form such that the positive root ξ of $f(X, 1)$ is >1; then, for $\mu = [\xi]$, we shall say that the strongly reduced form $f' = f \circ JT^\mu$ is the *Lagrangian transform*, or more precisely the *first Lagrangian transform* of f; it is thus given by

$$f'(X, Y) = f(\mu X + Y, X) .$$

The same procedure can then be iterated, i.e. applied to f', then to the transform f'' of f', etc., giving successive Lagrangian transforms f', f'', etc.; thus, in lemma 5, the forms $f_1, f_2, ..., f_{m-1}$ are the first $m-1$ successive Lagrangian transforms of f_0. For instance, in Chap.II, §XIII, the forms $(-1)^i F_{i-1}$, for $i = 1, 2, 3$, etc., are the successive Lagrangian transforms of the weakly reduced form $X^2 - NY^2$.

From these facts Lagrange deduces the solution of his problem, formulating it in terms of forms "reduced in the Lagrangian sense". As noted above, it can be formulated just as well (and more concisely and elegantly) in terms of "strongly reduced forms"; it is contained in the following theorem:

Theorem 2. Let $f = (a, b, c)$, $F = (A, B, C)$ be two equivalent indefinite forms, both strongly reduced; put $f^* = (c, -b, a)$, $F^* = (C, -B, A)$. Then either F or F^* is one of the successive Lagrangian transforms of either f or f^*.

Let S be a substitution in $\mathscr{G}$, such that $F = f \circ S$. Take first the case when S is in $\mathscr{G}_+$. Then, by theorem 1, we can write

$$S = T^{\mu_0} J T^{\mu_1} J ... J T^{\mu_m}$$

where all the μ_i are >0 except possibly the first and last one which are $\geqslant 0$. If $\mu_0 = 0$, $\mu_m > 0$, lemma 5, together with lemma 2, shows that F is the m-th Lagrangian transform of f. If μ_m were 0, we could write

$$SJ = S'JT^{\mu_{m-1}}$$

with $S' = J$ if $m = 1$, and anyway S' in $\mathscr{G}_+$. This gives

$$F \circ J = f \circ S'JT^{\mu_m-1};$$

here, as f and $F \circ J$ are weakly reduced, we can apply lemma 3 to f, $f \circ S'$ and $F \circ J$; it shows that $f \circ S'$ must be weakly reduced. Then lemma 2, applied to $f \circ S'$, shows that the form

$$F = (f \circ S') \circ JT^{\mu_m-1}J$$

cannot be strongly reduced, which contradicts our assumption; thus μ_m is >0. Now we show that μ_0 must be 0. In fact, if $\mu_0 > 0$, we can write $F = f' \circ S_0$ with $f' = f \circ J$, S_0 being given by

$$S_0 = JT^{\mu_0}J \dots JT^{\mu_m};$$

as f' is weakly reduced, lemma 5 shows that $f' \circ JT^{\mu_0}$ is strongly reduced if $m \geq 1$; the same is true by assumption if $m = 0$. Then lemma 2 shows that the positive root of $f'(X, 1)$ must be $>\mu_0$; but this cannot be, since that root is $1/\xi$ if ξ is the positive root of $f(X, 1)$, and we have $\xi > 1$ since f is strongly reduced. This proves our theorem for S in $\mathscr{G}_+$.

If S is not in $\mathscr{G}_+$, then, by lemma 1, it can be written as $ES'E'$ with S' in $\mathscr{G}_+$ and E, E' of the form $\begin{pmatrix} \pm 1 & 0 \\ 0 & \pm 1 \end{pmatrix}$. Call H the substitution $(x, y) \rightarrow (y, -x)$, i.e., in matrix form

$$H = \begin{pmatrix} 0 & 1 \\ -1 & 0 \end{pmatrix};$$

we have

$$H^2 = -1_2, \quad HJ = -JH = \begin{pmatrix} 1 & 0 \\ 0 & -1 \end{pmatrix}$$

where 1_2 is the identity in $\mathscr{G}$. Now each one of the substitutions E, E' can be expressed, either as $\pm 1_2$, or as $\pm HJ = \mp JH$. Thus lemma 1 shows that every substitution S in $\mathscr{G}$ can be expressed in one of the forms $\pm S'$, $\pm HS'$, $\pm S'H$ or $\pm HS'H$

with S' in $\mathcal{G}_+$. Consequently, if $F = f \circ S$, one of the forms F and $F^* = F \circ H$ is the transform, by a substitution S' in $\mathcal{G}_+$, either of f or of $f^* = f \circ H$. In view of what has been proved above, this completes the proof of our theorem. It should be noted that, if $f = (a, b, c)$ is weakly (resp. strongly) reduced, then so is $f^* = f \circ H = (c, -b, a)$.

To a result, equivalent in substance to theorem 2 (but expressed in terms of forms "reduced in the Lagrangian sense"), Lagrange adds one further remark. Since there are only finitely many reduced forms with a given discriminant, some form must occur twice in the sequence of the Lagrangian transforms of any given form f, so that such a sequence must be periodic, at least from a certain point onwards. Thus, as Lagrange observes (*Lag.*III.740–741), the sequence need not be continued beyond that point, so that the process is a finite one.

Taking our cue, however, from the solution of Pell's equation (*Lag.*II.494–496; cf. *ibid.*,pp.429–443 and 603–615, and above, Chap.II, §XIII), we can proceed one step further. Let f_0 be a strongly reduced form; let f_1, f_2, etc., be the sequence of its Lagrangian transforms, with $f_i = f_{i-1} \circ JT^{\mu_i}$ as before; call ξ_i the positive root of $f_i(X, 1)$; by lemma 5, we have

$$\mu_{i+1} = [\xi_i], \ \xi_{i+1} = \frac{1}{\xi_i - \mu_{i+1}}, \ \xi_0 = \mu_1 + 1/(\mu_2 + 1/ \cdots);$$

μ_1, μ_2, etc. are the successive integers occurring in the continued fraction for ξ_0; as the sequence (f_i) is periodic, at least from a certain point on, so is the sequence (μ_i). Now put $f_i^* = f_i \circ H$. In view of the easily verified identity

$$H(JT^\mu)^{-1}H^{-1} = -JT^\mu,$$

we have:

$$f_{i-1}^* = f_i^* \circ JT^{\mu_i}.$$

As all the f_i^* are strongly reduced, this shows that f_{i-1}^*, f_{i-2}^*, etc., are the successive Lagrangian transforms of f_i^*. Now, since the forms f_i must repeat themselves, assume that

$f_{j+n} = f_j$ for some j and some $n > 0$. Then we have $f^*_{j+n} = f^*_j$ and therefore $f^*_{j+n-1} = f^*_{j-1}$ etc., and finally $f^*_n = f^*_0$, which implies $f_n = f_0, f_{n+i} = f_i, \mu_{n+i} = \mu_i$ for all $i > 0$. That being so, since we have $f^*_n = f^*_0$, the successive Lagrangian transforms of f^*_0 are no other than $f^*_{n-1}, f^*_{n-2}, ..., f^*_1$ and again f^*_0. Consequently theorem 2 can be replaced by the following more precise statement: the strongly reduced forms equivalent to f_0 are $f_1, f_2, ..., f_{n-1}$, and $f^*_0, f^*_1, ..., f^*_{n-1}$ if these are different from $f_0, f_1, ..., f_{n-1}$.

Additional Bibliography and References

ĀRYABHAṬA: cf. K. Elfering.

L. Aubry, Solution de quelques questions d'analyse indéter-minée, Sphinx-Œdipe, 7e année (1912), pp. 81–84.

BACHET: Claude Bachet, sieur de Méziriac, Problèmes plaisants et delectables qui se font par les nombres, avec leur demonstration, Lyon, Pierre Rigaud 1612; 2e éd., 1624.

JAC. BERNOULLI: Jacobi Bernoulli . . . Ars Conjectandi, Opus Posthumum . . . Basileae 1713.

BOMBELLI: (a) L'Algebra, Opera di Rafael Bombelli da Bologna, divisa in tre libri . . . In Bologna, per Giovanni Rossi 1572, con licenza de' Superiori; (b) Rafael Bombelli da Bologna, L'Algebra, Prima edizione integrale, Feltrinelli, Milano 1966.

S. Chowla and W. E. Briggs, On discriminants of binary quadratic forms with a single class in each genus, Can. J. of Math. 6 (1954) 463–470.

B. Datta and A. N. Singh, History of Hindu Mathematics, A Source Book, 2 vol., Lahore 1935–1938 (vol.II, Algebra).

DIOPHANTUS: (a) Diophanti Alexandrini Arithmeticorum libri sex, et de numeris multangulis liber unus. Nunc primum Græcè et Latinè editi, atque absolutissimis Commentariis illustrati. Auctore Claudio Gaspare Bacheto Meziriaco Sebusiano V.C., Lutetiae Parisiorum, Sumptibus Hieronymi Drouart, via Jacobæa, Sub Scuto Solari

M.DC.XXI Cum Privilegio Regis; (b) Diophanti Alexandrini Arithmeticorum libri sex, et de numeris multangulis liber unus. Com Commentariis C.G.Bacheti V.C. et observationibus D.P. de Fermat Senatoris Tolosani. Accessit Doctrinae Analyticae inuentum nouum, collectum ex varijs eiusdem D. de Fermat Epistolis. Tolosæ, Excudebat Bernardus Bosc, è Regione Collegij Societatis Iesu, M.DC.LXX.

DIRICHLET-DEDEKIND: Vorlesungen über Zahlentheorie von P.G.Lejeune Dirichlet, herausgegeben und mit Zusätzen versehen von R.Dedekind, 4te Aufl., Braunschweig, F.Vieweg und Sohn, 1894.

K. Elfering, Die Mathematik des Āryabhaṭa I, Text, Übersetzung aus dem Sanskrit und Kommentar, W.Fink Verlag, München 1975.

ENCYCLOPÉDIE, ou Dictionnaire raisonné des Sciences, des Arts et des Métiers, par une Société de Gens de Lettres. Mis en ordre et publié par M. Diderot, de l'Académie Royale des Sciences et des Belles-Lettres de Prusse; et quant à la Partie Mathématique, par M. d'Alembert, de l'Académie Royale des Sciences de Paris, de celle de Prusse, et de la Société Royale de Londres (Tome Quatrième, à Paris, MDCCLIV, avec approbation et privilège du Roy).

G. Eneström: (a) Der Briefwechsel zwischen Leonhard Euler und Johann I. Bernoulli, Bibl. Math. 4 (1903) 344–388, 5 (1904) 248–291, 6 (1905) 16–87; Der Briefwechsel zwischen Leonhard Euler und Daniel Bernoulli, ibid. 7 (1906–07) 126–156; Zur Geschichte der unendlichen Reihen um die Mitte des siebzehnten Jahrhunderts, ibid. 12 (1911–12) 135–148; (b) Verzeichnis der Schriften Leonhard Eulers, Jahresb. d.D.M.V. (Ergänzungsband IV), 1910–1913.

A. Enneper, Elliptische Functionen, Theorie und Geschichte, Halle 1876.

EULER: Leonhardi Euleri Opera Postuma Mathematica et Physica, Anno 1844 detecta . . . edd. P.-H.Fuss et Nic. Fuss, Petropoli 1862 (Reprint, Kraus 1969).

L. Euler und Chr. Goldbach, Briefwechsel, edd. A.P.Juškevič und E. Winter, Berlin 1965.

FAGNANO: Produzioni Matematiche del Conte Giulio Carlo

di Fagnano, Marchese de' Toschi . . . In Pesaro, l'anno del Giubileo M.DCC.L . . . con licenza de' Superiori (2 vol. = *Fag*.I–II).

FERMAT: Varia Cpera Mathematica D.Petri de Fermat Senatoris Tolosani . . . Tolosae, Apud Joannem Pech . . . juxta Collegium PP. Societatis Jesu M.DC.LXXIX.

FRENICLE: Traité des Triangles Rectangles en nombres ("par M. Frenicle de Bessy", in Mémoires de l'Academie Royale des Sciences, tome V, 1729, pp. 127–206; "la première Partie du Traité . . . avoit été imprimée dès l'année 1676, in douze, et réimprimée avec la seconde en 1677 au Louvre").

E. GALOIS, Sur la théorie des nombres, Bull. de Férussac XIII (1830), p. 428–436 = Œuvres mathématiques d'Évariste Galois, Paris Gauthier-Villars 1897, pp. 15–23.

A. Genocchi, Démonstration d'un théorème de Fermat, Nouv. Ann. de Math. (III) 2 (1883) 306–310.

A. GIRARD: Les Œuvres Mathématiques de Simon Stevin . . . reveu, corrigé et augmenté par Albert Girard, Leyde, Elzevir 1634.

F. Grube, Ueber einige Euler'sche Sätze aus der Theorie der quadratischen Formen, Zeitschr. f. Math. u. Physik 19 (1874) 492–519.

H. Hasse, Kurt Hensels entscheidender Anstoss zur Entdeckung des Lokal-Global-Prinzips, Crelles J. 209 (1962) 3–4.

J. L. Heiberg und H. G. Zeuthen, Einige griechische Aufgaben der unbestimmten Analytik, Bibl.Math. 8 (1907–1908) 118–134.

HURWITZ: Mathematische Werke von Adolf Hurwitz, Basel, Birkhäuser, 2 vol. 1932–1933.

J. Itard, (a) Arithmétique et Théorie des Nombres, P.U.F. Paris 1973; (b) Sur la date à attribuer à une lettre de Pierre Fermat, Rev.Hist. des Sc. 2 (1948) 95–98.

W. Knorr, Archimedes and the measurement of the Circle, A new interpretation, Arch.Hist.ex.Sc. 15 (1975) 115–140.

J. LANDEN, An investigation of a general theorem for

finding the length of any arc of any conic hyperbola, by means of two elliptic arcs, with some other new and useful theorems deduced therefrom, Phil. Trans. LXV (1775) 283–289.

LEGENDRE: (a) Recherches d'Analyse Indéterminée, Par M. Le Gendre, in: Histoire de l'Académie Royale des Sciences, Année M.DCCLXXXV, avec les Mémoires de Mathématique et de Physique pour la même Année . . . A Paris, de l'Imprimerie Royale M.DCCLXXXVIII, pp. 465–559; (b) Adrien-Marie Le Gendre, Mémoire sur les Transcendantes elliptiques . . . lu à la ci-devant Académie des Sciences en 1792, A Paris, l'an deuxième de la République; (c) Essai sur la Théorie des Nombres; par A.M. Le Gendre, de l'Institut National, A Paris, Chez Duprat, Libraire pour les Mathématiques, quai des Augustins, An VI; (d) Adrien-Marie Legendre, Théorie des Nombres, troisième édition. Paris, Firmin-Didot, 2 vol. 1830.

MACLAURIN: Colin MacLaurin, A treatise of fluxions, Edinburgh, 2 vol. 1742.

D. Mahnke, Leibniz auf der Suche nach einer allgemeinen Primzahlgleichung, Bibl. Math. 13 (1912–1913) 29–61.

A. de MOIVRE: (a) The Doctrine of Chances, London 1718; (b) Miscellanea Analytica de Seriebus et Quadraturis . . . (1 vol. + Supplementum) Londini 1730.

O. Neugebauer and A. Sachs, Mathematical cuneiform texts, New Haven 1945.

B. PASCAL, Œuvres Complètes, éd. J. Chevalier, Bibl. de la Pléiade, Gallimard 1954 (pp. 97–108: Traité du triangle arithmétique; pp. 194–208: Histoire de la roulette).

K. Shukla, Ācārya Jayadeva the mathematician, Ganita 5 (1954) 1–20.

J. Steinig, On Euler's idoneal numbers, Elem.d.Math. 21 (1966) 73–88.

THEON: Theonis Smyrnaei . . . Expositio rerum mathematicarum ad legendum Platonem utilium, rec. E.Hiller, Lipsiae, Teubner 1878 (and: Theon of Smyrna, Mathematics useful for understanding Plato, transl. by R. and D. Lawlor . . . San Diego, Wizards Bookshelf 1979).

Index Nominum

Index Rerum

divisor (of a binary quadratic form; cf. prime): 217–218, 320, 332.
double equations: 28–29, 105–107, 138–139, 243, 253.

elastic curve (= elastica): 182, 244.
elliptic (functions, integrals): 1, 122, 130, 182–183, 243–251, 298–299, 325; (curves = curves of genus 1): cf. equations of genus 1.
equations: cf. diophantine; (of genus 0): 28, 104, 179–180, 233–239, 315, 317, 327–328, 341–344; (of genus 1): 28–29, 103–112, 130–134, 180, 242, 252–256, 296–307, 316.
equivalence (of forms, equivalent forms; cf. classes): 217, 318–319; (of equations): 339.
euclidean algorithm: 5, 7, 94, 126, 177, 193.
Euler-MacLaurin (formula, summation): 184, 257–261, 272.
Euler's theorem ($a^{\varphi(N)} \equiv 1 \bmod N$): 57, 177, 193.
eulerian product: 266–267.

factoring: cf. prime number decomposition.
Fermat's conjecture (on $2^{2^n} + 1$): 2, 58, 172.
Fermat's equation (= "last theorem": $x^n + y^n = z^n$): 49, 104, 173, 181; (case $n = 4$): 61, 77–79, 180, 181; (case $n = 3$): 61, 116–117, 181, 239, 242; (case $n = 5$): 335–338.
Fermat's theorem ($a^{p-1} \equiv 1 \bmod p$): 55–57, 175–177, 189, 193–194.
figurative numbers: 7.
form: cf. Gaussian, Lagrangian, quadratic.

Gaussian (classes, equivalence, forms): 319–320; (field, ring): 66, 125.
Goldbach's conjecture: 2, 170.

Hasse's (or: Hasse-Minkowski) principle: 60, 339, 340–342.

pythagorean triangles: 8–10, 13, 76–79, 82, 111.

quadratic (fields): 66, 125–129, 202; (forms): 80–81, 83–91, 128, 179, 204–205, 211–218, 222, 292–295, 316, 318–322; (reciprocity): cf. reciprocity; (residues): cf. residues; (rings): cf. quadratic fields.

real: cf. imaginary.
reciprocity (quadratic —, law of quadratic —): 62–63, 187, 209, 219, 287–291, 326, 328–330, 344–345.
recurrent (sequences): 183, 237.
reduced: 95; ("in the Lagrangian sense"): 321; (weakly, strongly —): 353.
reduction: cf. Lagrangian reduction.
residue (quadratic —, m-th power —): 61–66, 79, 177, 190–192, 195, 196, 209, 328.

side and diagonal numbers: 16, 90, 92.
submultiple numbers: 52–54.
sums of n squares (for $n = 2$): 10–11, 30, 34, 59, 66–74, 177–178, 210–211; (for $n = 3$): 30, 34, 59, 101–103, 292–294, 331; (for $n = 4$): 3, 30, 34, 102, 173, 174, 177, 178–179, 226–229, 294–295, 315–316.
sums (of three triangular numbers, four squares, five pentagonal numbers, etc.): 3, 61, 79, 101, 119, 173.

Wilson's theorem: 65–66, 201, 316.

The trouble with this explanation was that it assumed the Washington policy-makers to be just as naïve and gullible as newspaper readers. If Tony Imbert and his crowd were supposed to have had a veto power on the Guzmán plan, the Bundy mission was doomed from the outset, and the only problem is how anyone in the White House or the State Department could have thought otherwise. In fact, the Guzmán government was never intended to include the neo-*trujillista* elements tied up with Imbert, and the "broadly-based coalition" was not the policy Guzmán was supposed to represent but a totally new policy predicated on the failure of the Guzmán plan.

The "broadly-based coalition," which the State Department began to promote toward the end of May, was another in an already long list of political and diplomatic abortions. In effect, it envisaged the bringing together of both the Caamaño and the Imbert forces, or of the "moderate" elements in both camps, together with others not directly implicated in the struggle, into a working harmony of interests. The idea was launched at the worst possible time, in the wake of what the Boschists considered to be an odious betrayal of the Guzmán plan and in the midst of the Imbert forces' successful offensive against Caamaño's northern sector. While Secretary Rusk was talking about the "broadly-based coalition," Colonel Caamaño's territory was cut in half with either the connivance or toleration of U.S. troops, and the clean-up in the northern sector of the city made Imbert more determined than ever to finish off the now besieged Caamaño garrison with another violent blow.

Politically, however, the "coalition" concept was revealing. It might have made sense in the United States, with a solemn, centuries-tested constitution providing both first principles and rules for organic change. In such a system, deals and combinations by nominal political rivals may serve a useful purpose or at least may avoid any fatal disaster in times of stress, however much they may ruffle political purists. But in the Dominican Republic, the constitutional first principles and rules for organic change were lacking. These were precisely what the revolt was about. In such circumstances, the "broadly-based coalition" was a purely mechanical device, bereft of moral principle and the rule of law. It demonstrated once again how much greater was the political than the geographical distance between the United States and the Dominican Republic.

The struggle between contending forces and views in Washington and the mixed ideas and feelings of individual officials, prevented anything like a clear-cut decision. If there had been a consistent, unanimous pro-Bosch line in Washington, the Benoit and Imbert juntas would never have been set up; and if there had been a consistent, unanimous anti-Bosch line, the Guzmán formula would never have been entertained. Imbert's offensive against the northern sector of Caamaño's front was, to say the least, tolerated, but his planned follow-up against the southern sector was not. The "coalition" plan at first implied some kind of accommodation between Imbert and Caamaño and then turned into a search for someone outside either camp who could presumably draw some strength from both.

Thus, by June, a new Dominican politician was given the Washington buildup. He was Dr. Joaquín Balaguer, a well-known jurist, educator and historian, who had served the Trujillo dictatorship for almost thirty years as its most respectable intellectual front and had in return been rewarded with just about every honor it was in the power of the regime to bestow. He became Trujillo's vice-president in 1957, and was promoted to the Presidency in August 1960 to take some of the sting out of the O.A.S.'s investigation of the Trujillo regime. When Trujillo was assassinated in May 1961, Balaguer continued as President, but he realized much sooner than the other *trujillistas* that an epoch had come to an end. In the next year, Balaguer tried to ride out the post-Trujillo storm by making concessions to leftist demands and preparing the way for constitutional reform. Juan Bosch relates that Balaguer offered to turn over the presidency to him in December 1961.* This was not Bosch's idea of a democratic solution, and Balaguer instead formed a "Council of State," with himself as President and Antonio Imbert as one of its seven members. But penances and payoffs could not save Balaguer; the then dominant opposition, organized as the *Unión Cívica Nacional,* and backed by the United States, saw power in its grasp if it could destroy him politically. The UCN therefore waged a demagogically bitter campaign against him as the very incarnation of *trujillismo* and as Trujillo's political heir. Balaguer's military chieftain, General Pedro Rodríguez Echavarría, tried to save him by staging a military coup in January 1962, but the U.S. Chargé d'Affaires, John Calvin Hill, helped to break it up in 48 hours by threatening economic and possibly military sanctions. The "Council of State" was reorganized with the former Trujillo cabinet member and now UCN leader, Rafael F. Bonnelly, as the new President, and Donald Reid Cabral as Balaguer's replacement. Balaguer was literally driven into exile in March 1962 as an unreconstructed *trujillista,* and he was not able to return until June 1966. The man who could not live down his past in 1962 had become the United States' man of the future in 1965.

Yet Balaguer was a wily, old politician who knew his people, and his U.S. backers might

* Juan Bosch, *Crisis de la Democracia de América en la República Dominicana,* p. 52.

well have learned from him. In a statement on behalf of his *Partido Reformista* of September 12, 1965, Balaguer referred to the "popular movement of April 24"; he did not make the mistake of calling it a "Communist plot." He proffered the backing of his party to the next provisional government and expressed his faith in Provisional President García Godoy's moral and intellectual qualifications "to fulfill with absolute probity the mission which has been entrusted to him."

After all this, after the support of Wessin, the instigation of Benoit, the backing of Imbert, the doublecross of Guzmán, and the promotion of Balaguer, it should have been easy to predict whom the United States would in the end settle for—a reactionary puppet of a reactionary power, of course.

Instead, the man whom the United States backed as Provisional President of the Dominican Republic was Héctor García Godoy, the former Foreign Minister of the government of Juan Bosch, no puppet and no reactionary. He was able to take office for nine months on September 3, 1965, only because the United States decided to withdraw its financial support from Imbert's junta and effectively make known its will to the Dominican military. García Godoy's provisional regime was as full of contradictions and inconsistencies as all the rest, but at least a man of good will and democratic purposes was chosen for the job.

Barely a week later, General Wessin y Wessin was forced to leave Santo Domingo by, among others, the U.S. ambassador and commanding general. As General Wessin later told the unlikely story, a U.S. official visited his home towards the end of August, 1965. He pretended that he had seen an advertisement for the sale of the house which he offered to buy. General Wessin's wife asked him $50,000 for a house that was at most worth half as much, ostensibly to discourage him from coming back. But he returned at 1 P.M. that same day and talked to General Wessin himself. This time he offered General Wessin $8,000 immediately and the remainder when he had boarded the plane to take him out of the country. To this proposition, which was obviously a form of bribe, General Wessin allegedly replied that he was willing to sell his home for $50,000 but saw no need to leave the country. The U.S. official told him to "sacrifice" himself "so that peace could return to the country."

At midnight the following day, General Wessin received two other visitors—David Phillips of the C.I.A. and the army attaché Lt. Col. Joseph W. Weyrick. They renewed the offer, with the additional inducements that he would be a "guest of honor" in the Canal Zone in Panama, and make "pleasure trips" to various U.S. overseas military bases. No agreement was again reached. On September 8, President García Godoy sent for General

Wessin and again asked him to leave the country in the interests of "peace." General Wessin promised to answer the following day but, instead, went to his headquarters at the San Isidro base and there "alerted" his troops and ordered the return of his unit's tanks to the camp. These troop movements so alarmed the U.S. command that Wessin's tanks were stopped on their way by U.S. contingents and detained. General Wessin then went to see the U.S. divisional commander, Brig. Gen. John R. Deane, and apologized for having alerted the tanks without having given him previous notice. While they were talking, Lt. Gen. Palmer called and told General Wessin that he wanted to see him. The latter in effect replied that General Palmer could come to see him at his home.

It is clear from General Wessin's own recital of these events that he had made himself into an intolerable threat and nuisance to every one. He had made no secret of his contempt for his fellow-officers and was, in fact, partly responsible for the previous expulsion from the country of eight leading officers. He was an incorrigible, uncontrollable menace to any civilian government which did not meet with his approval; his black list included Joaquín Balaguer as well as Juan Bosch. He saw so many Communists everywhere that he was willing to credit the Dominican Communists with a "hard-core" membership of 20,000 besides many more thousands of sympathizers—a figure reached by the simple process of making every Boschist and many others into Communists. For the U.S. and Inter-American military command, his threatening movement of tanks and troops on September 8 was the last straw.

At 6 P.M. on September 9, Lt. Gen. Palmer put in an appearance at General Wessin's home. But he was not alone. He was accompanied, according to General Wessin, by his staff, by General Hugo Panasco Alvim of Brazil, commander-in-chief of the so-called Inter-American Peace Force, and his staff, by General Deane, commander of the U.S. 82nd Airborne Division. and his staff, the chiefs of the Dominican police, army and navy, and several jeeploads of troops. General Panasco Alvim acted as spokesman for the group and told him, as so many others had done before, that he had to leave the country for the sake of a peaceful settlement. To ease Wessin's way, General Panasco Alvim offered him the post of Dominican Consul in Miami, Florida, which Wessin accepted but, as he put it to **192** Jules Dubois, "with mental reservations." After demurring for some time against General Panasco Alvim's demand, General Wessin asked permission to bid farewell to his troops, and it was granted.

But the Inter-American commanders were taking no chances. After General Wessin had made his farewell, he was escorted to the military academy at the San Isidro base, where he was required to hand over his command to Colonel Elio Osiris Perdomo. Then, he was immediately hurried to a U.S. Army helicopter and put on a military transport plane which brought him to Fort Amador in Panama. Later, General Wessin complained bitterly that he had not been permitted to go home to change his clothes, get his passport and money, or say good-by to his family. From Panama, he was flown to Miami, where he soon wrote a blistering letter protesting his treatment. "Never would I have imagined," he complained, "that an army officer of my rank would have been taken to the airport in full uniform and tossed out of the country with a bayonet at his back." But, in his subsequent testimony, he explained that he had referred to the bayonet at his back merely as a "figure of speech." Everyone was always misunderstanding Elías Wessin y Wessin.*

---

* General Wessin's story is mainly taken from his testimony before the Senate Internal Security Subcommittee, pp. 162-75. He gave another and, in some details slightly different version of his removal from the Dominican Republic to Jules Dubois in the Chicago *Tribune*, October 6, 1965. The letter protesting his removal was published in the Chicago *Tribune*, September 17, 1965.

15

And quite as remarkably, Under Secretary Mann was able to say on October 12, 1965, in San Diego: "It has been suggested that non-intervention is thought by some to be an obsolete doctrine."

Now who in the world could have thought that? Could it possibly have been Under Secretary of State for Economic Affairs Thomas Mann who, in two interviews, had criticized both the O.A.S. and the U.N. charters for having been drawn up in "19th century terms"?* Could it have been that elder statesman Averell Harriman who was sent to Latin America to explain U.S. policy in the Dominican Republic and had told an audience in Montevideo that the principle of non-intervention was becoming "obsolete"?† Might

194

---

* New York *Times,* May 9, 1965, and *Look,* June 15, 1965.

† New York *Times,* May 7, 1965.

President Johnson have had that thought in mind when he observed on May 28: "The first reality is that old concepts and old labels are largely obsolete"? Or could it perchance have occurred to Representative Armistead I. Selden, Jr., the administration's most ardent defender in the House?

On September 20, 1965, the House of Representatives passed a resolution introduced by Representative Selden which inferentially vindicated the U.S. action in the Dominican Republic by the overwhelming majority of 312 to 52. This resolution was so ambiguously worded that it could mean anything or nothing. But it was generally interpreted as an effort to justify the unilateral use of force by any nation which considered itself threatened by "international Communism, directly or indirectly" in another nation.* Representative Selden said that the State Department "generally agrees with the objective of the proposed resolution," though it would have preferred a different wording based on the Punta del Este resolution of 1962. In any case, the State Department did not object to the resolution, the

---

* The key section of House Resolution 560, introduced by Representative Selden, read:

Resolved, That it is the sense of the House of Representatives that (1) any such subversive domination or threat of it violates the principles of the Monroe Doctrine, and of collective security as set forth in the acts and resolutions heretofore adopted by the American Republics; and

(2) In any such situation any one or more of the

intent of which was far more important than its language. As might have been foreseen, the Selden Resolution brought forth outbursts of protest from Latin American nations.

Mr. Mann's speech in San Diego on October 12 was in part intended to reassure the Latin American countries that the United States was still officially committed to the traditional doctrine of non-intervention. He knew of no Washington official, he said, who thought that the doctrine of non-intervention was obsolete. He himself, in fact, believed that it was the "keystone" of the inter-American system. That is why, he went on, the United States had refrained, in the first days of the revolt, from supporting either side in the Dominican Republic.

The following month, David Kraslow of the Los Angeles *Times* (November 20, 1965), to which the Senate Foreign Relations Committee's hearings had been made available, revealed that Mr. Mann himself had belied his San Diego position in his previous testimony before the committee. Mr. Kraslow compared

high contracting parties to the Inter-American Treaty of Reciprocal Assistance may, in the exercise of individual or collective self-defense, which could go so far as to armed force, and in accordance with the declarations and principles above stated, take steps to forestall or combat intervention, domination, control, and colonization in whatever form, by the subversive forces known as international communism and its agencies in the Western Hemisphere (*Congressional Record*, House, September 20, 1965, p. 23458).

what the news correspondents had been told in Santo Domingo and what Mr. Mann later admitted:

> U.S. officials on the scene, for example, told newsmen at the time that the U.S. had nothing to do with the creation of a civilian-military junta headed by General Antonio Imbert Barreras, one of the assassins of Dominican dictator Rafael Trujillo.
>
> Reporters quickly learned otherwise, and Mann confirmed in his testimony that Imbert was secretly backed by the U.S.
>
> Mann also confirmed what some newsmen had reported to the American people from Santo Domingo—that within about a week after the U.S. helped create the Imbert junta, Mann and other officials asked Imbert to surrender power because he lacked popular support.
>
> When newsmen checked on reports that the U.S. solicited a request from Dominican military leaders for Marines to be sent to Santo Domingo to protect Americans, officials there denied it. Yet Mann testified that this is precisely what happened.

One of the mysteries of this entire story is how an Under Secretary of State could participate in the moves which set up the Imbert junta, which tried to get rid of the Imbert junta, and which instigated the request for U.S. intervention from the Dominican military, admit all these actions in testimony before the Senate Committee, which was bound to be made public sooner or later, and then get up in San Diego and assure a fairly sophisticated audience that the United States had refrained from supporting either side in the Dominican Republic in the first days of the revolt. If I were of a more skeptical mind, and Mr. Mann's reputation for gravity were not so well established, I would almost be tempted to suspect that this allusion to the way we refrained from taking sides was some sort of private joke for the insiders in Washington who had read all the messages from Santo Domingo.

For about six months after April 1965, U.S. policy in the Dominican Republic was a kind of political phantasmagoria, dizzying to actors and onlookers alike.

Non-Americans are always likely to interpret this kind of extravaganza as a manifestation of diabolical malevolence; Americans are more likely to regard it as an exhibition of monumental incompetence. Every move was cancelled out by another move; every rationale was inferentially disavowed by a later one. The Guzmán formula was not acceptable in May, but a distant variant—the García Godoy formula—was put through in September. General Wessin y Wessin was a hero in April and a villain in September, though his Dominican version of the John Birch Society mentality had not changed at all, and he was just as apt to accuse Donald Reid Cabral and U.S. officials of serving the Communist cause as

Juan Bosch and Héctor García Godoy. It may be argued that this overabundance of disappointed suitors proves that the United States never played anyone's game and strictly maintained the neutrality which it pretended to observe. I rather think it merely proves that at different times the United States played everyone's game and was always unneutral to someone. A succession of inconsistencies does not add up to consistency any more than a collection of falsehoods is equivalent to the truth.

It had mattered little, indeed, who had represented the democratic alternative to the Imberts and the Wessins. The particular personality or political style of Juan Bosch was never the real issue. When Antonio Guzmán was proposed for the Presidency, he immediately received the same treatment Bosch had received, and then it was the turn of García Godoy—all very different personalities with different backgrounds and different styles. If one could imagine García Godoy followed by a Dominican counterpart of Lyndon Johnson, there is no doubt the air would again have rung with the same charges of "Communist influences." After all, the August 1965 issue of the organ of the John Birch Society had determined mathematically that the Communists influenced or controlled 60 to 80 per cent of economic and political affairs in the United States and that the landing of the U.S. marines in Santo Domingo was directed by "what often seems to be Communist headquarters in Washington—officially called the State Department." If this was lunacy, it was not, unfortunately, restricted to the United States.

200

16

In a deeper sense, then, the Dominican crisis was an expression of a crisis in the use and abuse of anti-Communism. Only some aspects of this larger problem can be touched on here.

It is no longer quite as clear as it used to be what Communism is. Nevertheless, in all its existing forms, it is still a system of political, intellectual, and social repression, based on a single dogma, and a single source of power, at best a party dictatorship, at worst a personal dictatorship. It has proven its ability to debauch the noblest ideals and to commit the most monstrous crimes. Yet, whatever form it still takes, Communism has its sacred books, its recognized national and international chiefs, and its tradition of faith and discipline.

Anti-Communism is not like that. It is merely a negation of Communism. The Communist world, however transitional it may be, is still a relatively restricted, prescribed order. The anti-Communist world is, by comparison, unrestricted and unprescribed. It has no sacred dogmas or too many, no leaders or too many, no causes or too many. It takes in the best and worst of humanity. The anti-Communism of a Hitler or a Trujillo is just as evil and repulsive as the Communism of a Stalin or an Ulbricht. As a result, anti-Communism by itself tells us nothing about whether a cause is worth fighting for. The Stalins always tell us that we need to fight with them against the Hitlers, and the Hitlers always tell us that we need to fight with them against the Stalins. Thus arise the ordeal and the glory of humane democratic anti-Communism—that it must usually fight on more than one front. What we must always ask is: What kind of anti-Communism do you stand for?

The problem of anti-Communism is complicated by an even more disturbing factor. There is no Chinese Wall between Communism and anti-Communism; almost every form of anti-Communism believes that other forms wittingly or unwittingly play into the Communists' hands. Hitlerism, to cite an extreme example, was not merely an evil in itself; it was the form of anti-Communism on which Stalinism fed the most. Once the choice could be reduced to Hitler or Stalin, thousands who might have chosen neither felt that they had to choose Stalin or Hitler. Conservatives think that liberal anti-Communism is really the anteroom to Communism; liberals think that conservatives seek to perpetuate the injustices and inequalities on which Communism thrives. There are those who thought that the late Senator Joseph McCarthy was the scourge of Communism; and those who were almost convinced that only a secret agent of Moscow could have sought to destroy trust in the U.S. Army's top leadership. The *National Review* thinks that the John Birch Society is causing grave damage to the anti-Communist cause; and the John Birch Society accuses everyone outside its orbit of selling out to the Communists.

Thus, many anti-Communists think that other anti-Communists are not serving the anti-Communist cause effectively enough: some also believe that this ineffectiveness positively helps the Communist cause or is even an integral part of an all-embracing "Communist conspiracy."

It is necessary to make these distinctions, in a consideration of the Dominican crisis, because many who were willing to admit that Juan Bosch was no Communist were also unwilling to see the pro-Bosch revolt succeed on the ground that Bosch's anti-Communism was too "soft." An editorial in *Life* (May 14, 1965) expressed this idea in a relatively genteel fashion: "The moment the rebel leadership was infiltrated by Castroite Communists, the return of former President Juan Bosch to the office he lost in a military coup two years ago ceased to be an acceptable solution to the crisis. Under *fidelista* auspices, Bosch's brand of liberalism and ineffectual, if well-meant, anti-Communism, would have lasted about as long as an icicle on the Avenida Independencia." Of course, this implied that the return of Bosch would have been an acceptable solution before the alleged infiltration of the Castroite Communists. And what "infiltration" meant, the editorial never made clear. But the general idea was clear enough and was repeated, in far more vulgar and offensive terms, in hundreds of other editorials.*

And so actual Communism and "softness on Communism" imperceptibly tend to melt into

---

* The readers of *Time* must have been surprised to find the following explanation of the Latin American Communists' "new" tactics in the issue of August 6, 1965: "They are best exemplified by the Dominican Republic, where the Communists resorted to the old 'popular front' strategy, muscling into a legitimate non-Communist rebel movement with hopes of duping its idealistic leader, Juan Bosch." But *Time's* readers had previously been led to believe that the Communists had not merely "hoped" to muscle in but had actually taken over, and that U.S. intelligence had "flatly reported" Bosch's pre-revolt collusion with the Communists. Still, we must be grateful for small favors. By August 1965, at any rate, *Time* had discovered that the rebel movement had been legitimately non-Communist and that Juan Bosch was an "idealistic leader."

each other or, in practice, are virtually reduced to the same thing. The only question, then is: who is to decide who else is "soft on Communism"?

A book by the late deLesseps S. Morrison devoted pages to the proposition that the "democratic Left" in Latin America is vulnerable to "Communist infiltration," becomes an "easy target" for the Communists, or "opens the door" to the Communists. Mr. Morrison gives Bosch's 1963 regime as the horrible example of this "democratic Left" infirmity. He cites former Costa Rican President José Figueres as the organizer of the tendency and includes in it, besides Bosch, such diverse figures as former Guatemalan President Juan José Arevalo, former Puerto Rican Governor Luis Muñoz Marín, former Venezuelan President Rómulo Betancourt, Peru's Victór Haya de la Torre, Costa Rican President Francisco J. Orlich, former Bolivian President Victór Paz Estenssoro, former Honduran President Ramón Villeda Morales, and the earlier Fidel Castro. Obviously, the problem extends far beyond Bosch personally, and any one of these figures, lumped together so indiscriminately, might have found himself in the same position and have been subjected to the same kind of treatment. Mr. Morrison was not, to be sure, one of the brightest stars in the U.S. diplomatic firmament—but he was U.S. Ambassador to the Organization of American

205

States from 1961 to 1964. The views that he and his "editor," Gerold Frank, put into his book were, and are, common currency in U.S. diplomatic and political circles.*

Another former and far more distinguished U.S. diplomat, Ellis Briggs, used the same technique to justify the ouster of Premier George Papandreou in the Greek crisis of 1965. Papandreou is no Communist, he wrote, but he is "incompetent, does not answer his doorbell," and "has never learned not to play with Communists."† Mr. Briggs was U.S. Ambassador to Greece from 1959 to 1962. Indeed, there were startling similarities between the Greek crisis and the Dominican crisis.

I do not wish to oversimplify the problem. I believe that there is a sense in which it is right and natural that some anti-Communists may question the effectiveness of other anti-Communists and even conscientiously believe that some anti-Communisn does more harm than good.

But I am convinced that the anti-Communism which has made a paranoid slogan or a political racket of the terms "infiltration" and "softness" has led us onto the path of disgrace and disaster.

206

---

* deLesseps S. Morrison (and Gerold Frank), *Latin American Mission*, Simon & Schuster, especially Ch. 18.

† New York *Times*, August 6, 1965.

The concept of "infiltration" is obviously far from simple. Communists do not bother to "infiltrate" their own organizations or governments; it must mean that they seek to gain influence in, or take over, non-Communist or anti-Communist movements. It also makes a difference whether they come in anonymously or openly, in the rank and file or in the top leadership. How much "infiltration" is unavoidable, how much dangerous, how much intolerable? To what extent can a democratic organization inhibit or prohibit such tactics? Are the Communists the only political group that uses them? Questions such as these might well occupy some of the time and energy of political scientists.

But when the term "Communist-infiltrated" is flung about wantonly, irresponsibly, indiscriminately, it is nothing but a swindle, blackmail, or an incantation. I have seen hundreds, perhaps thousands, of references to Juan Bosch's "Communist-infiltrated government," and I have yet to see a single member of that government identified as a Communist. The line between "Communist," "Communist-controlled," Communist-infiltrated," and "Communist-influenced," has been all but obliterated in what has become very common usage, and repetition has given these terms a social sanction which is not only undeserved but which may wreak social havoc if improperly employed. Anyone can play this game of "hardness" and "softness" on Communism until everyone is outbid by the John Birch Society. If an anti-Communist policy is short-sighted or stupid, it matters little whether it is "hard" or "soft." If U.S. policy in the Dominican crisis demonstrated nothing more, it would still be worthy of continued study and reflection.

On the following pages can be found details of other
COMMENTARY REPORTS.
For information write to

COMMENTARY REPORTS
165 East 56th Street
New York, N.Y. 10022

DATE DUB			
MAR 21 22			
MAY 21 22			
GAYLORD M-2			PRINTED IN U.S.A.